Fungi

Experimental Methods In Biology

Second Edition

MYCOLOGY SERIES

Editor
J. W. Bennett
Professor
Department of Plant Biology and Pathology
Rutgers University
New Brunswick, New Jersey

Founding Editor
Paul A. Lemke

Fungi
Experimental Methods In Biology
Second Edition

Ramesh Maheshwari

Indian Institute of Science
Bangalore, India

CRC Press
Taylor & Francis Group
Boca Raton   London   New York

CRC Press is an imprint of the
Taylor & Francis Group, an **informa** business

Front cover: The tube-like hyphae of the pink-orange mold *Neurospora crassa* are divided into multinuclear compartments by cross walls. The fungus was transformed with a gene from a jellyfish that encodes a green-fluorescent protein. Nuclei in living hyphae tagged by fusion of GFP to histone H1 gene were visualized by confocal microscopy. The nuclei become spindle-shaped (inset) as they move from one compartment to another compartment of branched hyphae through centrally perforated cross walls. This image was provided by Nick D. Read of the University of Edinburgh and is used with permission from Elsevier.

Back cover: The plant-parasitic fungus *Puccinia graminis tritici* has been a scourge of mankind since ancient times. It infects stems and leaves of the wheat plant and forms rust-colored pustules. Each pustule contains countless numbers of wind-disseminated urediospores shown at top right. The fungus survives in stubble in the form of thick-walled bi-celled teliospores. The photographs were provided by Jacolyn A. Morrison of Cereal Rust Research Laboratory at the University of Minnesota.

CRC Press
Taylor & Francis Group
6000 Broken Sound Parkway NW, Suite 300
Boca Raton, FL 33487-2742

First issued in paperback 2016

© 2012 by Taylor & Francis Group, LLC
CRC Press is an imprint of Taylor & Francis Group, an Informa business

No claim to original U.S. Government works

ISBN 13: 978-1-138-19925-5 (pbk)
ISBN 13: 978-1-4398-3903-4 (hbk)

Visit the Taylor & Francis Web site at
http://www.taylorandfrancis.com

and the CRC Press Web site at
http://www.crcpress.com

Dedication

David D. Perkins
(1919–2007)

Contents

Part III
Gene Silencing

Part VI
Populations

Preface to the First Edition

Fungi are organisms generally composed of tubes that are invisible to the naked eye. The cells of these tubes are multinucleate and in cytoplasmic continuity. Fungi are among the oldest and largest living organisms, rivaling the mass of a California redwood tree or a blue whale. As the chief agents of decomposition of organic matter, fungi contribute to the sustenance of the carbon cycle. As mycorrhizal partners of roots, they provide the primary mechanism for the capture of nutrients used by plants, thereby contributing to the green cover on earth. Some fungi occur as endophytes in plants or as symbiotic partners with algae, allowing the mutualistic partners to tolerate and grow in harsh conditions that they could not do otherwise. As virulent pathogens of plants, fungi are a constant threat in agriculture and forestry. Since antiquity, fungus has been exploited either unwittingly or intentionally for the conversion of grape juice into ethanol in wine. As producers of antibacterial compounds, fungi are sources of life-saving drugs. They are the only eukaryotic organisms that can thrive at temperatures beyond which no plant or animal can live. Though potentially immortal, a few fungi have a limited life span, providing valuable models for investigating the mechanisms in aging and death. Fungi are now at the forefront of research on mechanisms in gene silencing, biological rhythm, mating processes, biogenesis of intracellular organelles, adaptations to hostile habitats, structure of natural populations, and speciation. Because of their small genomes, fungi are being used in "systems biology" to understand the connections between genes, proteins, and metabolic and signaling pathways.

This book on fungi is an outcome of my association with graduate students in biochemistry. The majority of these students had little or no previous exposure to fungi and a few not even to biology. I found that students became interested in fungi if an attempt was made to demonstrate fungi in natural situations and how they could be used to understand complex biological questions, in particular if the design of the experiments that were done to obtain information were described. Today's accelerated pace of research, aided by new instruments and techniques combining the approaches of genetics, biochemistry, and cell biology, has changed the character of mycology, necessitating a new approach for the organization of the subject matter and learning about the fungi.

This book should be useful both for a beginning research worker and a professional. The subject matter is divided into six parts, comprising 14 chapters. Each chapter is self-contained and written in a style that enables the reader to progress from elementary concepts to current thinking on a topic. Throughout, attention is drawn to unsolved questions. References are given only to selected publications, primarily for details on the design of experiments that were used to obtain information, and for the identification of some of the key players. Finally, an Appendix gives the principles in naming fungi, estimated to comprise more than 1.5 million species, and of their broad classification. Many authors include the slime molds in the fungi. Since they consist of naked cells, move in amoeboid fashion, and ingest particulate food, I have excluded them.

Ramesh Maheshwari

Preface to Second Edition

The ease with which yeasts and molds can be cultivated in simple nutritive media has made these eukaryotic organisms the choice material for basic and applied research. *Fungi: Experimental Methods in Biology* (FEMIB) gives an account of real experiments that have been carried out on the diverse lifestyles of these organisms. The subject matter here is grouped in six sections. In addition to recent information, each chapter includes a historical perspective and illustrations that enable the reader to progress from elementary concepts to advanced research. The summary of each chapter focuses on unsolved problems. A glossary of mycological and interdisciplinary terms serves as a compilation of the concepts introduced in the chapters.

For the first edition of FEMIB, Dr. Amitabha Chaudhuri acceded to my request to contribute a chapter on yeast. Amitabha has revised this chapter for the second edition, FEMIB 2, by incorporating new advances.

The photographs on the front and the back cover were kindly provided by Nick D. Read and Jacolyn A. Morrison. Among the other colleagues who provided photos/illustrations are: Susan E. Anagnost, K. Mahalingeshwara Bhat, Nicholas Brazee, Jan Dijksterhuis, Jean Paul Ferrero/Ardea, Kenneth Hammel, M. Kapoor, Mark R. Marten, (late) P. Maruthi Mohan, Kurt Mendgen, Erich-Christian Oerke, Namboori B. Raju, William Sanders, Monika Schmoll, Matthew S. Springer, Ewald Srebotnik, Richard C. Staples, W. Dorsey Stuart, T. Suryanarayana, and Polona Znidarsic-Plazl. The organization of text matter was reviewed by Rowland H. Davis, and Namboori B. Raju reviewed Chapter 10.

I was fortunate to discover the artistic talents of young Subhankar Biswas of Bangalore. All unaccredited illustrations are by him. Although not so mentioned, he was my local production editor for FEMIB 2.

FEMIB 2 is dedicated to the memory of David D. Perkins for his mentorship of genetic approach to fungal biology. My wife, Manjuli, and son, Govind, have provided all possible help while writing this book. I thank John Sulzycki, my editor at Taylor & Francis, for his wise counsel.

Ramesh Maheshwari
Bangalore
June 2011

About the Author

Ramesh Maheshwari received his PhD degree from the University of Wisconsin–Madison. He did his postdoctoral work at the University of Michigan–Ann Arbor and at Stanford University, California. He has held academic appointment as a professor of biochemistry at the Indian Institute of Science in Bangalore. He is the author of *Fungi: Experimental Methods in Biology* and over 100 scientific papers on fungal and plant physiology. The thrust of his research over a span of 35 years was on the physiology and enzymes of thermophilic fungi and the genetics and ecology of *Neurospora*.

PART I

The Unique Features of Fungi

CHAPTER 1

The Hyphal Mode of Life

Considering their ubiquity, diversity, antiquity, longevity, and ability to colonize vast land areas, the fungi composed of a system of interconnected tubes can be regarded as the most successful organisms.

 With the exception of the yeast fungi, the large majority of fungi are composed of microscopic tube-like structures called the hyphae (sing., hypha). The hyphae spread and penetrate into substratum such as leaf litter, a fallen log, herbivore dung, or an artificially made agar medium. A hypha is shaped as a cylindrical tube with an average diameter of 4–6 µm and has a tapering tip that branches subapically (Figure 1.1). Each branch has a tip of its own. By iteration of this hyphal modular unit, a radially expanding mycelium is formed (Figure 1.2). At places, the branched hypha is bridged by short lateral outgrowths, bringing the entire mycelium into a protoplasmic continuity. The mycelium spreads over and penetrates into the substratum, secreting digestive enzymes that decompose the polymeric constituents of the substratum and absorbs the solubilized carbon, nitrogen, phosphorus, potassium, and sulfur compounds for its growth. This mode of fungal growth is inferred from examination of cultures grown on nutrient medium solidified with agar and based on observations of fungi growing in litter and in liquid-shake cultures wherein the mycelium grows as dispersed mycelium or assumes the outline of spherical pellets.

1.1 FEATURES OF HYPHAE

1.1.1 Apical Extension

For a fungal hypha, life is at its tip.

—G. W. Gooday and N. A. R. Gow (1990)

 Early in the last century, measurements were made of the distance between the growing hyphal tip and the first septum, and the distances between the successive demarcated segments. While the former increased, the latter remained constant, from which it was concluded that the growth of hypha is confined to the tip. When ^{14}C-labeled N-acetylglucosamine, a radioactive precursor of chitin (a structural component of the fungal cell wall), was fed to a growing mycelium and the site of its incorporation in hyphae determined by autoradiographic imaging of the whole mycelium, the incorporation of the label was observed only in the terminal region of the hypha (Wösten *et al.*, 1991). All vital cellular organelles are present in about 100 µm of the hyphal tip. The fungal strategy of growth is to keep the terminal region extending by the active forward movement of the protoplasm, although the mechanism by which the protoplasm is drawn toward the tip, leaving the empty tube behind, remains a mystery. To determine the mechanisms in polarized growth of hypha,

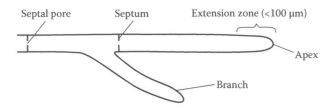

Figure 1.1 A hyphal modular unit (diagrammatic).

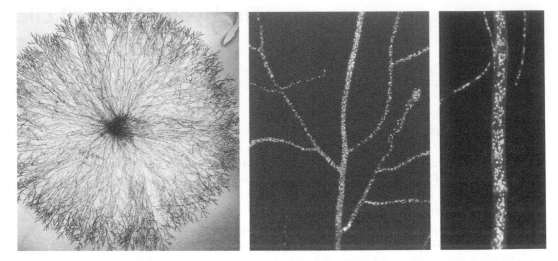

Figure 1.2 Left, a two-day-old colony of fungus *Neurospora crassa* grown from a small amount of mycelium placed on the center of an agar medium in a Petri dish. Middle, a small part of mycelium stained with a DNA-binding fluorescent dye. Right, an enlarged view of a multinucleate hyphal compartment.

a variety of approaches are being used: video-enhanced microscopy of the movement of green fluorescent protein (GFP) tagged organelles in living hypha; fluorescence imaging of distribution of ions; micro-electrode measurement of pH along hypha; patch-clamp detection of ion channels; immunofluorescent detection of enzyme distribution; and measurement of turgor by the plasmolysis method. Since the unicellular fungus yeast shows polarized bud formation and its genome sequence is now known, homologues of budding in the filamentous fungi may provide clues to the mechanism of polarized growth of fungal hypha. With a system of main and branched hyphae extending outward to secure food, a mycelium colony can spread over astonishingly large areas.

1.1.2 Spread and Longevity

For example, in Michigan, a tree-root-infecting fungus *Armillaria* sp. (Figure 1.3) that had colonized an area about the size of a football field was discovered in a forest (Smith *et al.*, 1992). This fungus forms honey-colored, mushroom-like fruit bodies at the base of the tree trunk; hence it is commonly known as the honey fungus. The question was asked if the large area of the forest was colonized by a single individual that had grown from a single spore or from genetically different individuals, each originating from a different reproductive propagule (Figure 1.4), even though it could not be directly seen as a single colony. Microscopic tracking of mycelium was precluded because soil is an opaque medium and the covered area could be very large. Hence a microbiological-cum-molecular approach was taken. A feature of *Armillaria* is a longitudinal aggregation of hyphae into dark-colored rhizomorphs that grow beneath the bark of a living tree (see Figure 1.18) to tap nutrients

Figure 1.3 Postage stamp showing fruiting bodies of *Armillaria mellea* (syn., *Armillariella*). This species is related to *Armillaria bulbosa*. The Armillaria fungi attack almost all species of hardwoods and conifers and live as saprobes in the soil on dead wood.

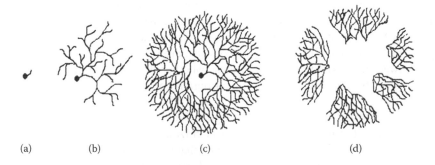

| (a) | (b) | (c) | (d) |

Figure 1.4 Diagram of stages in development of an individual colony of a fungus. (a) Germinating spore (black dot) that has produced a short hypha. (b) Branching to form a radially expanding mycelium. (c) Interconnected hyphal network formed by fusion of hyphae. (d) Disconnected mycelium of an individual fungal colony.

and extend through soil until they find a new tree. Since the rhizomorphs have a large food base as they come out from the stump of a dead tree, they are capable of extending through nonsupportive soil terrain and infecting the roots of healthy trees (Figure 1.5). Samples of rhizomorphs taken from a forested area were placed on a nutrient agar medium and the mycelium allowed to spread out. To determine if the mycelium thus sampled from a large area belonged to an individual fungus that had originally developed from a single spore, a compatibility test was performed. Mycelial isolates were oppositely paired in all combinations on nutrient medium in Petri dishes, the rationale being that if the mycelial growth intermingled to form a continuous mat it would indicate that the isolates were genetically related, i.e., they belonged to an individual fungus. On the other hand, if the isolates showed a zone of aversion, i.e., area between the paired mycelia not penetrated by hyphae (termed "barrage," an incompatible reaction), it would indicate that these belonged to different individuals. Results suggested that all mycelium isolates were those of a single individual.

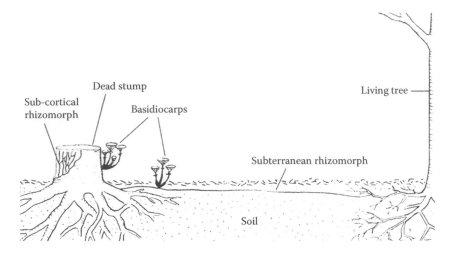

Figure 1.5 Hyphal strategy of hyphal spread and infection of forest trees. (From Ingold, C.T. and Hudson, H.J., *The Biology of Fungi,* 1993, Chapman and Hall, London. With kind permission from Springer Science and Business Media.)

The genetic relatedness of the mycelial isolates was confirmed by DNA fingerprinting method using the random amplification of polymorphic DNA (RAPD). In this technique the DNA extracted from mycelial isolates was cut with bacterial restriction endonuclease. The product was separated on an agarose gel by electrophoresis and the pattern of DNA fragments served as a fingerprint of a DNA molecule. An arbitrarily designed 10-base-pair sequence was annealed to DNA isolated from the strains, and a polymerase chain reaction (PCR) was carried out. The DNA fingerprints of fungal isolates from an extensive area were identical (Figure 1.6, clone 1). The results showed that a single genet, regardless whether its mycelium had remained intact or become fragmented because of physicochemical regions or predators in soil, had colonized an approximately 15-hectare area.

Having ascertained the individuality of this giant fungus colony, the question arose of determining its total mass. From the average weight of rhizomorph samples and their numbers in a representative area of the forest soil, it was estimated that this *A. bulbosa* genet was at least 10,000 kg of biomass. Furthermore, from the growth rate of the fungus on wooden stakes buried in soil, the age of the colony was estimated to be around 1500 years. This large and aged living organism may still be growing. Subsequently, a clone of *Armillaria ostoyae* was discovered in Oregon, with a mycelium spread over 890 hectares and estimated to be 2400 years old. It is generally thought that hyphae are capable of potentially unlimited growth and that fungi are immortal organisms. However, as we shall note in Chapter 16, senescing strains of fungi have been discovered.

1.1.3 Large Surface Area

Hyphae are generally 3–4 μm in diameter. Older hyphae can be up to 10 μm wide. A large surface area in relation to the total mass of protoplasm maximizes hyphal contact with the environment for uptake of raw materials for biosynthesis, for gas exchange, and for the release of metabolic by-products. A large surface area is advantageous in other ways too. For example, the space between the plasma membrane and the cell wall (intramural space) is used for storage of carbon compounds as well as for enzymes required to utilize it in times of need, viz., for maintenance of cellular structure when an energy source is not available. This notion is supported by the fact that washed mycelia exhibit a high endogenous rate of respiration for several hours.

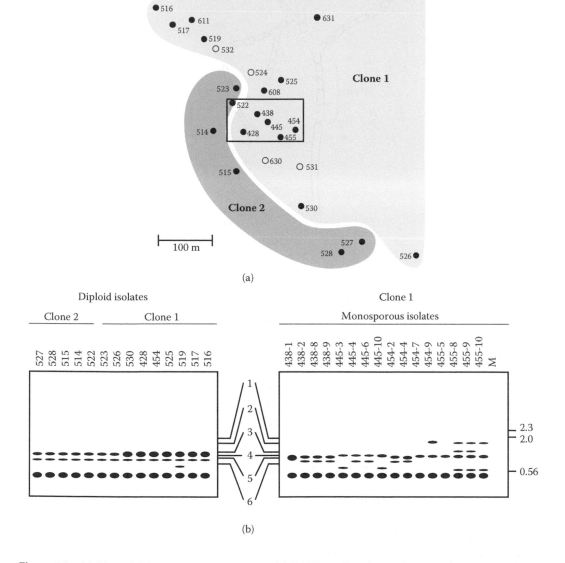

Figure 1.6 (a) Map of rhizomorph sampling area. (b) RAPD profile of mycelia grown from sampled rhizomorphs. Clone 1 is an individual genet that had colonized from rhizomorph mycelium or basidiospores. (Data from Smith, M., Bruhn, J.N., and Anderson, J.B., *Nature* 356, 428–431, 1992.)

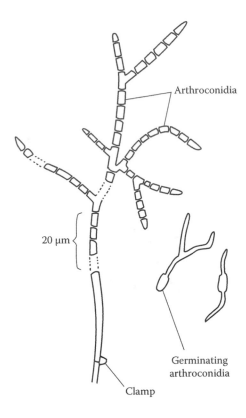

Figure 1.7 Arthroconidia. (From Ingold, C.T. and Hudson, H. J. *The Biology of Fungi,* 1993, Chapman and Hall, London. With kind permission from Springer Science and Business Media.)

The large surface area of the hypha is not without its disadvantages. Because of the single-cell thickness of the hypha, the environment has a direct effect on it, rendering the thin-walled hypha vulnerable to desiccation. Fungi therefore grow optimally either in aqueous medium or in a very humid atmosphere. In adverse conditions the hyphal tip perceives a signal and apparently produces a conidiation-inducing factor that diffuses behind it, inducing formation of double septa along the length of the hypha. The cells disarticulate and function as propagules called arthroconidia (Figure 1.7). Arthroconidia are the simplest type of spores formed by hypha.

1.2 HYPHAL STRUCTURE

The first electron microscopic studies of hyphae used OsO_4 for fixation. Subsequent studies showed that many structural components are better preserved by freeze substitution. Since the prime motivation was to find the reasons for the strictly polarized growth of the hypha, the apical cell has been the focus for ultrastructure studies (Figure 1.8).

1.2.1 Cell Wall

Survival of fungi involves hyphae that penetrate into living or dead tissues and absorb nutrients, spread into soil, or grow into air for dissemination of spores. This demands that the fungal protoplasm is encased in a cell wall that is flexible and yet of high mechanical strength and carries

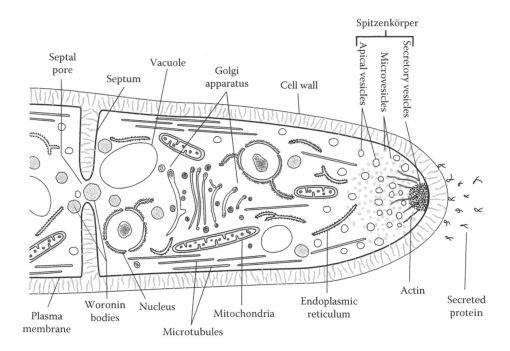

Figure 1.8 Diagram of cellular structures in apical cell.

with it a wide range of hydrolytic enzymes on its surface for digestion and utilization of exogenous insoluble organic matter (litter) for energy and biosynthesis. The cell wall not only determines the morphology of a fungal colony but is also involved in mating and interaction of fungi with their plant or animal hosts. A hypha must have adhesive properties and the ability to sense surface configurations for orienting direction of its growth (Chapter 6). Additionally a hypha must have a strategy for water retention such that it can develop the high turgor pressure needed for cell separation in litter and forcing entry. Hyphal cell wall architecture with hydrolytic enzymes attached to plasma membrane below a porous or semi-porous cell wall is thought to cleave linkages in the substratum as the hypha extends.

The cell wall is generally a multilayered structure (Figure 1.9) with an inner α-glucan layer (a glucan is a polymer of glucose), a middle β-glucan layer, and an outer layer of glycoprotein, in addition to chitin, which occurs in some cell walls. Howard and Aist (1979) reported that the hyphal apex of *Fusarium* exhibits a four-layered cell wall when a cryofixation method was used but exhibited a two-layered wall in the conventional chemical-fixed material, reminding us that knowledge of the structure of hyphae may improve with newer histochemical and microscopic techniques.

1.2.2 Glucan

Purified cell wall fragments, chiefly of *Phytophthora infestans,* of *Neurospora crassa* and of *Schizophyllum commune,* were extracted with hot water, with detergent (sodium dodecyl sulfate), or with hot hydrochloric acid or alkali (KOH or NaOH) and were characterized following successive digestion with specific polymer-degrading enzymes (chitinase, cellulase, laminarinase, pronase). The images of shadowed preparations viewed by electron microscopy provided evidence of a reticulum of chitin microfibrils (Bartnicki-Garcia, 1966; Hunsley and Burnett, 1970; Sietsma and Wessels, 1979). The hyphal apex is comprised of a more or less coaxial sequence of polymers:

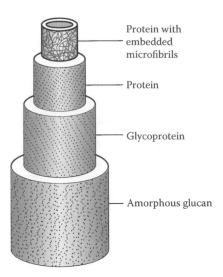

Protein with embedded microfibrils

Protein

Glycoprotein

Amorphous glucan

Figure 1.9 A diagram of hyphal wall layers (Adapted from Burnett, J.H., *Fundamentals of Mycology,* Edward Arnold, London, 1976, chap. 3.)

glucan outside, protein and chitin (or cellulose) within. Freeze-dried solubilized material was also studied by x-ray diffraction (Burnett, 1976; Schoffelmeer *et al.*, 1998; de Groot *et al.*, 2009). The yeast cell wall is bilayered, but most filamentous fungi have a third outermost electron-dense glycoprotein layer that binds lectin and accounts for 50–60% of the total mass of the wall, suggesting that glycoproteins form an external layer covering an inner layer composed of chitin and glucan. Since the glucan layer is removed by laminarinase treatment, it is considered to be principally composed of α-(1→3)- and α-(1→6)-linked anhydroglucose units. The cell walls of fungi that are now classified under Straminipila (see Appendix) have cellulose too. Many of the cell wall proteins are glycoside hydrolases or transglycosidases. These are covalently attached to the plasma membrane via glycosylphosphatidylinositol (GPI) linkages (Figure 1.10) and are postulated to be involved in constant remodeling of cell wall (de Groot *et al.*, 2009). This layer normally masks the inner electron-transparent layer of interwoven microfibrils of chitin. The polymers are apparently cross-linked, consistent with the hyphal wall withstanding enormous turgor pressure. The fact that chitin is present in fungal cell walls but is absent in animal cells suggests that anti-chitin compounds can be prepared for control of fungal infections of humans and animals.

The composition and structure of cell walls vary with age and environmental conditions. It is different in growing and nongrowing regions of the single hypha or when the hypha growing along an aqueous surface grows into air to form and discharge spores. A challenging question is how an insoluble, three-dimensional structure mainly composed of protein and polysaccharides is assembled external to the plasma membrane. Membrane preparations have shown β-(1,3)-glucanosyltransferase attaches to membrane through a glycosylphosphatidylinositol, suggesting that oligosaccharides pass through plasma membrane. At the plasma membrane, GPI-anchored enzymes transfer other polymers (chitin, galactomannan) onto the nonreducing ends of β-(1,3)-glucan, generating a branched structure.

Fungal colonies have three main types of mycelial morphology (Garnjobst and Tatum, 1967): (i) spreading, (ii) restricted button-shaped colonies, and (iii) colonial, consisting of roughly parallel hyphae. The changes in morphology of a colony are due to changes in its chemical composition, which in turn is due to altered enzyme levels leading to altered levels of glucan and chitin. The chemistry of the wall apparently varies in different regions of a single colony. Since

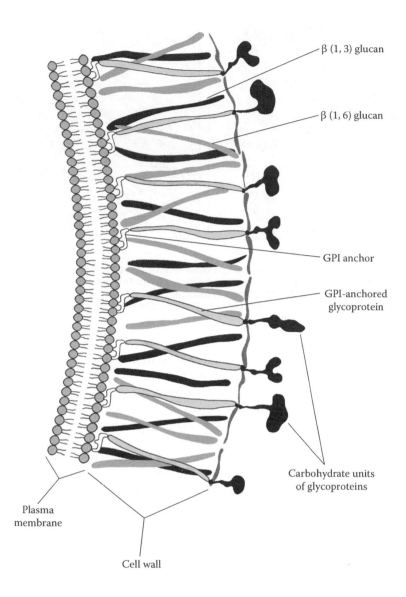

β (1, 3) glucan

β (1, 6) glucan

GPI anchor

GPI-anchored
glycoprotein

Carbohydrate units
of glycoproteins

Plasma
membrane

Cell wall

Figure 1.10 Diagram of hyphal cell wall.

a hypha is single-cell thick and the cell wall is in direct contact with the environment, breaking and remaking of walls must be a constant process. The cell wall is to be regarded as a highly dynamic structure.

1.2.3 Hydrophobin

Terrestrial fungi grow in tight contact with the substratum to perceive microscopic surface signals to orient the direction of their growth or to lower the water surface tension to be able to grow into air for formation and dissemination of spores (Talbot, 1997). The molecule that determines such growth of the hyphae is hydrophobin—a hydrophobic protein that is extractable by strong trifluoroacetic acid or formic acid. Hydrophobins have a high proportion of nonpolar amino acids, which allows them to self-aggregate to form a film that appears as bundles of mosaic rodlets in

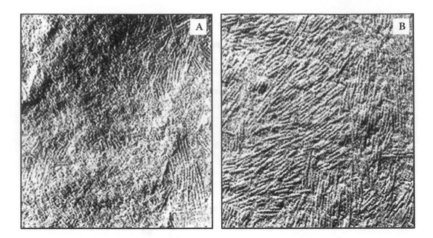

Figure 1.11 Hydrophobin. (A) Electron micrograph of surface of aerial hyphae of *Schizophyllum commune* after freeze-fracturing and shadowing showing layer of hydrophobin rodlets. (B) Electron micrograph of rodlets after drying solution of extracted hydrophobin. (Reprinted from *Trends Plant Sci* 1, J.G.H. Wessels, Fungal hydrophobins: Proteins that function at an interface, 9–15, 1996, with permission from Elsevier.)

electron micrographs (Figure 1.11). Hydrophobin was discovered by Joseph Wessels and his co-workers. It is estimated to constitute 10% of total proteins in fungi (Wessels, 1996).

Blondel and Turian (1960) were among the first to undertake studies of cytoplasmic organization within fungi. Using fixed hyphae of an aquatic mold, *Allomyces macrogynus*, it was demonstrated that RNAse-sensitive ribonucleoprotein particles are distributed in the cytoplasm. Other components demonstrated were nuclei, nucleolus, and mitochondria. In subsequent years with the use of cryofixation, live-cell imaging using markers to selectively tag the actin, the microtubule, the endoplasmic reticulum (ER), the mitochondria, or the vacuole have constructed a finer picture of hyphal compartment. A revolution is occurring in our knowledge of the fine structure of fungal cells, chiefly imaging of live cells by confocal microscopy. Table 1.1 summarizes information that has come using vital stains or by using transgene strains. The gene encoding the green fluorescent protein is fused to a gene whose protein product is to be determined, and the GFP-gene construct is transformed into a fungal host that is used for live-cell imaging. As an example, endocytosis in animal cells is an important process for signal transduction, reconstruction of cell polarity, acquisition of nutrition from cell exterior, and the uptake of the plasma membrane and membrane proteins into cells. Plasma membrane proteins and extracellular molecules are packaged into vesicles that are formed by invagination of the plasma membrane and thereafter transported to endosomes. Whether endocytosis occurs in the fungi is debated.

1.3 INTERNAL STRUCTURE

The picture that has emerged of the hyphal apex from examination of living and fixed material is diagrammed in Figure 1.8 in the form of a longitudinal section.

1.3.1 Microtubules and Actin Filaments

Just as the whole hypha is polarized, so are its cytoskeleton elements. Staining of hyphae with an anti-tubulin antibody bound to a fluorescent dye, fluorescein isothiocyanate (FITC), or with GFP-tagged tubulin revealed an extensive array of green-stained, long pipe-like structures, called

Table 1.1 Fine Structure of Hypha in Various Fungi

Subcellular Structure	Fungus	Method	Key Finding	Selected Reference(s)
Cell wall	*Neurospora crassa*	Fixation by OsO_4, staining by uranyl acetate.	Chitinous wall containing fibrils; endoplasmic reticulum associated with plasma membrane.	Shatkin and Tatum (1959)
	Schizophyllum commune, Phytophthora infestans	Pronase, chitinase, cellulose, laminarinase-treated cell walls shadowed and examined by electron microscopy (EM).	Presence of glucan and chitin. Mechanical rigidity due to microfibril component. Cellulose present in *P. infestans*.	Hunsley and Burnett (1970)
	Fusarium acuminatum	Freeze substitution (FS) and conventional glutaraldehyde (CS), staining by uranyl acetate, infiltration in epoxy resin before sectioning.	FS-fixed tip cell is 4-layered and 2-layered in CS. Total width 60 nm in FS.	Howard and Aist (1979)
	Yeasts	Various. Extraction of purified cell walls. Enzymatic and mass spectrometric analysis of solubilized products, electron microscopy.	Internal layer containing the shape-conferring, stress-bearing polysaccharides and a fibrillar outer layer consisting of galactomannoproteins that extend away from the cell surface.	Klis *et al.* (2006)
	Aspergillus nidulans	Extraction of purified cell walls by salt, mild alkali, mercaptoethanol, enzymes, and mass spectrometric analysis of solubilized products.	Cell wall contains α-glucan, β-(1,6)-glucan linked to β-(1,3)-glucan, chitin, and hydrophobins. GPI-linked proteins are the largest class of cell wall proteins.	De Groot *et al.* (2009)
Nucleus	*Allomyces macrogynus*	Transmission electron microscopy (TEM) of OsO_4 fixed cells.	Variable shape and size 1.5—4 μm. Double membrane pierced by pores.	Blondel and Turian (1960)
	Neurospora crassa	Confocal microscopy of live histone H1-GFP tagged hyphae.	Nuclei oval or pear-shaped, absent in first 20–25 μm tip. One spindle pole body, position of which indicative of the direction in which nucleus moves, and perhaps of the direction of elongation of hypha.	Freitag *et al.* (2004)
Nucleolus	*Allomyces macrogynus*	EM of OsO_4—fixed cells.	Dense granular.	Blondel and Turian (1959)
Mitochondria	*Allomyces macrogynus*	TEM of OsO_4/uranyl acetate-fixed cells.	Mostly elongate, double membrane with inner membrane (cristae) either penetrating deep or not, randomly distributed among ribonucleoprotein particles close to nucleus.	Blondel and Turian (1959)

Continued

Table 1.1 (Continued) **Fine Structure of Hypha in Various Fungi**

Subcellular Structure	Fungus	Method	Key Finding	Selected Reference(s)
Ribosomes	*Allomyces macrogynus*	TEM of OsO₄/uranyl acetate-fixed cells.	150 to 250 Å particles distributed in cytoplasm, giving it coarse appearance.	Blondel and Turian (1959)
	Fusarium acuminatum	Freeze substitution.	Cell wall 4-layered; cisternae and microtubules extend into apex, 70–90 nm spherical or 90–110 oblong vesicles.	Howard and Aist (1979)
	Sclerotium rolfsii	Freeze substitution.	Aligned on the outer surface of mitochondria.	Roberson and Fuller (1988)
Golgi	*Saprolegnia ferax*	TEM of freeze-substituted cells.	Up to 2.5 μm elongated vesicular structures adjacent to plasmalemma and apparently fusing with it.	Heath *et al.* (1985)
	Sclerotium rolfsii	TEM of freeze-substituted cells.	Golgi throughout apical region.	Roberson and Fuller (1988)
Vesicles	*Fusarium acuminatum*	TEM of freeze-substituted cells.	Spherical (70–90 nm).	Howard and Aist (1979)
	Sclerotium rolfsii	TEM of freeze-substituted cells.	Spitzenkörper is a dense mass of vesicles.	Roberson and Fuller (1988)
Microtubules	*Aspergillus nidulans*	Confocal microscopy of GFP–tubulin stained living hyphae.	Tip-high gradient of microtubule subunits sliding tipwards.	Sampson and Heath (2005)
	Sclerotium rolfsii		Microtubules traverse Spitzenkörper, terminating at plasmalemma.	Roberson and Fuller (1998)
	Fusarium acuminatum	Compared freeze substituted and conventional chemical fixation.	Microtubules in 1 μm tip region in freeze-substituted hypha not seen in conventionally fixed material. Nonetheless, microtubule cytoskeleton is important for nuclear positioning and vesicle delivery; hence for polarized growth of hypha.	Howard and Aist (1979)
Actin	*Ashbya gossypii, Aspergillus nidulans, Fusarium acuminatum, Sclerotium rolfsii, Ustilago maydis*	Staining with rhodamine phalloidin; immunofluorescence microscopy; TEM immunogold labeling; computer-aided, three-dimensional reconstruction analysis and immunoblot identification.	Actin arrays below the apex. Latrunculin B disrupts actin polymerization, causing tip swelling. Postulated to play role in exocytosis, tip morphogenesis.	Roberson and Fuller (1988) Gupta and Heath (1997)

	Species	Method	Finding	Reference
Spitzenkörper/Vacuole	Sclerotium rolfsii	Freeze-substituted hyphae brought to room temp, embedded in epon-araldite, sectioned, stained with uranyl acetate and lead acetate.	Cluster of single unit-membrane-bound vesicles associated with Golgi cisternae and microtubules.	Roberson and Fuller (1998)
Vacuole	Gigaspora margarita Pisolithus sp. (ectomycorrhizal basidiomycete)	Confocal microscopy of live hypha stained with Oregon Green 488 carboxylic acid diacetate.	Longitudinally oriented motile tubular vacuoles. Suggested to play a role in longitudinal transport of P-containing substances via bulk flow.	Ashford (1998) Uetake et al. (2002)
Endoplasmic reticulum	Fusarium acuminatum	TEM of freeze-substituted hypha.	Smooth or rough.	Howard and Aist (1979)
	Aspergillus oryzae Sclerotium rolfsii	BipA-EGFP recombinant strain. Three-dimensional analyses using confocal microscopy.	A network of ER with increasing content toward the hyphal tip. Discontinuity of the ER network distribution across the first septum.	Srinivasan et al. (1996), Roberson and Fuller (1988)
	Aspergillus oryzae	ER network distribution specialized for apical growth was reconstructed in the intrahyphal hypha.		Maruyama et al. (2006)
Peroxisome	Aspergillus oryzae Yeasts	EM of glutaraldehyde/OsO$_4$ fixed hypha.	Abundant in fungus grown on oleate or methanol.	Escano et al. (2009)
Free calcium	Aspergillus nidulans	Confocal microscopy of recombinant GFP–calmodulin strain.	Concentrated at extreme hyphal apex, co-localized with Spitzenkörper; depletion using anti-CaM and anti-motor protein KipA caused defects in septation and anti-polarized growth; tip localization depends on actin filaments and motor protein.	Chen et al. (2010)
Polarisome	Aspergillus niger Aspergillus nidulans Neurospora crassa	Confocal microscopy of recombinant yeast SPA-2-GFP.	Co-localized with Spitzenkörper.	Meyer et al. (2008) Araujo-Palomares et al. (2009)

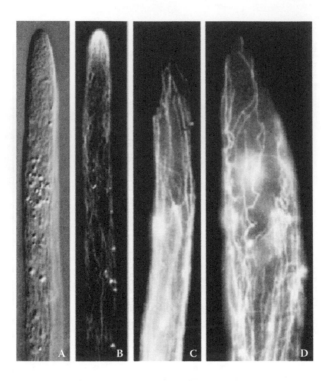

Figure 1.12 Hypha of *Saprolegnia ferax*. (A) Differential interference contrast image of tip. (B) Stained with rhodamine phalloidin showing actin concentrated at the tip. (C) and (D) Stained with anti-tubulin to show subapical microtubules. Few microtubules extend to the tip. (Reprinted from *Fung. Genet. Biol.* 30, Heath, I.B., Gupta, G., and Bai, S. Plasma membrane-adjacent actin filaments, but not microtubules, are essential for both polarization and hyphal tip morphogenesis in *Saprolegnia ferax* and *Neurospora crassa*, 45–62, 2000, with permission from Elsevier.)

microtubules, generally arranged parallel to the long axis of hyphae. Another cytoskeleton element visualized by staining with rhodamine phalloidin is a microfilament composed of the protein actin (Figure 1.12). Cytoskeleton in fungi is highly dynamic, assembling and disassembling in response to changing cellular needs, and functions as a track for migration and positioning of organelles (Riquelme *et al.*, 1998; Mouriño-Perez *et al.*, 2006).

1.3.2 Spitzenkörper and Polarisome

An apical cluster of small vesicles with no clear boundary has been discovered directly beneath the plasma membrane of the hyphal tip in living hyphae by phase contrast microscopy and vital staining with a membrane-selective fluorescent dye. This apical body is called *Spitzenkörper* in German. Video microscopy and image analysis of living hyphae showed a close correlation of Spitzenkörper trajectory and the direction of growth of the hypha (Figure 1.13) (Riquelme *et al.*, 1998). From its position and behavior, it has been suggested that Spitzenkörper is a collection center of vesicles containing enzymes and preformed polysaccharide precursors for cell wall synthesis, allowing their localized delivery to the hyphal apex (Bartnicki-Garcia *et al.*, 1995). Once the Spitzenkörper has discharged its contents, a new Spitzenkörper is reformed. This cycle is consistent with growth of fungal hypha occurring in pulses (Lopez-Franco *et al.*, 1994). Further, the association of the Spitzenkörper with a meshwork of microtubules and microfilaments suggests that its polarized trajectory is determined by the growing scaffolding of microtubules. Support for this view is the observation that benomyl, an inhibitor of microtubule assembly, markedly disturbed

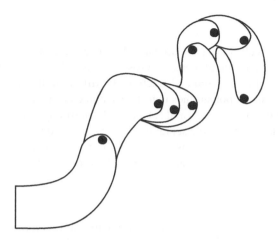

Figure 1.13 Tracings of video-images of Spitzenkörper trajectory in hypha during 9-min interval. (From Bartnicki-Garcia, 2002.)

the directionality of hyphal growth of wild-type *Neurospora crassa*, whereas a benomyl-resistant mutant was not affected.

The evidence for Spitzenkörper as a collection discharge center for vesicles containing membrane and cell wall precursors is circumstantial. However, their polarized delivery and insertion into the plasma membrane at the tip can explain the generation of hyphal shape and polarity. The vesicle membrane and plasma membrane at the tip may have specific proteins that tether the two communicating membranes for docking and fusion, as postulated in animal cells. At the core of pairing between the fusions of the vesicle with its target membranes lies an interaction between the homologous vesicle and target membrane proteins, called v- and t-SNARES, discovered in animal cells. Using specific antibodies, a tip-high gradient of t-SNARES in *Neurospora* hypha has been demonstrated (Gupta and Heath, 2002). The location of Spitzenkörper at the tip and of SNARES may together bolster rapid apical growth of the hyphal tip.

Another structure at the apex of the hypha, the polarisome, co-localizes with Spitzenkörper. Polarisome is thought to mediate the nucleation of actin cables. The effects on hyphal morphogenetic and growth patterns, using laser manipulation of cell components, is modulated by the balance between internal (cellular) parameters—concentration of vesicles, diameter of the filaments, growth rate, and turgor pressure—as well as external (laser) parameters—e.g., power, focus, and velocity of the trap movement. The manipulation of the apical cluster of secretory vesicles in order to disturb the pattern of tip growth and branch formation could provide a better understanding of the tip-growth-related processes and how these processes are regulated.

1.3.3 Calcium

As in the other tip-extending cells—for example, the plant root hair, pollen tube, rhizoid cell of the alga *Fucus*—the fungal hypha too contains a tip-high gradient of calcium ions (Levina *et al.*, 1995). In hyphae of *Saprolegnia ferax* (an aquatic mold) and *Neurospora crassa* (a terrestrial mold), the cytosolic calcium was measured by ratio-imaging emission intensities of Ca^{2+}-sensitive fluorescent dyes fluo-3 and Fura Red by confocal microscopy (Hyde and Heath, 1997; Levina *et al.*, 1995; Silverman-Gavrila and Lew, 2000, 2001). Fluorescence emission was localized in the 10-μm terminal region of the tip in the growing hyphae but not in the nongrowing hyphae. Hyphal elongation was inhibited by microinjection of Ca^{2+} chelators, suggesting that the tip-high calcium is required for tip growth.

1.3.4 Vacuole

Easily seen in the older cells of hyphae are vacuoles. These membrane-bound, mostly spherical organelles have been considered analogous to lysosomes of animal cells (Klionsky *et al.*, 1990). Vacuoles have been isolated from lysed cell fractions by differential centrifugation. They contain basic amino acids, nitrogen-rich polyamines that stabilize nucleic acids, and polyphosphate. Hence vacuoles serve as storage organelles for N and P. Confocal microscopy of living mycorrhizal hyphae has shown tubular vacuoles spanning the septal pore over an extended length, forming an interconnected network (Darrah *et al.*, 2006). This might suit the role of mycorrhizal hyphae as conduits interconnecting roots of different plant species over fairly long distances.

1.3.5 Septa and Woronin Body

Evenly spaced septa divide the hypha into compartments. In *A. nidulans* the apical cell (distance from the hyphal tip to the closest septum) is variable, whereas the intercalary compartments have a uniform length of 38 µm (Wolkow *et al.*, 1996), suggesting that some cellular mechanisms determine the site of placement of the septa. Septa are formed by centripetal growth of the cell wall and have a perforation through which cytoplasmic organelles, including nuclei, can pass. In Ascomycotina and Basidiomycotina, an electron-dense protein-body, called a Woronin body (a peroxisome-derived dense core), is present on either side of the septa, suggesting that these function as a plug for the septal pore. By regulated closing or opening of the septal pores, the movement of protoplasm may be redirected to any region in the mycelium, changing the direction of its extension for productive exploration of nutrient-rich surrounding areas. Closing of septal pores may reduce leakage when the hypha is ruptured. Another view is that septa contribute to the rigidity of hypha as the rungs in a ladder; for example, the *cross-wall-less* (*cwl*) mutants of *Neurospora crassa* are very weak, probably because the contents of a long hyphal cell ooze out as there are no cross walls to prevent "bleeding." Yet another function of septa could be to provide a reduced environment for redox-sensitive enzymes. In some higher fungi, such as the agarics, the septal pore is of elaborate construction with thickened sidewalls. Septa may allow spatial regulation of branch sites and in development of reproductive structures by redistribution of nutrients. By sealing septal pores, the translocation of nutrients can be redirected toward a developing fruit body. For example, the spores of *Coprinus sterquilinus* that fall on vegetation are taken in by herbivores and voided with dung, where they germinate, and subsequently the hyphae fuse and form a single three-dimensional interconnected mycelial network that operationally acts as a unit for maximizing extraction of nutrients from the substrate and their translocation to a developing basidiocarp (fruiting body). "Cooperation by fusing mycelia, rather than competition by individual colonies, is the general feature of fungal growth in nature" (Burnett, 1976).

1.4 NETWORKING

1.4.1 Branching

Branches arise subapically in acropetal succession, generally in close proximity to the septum. Hyphae tend to avoid the neighboring hyphae and grow outwardly from the center. The pattern of branching has been compared with the pattern of branching in a Christmas tree, with a main hypha and a series of branches borne alternately in two dimensions, suggesting a marked apical dominance. As the leading hyphae diverge from one another, apical dominance becomes weaker. Branching allows the hyphae to effectively colonize and exploit the surrounding area.

The mechanisms involved in the cellular organization of hypha are beginning to be studied by genetic methods using conditional mutants. Mutants with defects in branching are easily recognized because, in contrast to the wild type, they form compact (colonial) colonies. The *cot* (*colonial* temperature sensitive) mutant was identified as having a temperature-sensitive defect in hyphal growth: It produces colonial (tight, button-shaped) colonies at 32°C, but normal wild-type colonies at 25°C. In a nuclear distribution mutant of *Aspergillus nidulans*, branching intensity—which was determined as hyphal growth unit length (the ratio between the total hyphal length and the number of tips)—was higher in nutrient-rich medium than in nutrient-limiting medium (Dynesen and Nielsen, 2003). The cytoplasmic volume per nucleus is unaffected by substrate availability, suggesting that branching and nuclear division are co-coordinated to maintain this ratio. An ecological implication of this is that in nature the fungus can explore its surroundings with thin, sparsely branched hyphae (minimum formation of biomass).

1.4.2 Hyphal Fusion

As the different parts of a mycelium extend, neighboring hyphae become physically interconnected by cell fusion, bringing all branches into a cytoplasmic continuity. The process of hyphal fusion has been investigated in living hyphae by time-lapse imaging (Hickey *et al.*, 2002). Short lateral branches arise and redirect their growth toward each other by a remote sensing mechanism and undergo tip-to-tip hyphal self-fusion, but the nature of the signaling molecules is not yet known. Since the Spitzenkörper is observed where hyphae meet, this structure presumably delivers cell adhesion molecules and enzymes required for dissolution of cell walls at the point of contact of the hyphal tips to allow for cytoplasmic continuity (Figure 1.14). Formation of interconnected hyphae

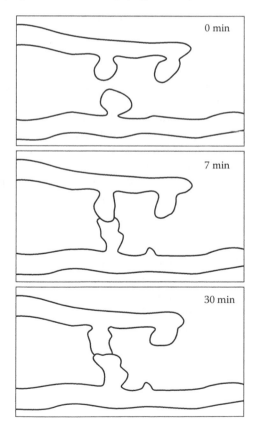

Figure 1.14 Hyphal fusion. (Redrawn from Hickey, P.C., Jacobson, D.J., Read, N.J., and Glass, L., *Fung. Genet. Biol.* 37, 109–119, 2002.)

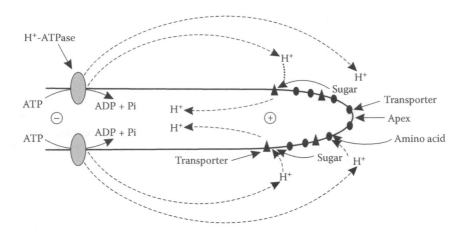

Figure 1.15 Diagram of proton pump and symport in hypha. (Based on Harold, F.M., *Fung. Genet. Biol.* 27, 128–133, 1999.)

allows the mycelium to forage in space and in time. By opening or plugging the septal pores, the mycelium redirects growth toward a nutritionally rich area; the cytoplasmic streaming enables the mycelium to store nutrients for translocation and packaging in spores.

1.5 MAIN FUNCTIONS

1.5.1 Nutrient Uptake

In 1974, a miniaturized microelectrode was used to measure voltage along germinating zygotes of brown algae, growing pollen tubes and root hair cells (Jaffe and Nuccitelli, 1974). It was found that these cells generate a current—a longitudinal pH gradient—in the surrounding medium due to an extracellular flow of proton (positive charge) that enters the tip and exits from behind. This observation was extended to fungi. Measurement of the pH profile along the hypha of *Neurospora* showed that the apical zone (200 to 300 µM) was relatively alkaline and the distal zone relatively acidic (Kropf *et al.*, 1984; Harold, 1999). The fungal hypha is electrically polarized because the energy of a proton gradient provided by the plasma membrane H$^+$-ATPase extrudes protons (H$^+$ ions), produced because of oxidative metabolism of glucose, from the rear region of the hypha, making this region of the hypha relatively acidic (Figure 1.15). The extruded protons re-enter from the apex, making it relatively alkaline. The hypha thus drives a current of protons through itself with an inward flow of protons from the tip and their efflux from the distal region.

The hypha secretes a variety of enzymes that break down the polymeric constituents of the substratum into simple forms by means of extracellularly secreted enzymes. A plasma membrane H$^+$-ATPase pumps out protons derived from the metabolism of glucose. The energy of a proton gradient is coupled to the uptake (symport) of ions, sugars, and amino acids. The rapid uptake of solubilized nutrients is the basis of the absorptive mode of nutrition of fungi. The spatial separation of H$^+$ pump and nutrient transporters suggests that the hypha is not only cytologically but also physiologically polarized.

1.5.2 Protein Secretion

Insoluble substratum, such as the plant litter on which fungi grow, generally contains polymeric cellulose, hemicellulose, starch, and lignin. The enzymes required for depolymerizing these

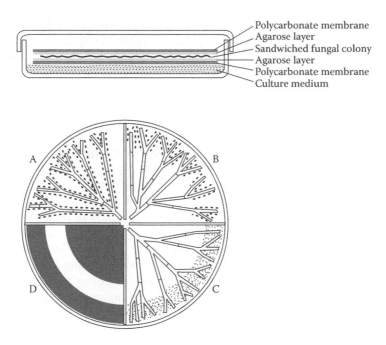

Figure 1.16   Protein secretion at growing hyphal apices. Above, diagram of method of culturing *Aspergillus niger* as sandwiched colony between two perforated polycarbonate membranes placed on starch medium. Below, (A) Diagram of autoradiograph image of mycelium for monitoring protein synthesis. (B) Diagram of autoradiograph showing chitin (cell wall) synthesis. (C) Diagram of immunogold labeling showing site of glucoamylase secretion. (D) Diagram of zone of starch degradation by I$_2$-KI staining. (Based on data from Wösten, H.A.B., Moukha, S.M. Sietsma, J.H., and Wessels, J.G., *J. Gen. Microbiol*. 137, 2017–2023, 1991.)

polymers are secreted from the hyphal tips. This was inferred from an experimental setup in which a colony of *Aspergillus niger* was grown sandwiched between two perforated polycarbonate membranes placed on a starch medium (Wösten *et al*., 1991). The sandwiched fungal colony could be lifted and exposed to labeled compounds, *N*-acetyl [^{14}C] glucosamine or [^{35}S] sulfate, washed and used for imaging by autoradiography (Figure 1.16) for monitoring, respectively, the site of chitin and new protein synthesis. This setup allowed simultaneous visualization of hyphal growth and the site of secretion of glucoamylase by immunogold labeling, as well as monitoring the zone of starch-degrading activity by I$_2$-KI staining. The results showed that cell wall synthesis is limited to the growing edge of hypha, whereas protein synthesis occurred throughout the hypha. However, Western blotting showed that secretion of glucoamylase occurred only from the hyphal apices. This suggests that the apical region is porous compared to the rest of the hypha, and that breakdown of the polymeric compounds is closely connected with apical growth.

1.6 MORPHOGENESIS

On an agar surface the wild-type *Neurospora* grows as a spreading mycelium, whereas several single-gene mutants showed bizarre morphologies. In each case biochemical analysis in Edward L. Tatum's laboratory correlated altered morphology to an increase in the level of glucan, a decrease in β-(1,3)-glucan, or both, suggesting that the morphological changes were defective in enzymes of carbohydrate metabolism, revealing that the cell wall is the principal determinant of the fungus's morphology (see Figure 1.11). In particular exogenous sorbose greatly restricted mycelium spread on colony morphology, with colonies appearing as "buttons" (Mishra and Tatum, 1972). Sorbose effect is due to inhibition of activities of β-(1,3)-glucan and glycogen synthetase. A sorbose plating medium was devised and used in counting survival rate of conidia after UV irradiation in mutation and genetic experiments. The main conclusion from these observations is that the cell wall/plasma membrane has a crucial role in morphogenesis.

1.6.1 Hyphal Differentiation

Within a single radially extending mycelial colony, one zone of hyphal differentiation may occur, even in cultures grown in liquid shake cultures, as shown in Figure 1.17. This is a section of pellet of the penicillin-producing fungus stained with cresol violet (Wittler *et al.*, 1986). The high content of cytoplasm in the outer L1 layer suggests that mycelia grow mainly in this region. At the inner boundary of L1 autolysis occurs with loss of cytoplasm. The next layer (L2) is not well stained, since the hyphae have already lost their cytoplasm and the remaining cell walls are slowly dissolved by exoenzymes of the fungus. The next layer (L3) again contains cytoplasm. In the center of the pellet (L4), the hyphae structure has already been completely dissolved. Using a needle sensor (micro-probe), it was shown that dissolved oxygen tension in the center was very low. The differentiation of mycelial layers within a pellet suggests that complex factors of molecular diffusion, convective flow, and transfer mechanisms are involved in colony differentiation.

The radiating system of hyphae in a single mycelial colony may actually comprise different hyphal types (Vinck *et al.*, 2005). When a colony of *Aspergillus niger* that had been transformed with a green fluorescent protein (GFP)–glucoamylase (GLA) was grown on an agar medium, nearly half of the hyphae at the periphery fluoresced strongly, whereas the other hyphae containing GLA::GFP fusion protein fluoresced less, showing that hyphae within the colony differed in the expression of the gene. This observation has implications in the use of fungal hyphae as biofactories wherein the

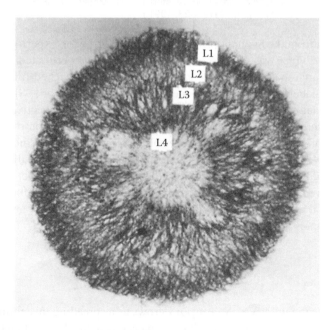

Figure 1.17 A section of cutting through a pellet of *Penicillium chrysogenum* grown in aerated culture and stained with cresol violet showing four layers (light microscopy). (From Wittler, R., Baumgartl, H., Lübbers, D.W., and Schügerl, K., *Biotechnol. Bioeng.* 28, 1024–1036, 1986. Copyright Wiley-VCH Verlag Gmbh & Co. KGaA. Reproduced with permission.)

Figure 1.18 Rhizomorph of *Armillaria ostoyae* on trunk of white ash *Fraxinus americana*. (Courtesy of Nicholas J. Brazee, University of Massachusetts.)

mycelium under fermentation conditions grows as pellets and certain genes may be activated only in some parts of the colony.

1.6.2 Rhizomorph

A mycelial habit of some mycorrhizal fungi belonging to the phylum Basidiomycotina is the formation of cords or strands by aggregation and intertwining of thousands of hyphae into a structure, the rhizomorph. The rhizomorph grows beneath the bark of a tree, tapping xylem vascular tissue for water (Figure 1.18), and the phloem vascular tissue for organic compounds synthesized by leaves serves as a device for conducting nutrients from a parasitized living tree. A rhizomorph extends over considerable distances across inhospitable terrain until it finds a new tree. Once again the process is initiated as the rhizomorph spreads up the trunk, beneath the bark, to tap the phloem of the victim. This habit of mycelial growth is undoubtedly the chief reason for the large spread of rhizomorph-forming fungi being among the largest and oldest living organisms.

1.6.3 Mushroom Fruiting Body

Common sights on litter and on decaying wood are the fruiting bodies of mushrooms (see Figure 4.1). A basidiomycete fruiting body, over 5.5 feet long, 284 kg in 1996, growing in a corner of the Royal Botanic Gardens, England, is mentioned in *Guinness World Records*. The development of the fruiting body is made possible by the synchronized growth toward each other of thousands of hyphae, their branching and interweaving, and their thickening and gluing together by mucilaginous β-1→6, β-1→3-linked glucan. Three principal types of hyphae involved in the construction of the hard fruiting bodies of enormous strength are thick-walled, unbranched, aseptate hyphae; thick-walled branched hyphae; and those with binding (gluing) properties. Synchronization of the activities of hyphae is a feature of fungal morphogenesis, but "the how" of this process has not been studied.

1.7 AUTOPHAGY

In *Penicillium chrysogenum* rapid changes in the nucleic acid and protein content of the fungus occurred when glucose feed was stopped. After five days the contents were about 25% of their original values, at which time electron microscopy revealed that though the older hyphae became empty; nevertheless the hyphal tip contained cytoplasm. Whereas nuclei in hyphal tips stained densely, the nuclei in the older hyphal compartments stained poorly. These degenerative changes were attributed to autophagy—the turnover of multilayered cell walls, of mitochondria, of ribosomes, and of the supernumerary nuclei. This process is mediated by lysosome-like structures (autophagosomes) for deriving C, N, and P under starvation conditions with regrowth of hyphae within some of the older hyphae ("intrahyphal hyphae"). The growth of some hyphae (or compartments) at the expense of others is a fungal strategy of keeping the tip alive and foraging even when sources of nutrients are limited.

1.8 CONCLUDING REMARKS

The interconnecting, intercommunicating network of fungal hyphae is a highly effective device for long-distance exploration and exploitation of the substratum for absorption of nutrients. Several questions pertaining to hyphal growth have come to the fore. How is the protoplasm drawn from the older compartments toward the tip cell, keeping it alive? How are the sites of initiation of branches determined? What mechanisms determine that cytokinesis in yeast-like fungi results in complete separation of mother and daughter cells but septation in filamentous fungi leads to generation of individual compartments within the hypha? How does the cell wall extend, and what determines the apical secretion of proteins? How do hyphae perceive environmental cues? How is the activity of thousands of hyphae synchronized that must grow toward each other for construction of a mushroom fruiting body? In the subapical region, is endocytosis of the plasma membrane required in the maintenance of polarized hyphal growth?

REFERENCES

Adams, D. J. (2004). Fungal cell wall chitinases and glucanases. *Microbiology* 150:2029–2035.

Allaway, W. G. and Ashford, A. E. (2001). Motile tubular vacuoles in extramatrical mycelium and sheath hyphae of ectomycorrhizal systems. *Protoplasma* 215:218–225.

Araujo-Palomares, C. L., Riquelme, M., and Castro-Longoria, E. (2009). The polarisome component SPA-2 localizes at the apex of *Neurospora crassa* and partially colocalizes with the Spitzenkörper. *Fung. Genet. Biol.* 46:551–563.

Ashford, A. E. (1998). Dynamic pleiomorphic vacuole systems: Are they endosomes and transport compartments in fungal hyphae? *Adv. Bot. Res.* 28:119–159.

Bae, T.-S. and Knudsen, G. R. (2000). Cotransformation of *Trichoderma harzianum* with β-glucuronidase and green fluorescent protein genes provides a useful tool for monitoring fungal growth and activity in natural soils. *Appl. Environ. Microbiol.* 66:810–815.

Bartnicki-Garcia, S., Bartnicki, D. D., Gierz, G., Lopez-Franco, R., and Bracker, C. E. (1995). Evidence that Spitzenkörper behavior determines the shape of a fungal hypha: A test of the hyphoid model. *Exp. Mycol.* 19:153–159.

Bartnicki-Garcia, S. (1966). Chemistry of hyphal walls of *Phytophthora*. *J. Gen. Microbiol.* 42:57–69.

Blondel, B. and Turian, G. (1960). Relation between basophilia and fine structure of cytoplasm in the fungus *Allomyces macrogynus* Em. *J. Biophys. Biochem. Cytol.* 7:127–133 (with 57 plates).

Boddy, L. (1999). Saprotrophic cord-forming fungi: Meeting the challenge of heterogeneous environments. *Mycologia* 91:13–32.

Brasier, C. M. (1984). Inter-mycelial recognition systems in *Ceratocystis ulmi*. In: D. H. Jennings and A. D. M. Rayner, eds. *The Ecology and Physiology of the Fungal Mycelium*. Cambridge University Press, Cambridge, pp. 451–497.

Brown, D. H., Jr., Giusani, A. D., Chen, X., and Kumamoto, C. A. (1999). Filamentous growth of *Candida albicans* in response to physical environmental cues and its regulation by the unique *CZF1* gene. *Mol. Microbiol.* 34:651–662.

Burnett, J. H. (1976). *Fundamentals of Mycology*. Edward Arnold, London, chap. 3.

Carlile, M. J. (1995). The success of hypha and mycelium. In: N. A. R. Gow and G. M. Gadd, eds., *The Growing Fungus*. Chapman and Hall, London, pp. 3–19.

Chen, S., Song, Y., Cao, J., Wang, G., Wei, H., Xu, X., and Lu, L. (2010). Localization and function of calmodulin in live-cells of *Aspergillus nidulans*. *Fung. Genet. Biol.* 47:268–278.

Darrah, P. R., Tlalka, M., Ashford, A., Watkinson, S. C., and Fricker, M. D. (2006). The vacuole system is a significant intracellular pathway for longitudinal solute transport in basidiomycete fungi. *Euk. Cell* 5:1111–1125.

Delgado-Álvarez, D. L., Callejas-Negrete, O. A., Gómez, N., Freitag, M., Roberson, R. W., Smith, L. G., and Mouriño-Pérez, R. R. (2010). Visualization of F-actin localization and dynamics with live cell markers in *Neurospora crassa*. *Fung. Genet. Biol.*, doi: 10.1016/j.fgb.2010.03.004.

de Groot, P. W. J., Brandt, B. W., Horiuchi, H., Ram, A. F. J., de Koster, C. J., and Klis, F. M. (2009). Comprehensive genomic analysis of cell wall genes in *Aspergillus nidulans*. *Fung. Genet. Biol.* 46:S72–S81.

Dynesen, J. and Nielsen, J. (2003). Branching is coordinated with mitosis in growing hyphae of *Aspergillus nidulans*. *Fung. Genet. Biol.* 40:15–24.

Escaño, C. S., Jurvaadi, P. R., Jin, F. J. *et al.* (2009). Disruption of the *Aopex 11-1* gene involved in peroxisome proliferation leads to impaired Woronin body formation in *Aspergillus oryzae*. *Euk. Cell* 8: 296–305.

Fischer-Parton, S., Parton, R. M., Hickey, P. C., Dijksterhuis, J., Atkinson, H. A., and Read, N. D. (2000). Confocal microscopy of FM-4-64 as a tool for analysing endocytosis and vesicle trafficking in living fungal hyphae. *J. Microscopy* 198:246–259.

Freitag, M., Hickey, P. C., Raju, N. B., Selker, E. U., and Read, N. D. (2004). GFP as a tool to analyze the organization, dynamics and function of nuclei and microtubules in *Neurospora crassa*. *Fung. Genet. Biol.* 41:897–910.

Garnjobst, L. and Tatum, E. L. (1967). A survey of new morphological mutants in *Neurospora crassa*. *Genetics* 57:579–604.

Gooday, G. W. and Gow, N. A. R. (1990). Enzymology of tip growth in fungi. In: I. B. Heath, ed., *Tip Growth in Plant and Fungal Cells*. Academic Press, New York, pp. 31–58.

Gupta, G. D. and Heath, I. B. (1997). Actin disruption by Latrunculin B causes turgor-related changes in tip growth of *Saprolegnia ferax* hyphae. *Fung. Genet. Biol.* 21:64–75.

Gupta, G. D. and Heath, I. B. (2000). Tip-high gradient of a putative plasma membrane SNARE approximates the exocytotic gradient in hyphal apices of the fungus *Neurospora crassa*. *Fung. Genet. Biol.* 29:187–199.

Gupta, G. D. and Heath, I. B. (2002). Predicting the distribution, conservation, and functions of SNAREs and related proteins in fungi. *Fung. Genet. Biol.* 36:1–21.

Hamasaki, M., Noda, T., Baba, M., and Ohsumi, Y. (2005). Starvation triggers the delivery of the endoplasmic reticulum to the vacuole via autophagy in yeast. *Traffic* 6:56–65.

Harispe, L., Portela, C., Scazzocchio, C., Peñalva, M. A., and Gorfinkiel, L. (2008). Ras GTPase-activating protein regulation of actin cytoskeleton and hyphal polarity in *Aspergillus nidulans*. *Euk. Cell* 7:141–153.

Harman, G. E. (2006). Overview of mechanisms and uses of *Trichoderma* spp. *Phytopathology* 96: 190–194.

Harold, F. M. (1999). In pursuit of the whole hypha. *Fung. Genet. Biol.* 27:128–133.

Harris, S. D. (2008). Branching of fungal hyphae: Regulation, mechanisms and comparison with other branching systems. *Mycologia* 100:823–832.

Harris, S. D. and Momany, M. (2004). Polarity in filamentous fungi: Moving beyond the yeast paradigm. *Fung. Genet. Biol.* 41:391–400.

Heath, I. B., Gupta, G., and Bai, S. (2000). Plasma membrane-adjacent actin filaments, but not microtubules, are essential for both polarization and hyphal tip morphogenesis in *Saprolegnia ferax* and *Neurospora crassa*. *Fung. Genet. Biol.* 30:45–62.

Heath, I. B. and Kaminskyz, S. G. W. (1989). The organization of tip-growth related organelles and microtubules revealed by quantitative analysis of freeze-substituted oomycete hyphae. *J. Cell Sci.* 93:41–52.

Heath, I. B., Rethoret, K., Arsenault, A. L., and Ottensmeyer, F. P. (1985). Improved preservation of the form and contents of wall vesicles and the Golgi apparatus in freeze substituted hyphae of *Saprolegnia*. *Protoplasma* 128:81–93.

Hickey, P. C., Jacobson, D. J., Read, N. J., and Glass, L. (2002). Live-cell imaging of vegetative hyphal fusion in *Neurospora crassa*. *Fung. Genet. Biol.* 37:109–119.

Higuchi, Y., Nakahama, T., Shoji, J.-Y., Arioka, M., and Kitamoto, K. (2006). Visualization of the endocytic pathway in the filamentous fungus *Aspergillus oryzae* using an EGFP-fused plasma membrane protein. *Biochem. Biophys. Res. Com.* 340:784–791.

Horio, T. and Oakley, B. R. (2005). The role of microtubules in rapid hyphal tip growth of *Aspergillus nidulans*. *Mol. Biol. Cell.* 16:918–926.

Howard, R. J. and Aist, J. R. (1979). Hyphal tip cell ultrastructure of the fungus *Fusarium*: Improved preservation by freeze-substitution. *J. Ultrastructural Res.* 66:224–234.

Hunsley, D. and Burnett, J. H. (1970). The ultrastructural architecture of the walls of some hyphal fungi. *J. Gen Microbiol.* 62:203–221.

Hyde, G. J. and Heath, I. B. (1997). Ca^{2+} gradients in hyphae and branches of *Saprolegnia ferax*. *Fung. Genet Biol.* 21:238–251.

Ingold, C. T. and Hudson, H. J. (1993). *The Biology of Fungi*. Chapman and Hall, London.

Jaffe, L. F. and Nuccitelli, R. (1974). An ultrasensitive vibrating probe for measuring steady electrical currents. *J. Cell Biol.* 63:614–628.

Jedd, G. and Chua, N. (2000). A new self-assembled peroxisomal vesicle required for efficient resealing of the plasma membrane. *Nature Cell Biol.* 2:226–231.

Klionsky, D. J., Herman, P. K., and Emr, S. D. (1990). The fungal vacuole: Composition, function, and biogenesis. *Microbiol. Rev.* 54:266–292.

Klis, F. M.., Boorsma, A., and De Groot, P. W. J (2006). Cell wall construction in *Saccharomyces cerevisiae*. *Yeast* 23: 185–202.

Kropf, D. L., Caldwell, J. H., Gow, N. A. R., and Harold, F. M. (1984). Transcellular ion currents in the water mold Achlya: Amino acid proton symport as a mechanism of current entry. *J. Cell Biol.* 99:486–496.

Levina, N. N., Lew, R. R., Hyde, G. L., and Heath, I. B. (1995). The roles of Ca^{2+} and plasma membrane ion channels in hyphal tip growth of *Neurospora crassa*. *J. Cell Sci.* 108:3405–3417.

Liu, F., Ng, S. K., Lu, Y., Low, W., Lai, J., and Judd, G. (2008). Making two organelles from one: Woronin body biogenesis by peroxisomal protein sorting. *J. Cell Biol.* 180:325–339.

Lopez-Franco, R., Bartnicki-Garcia, S., and Bracker, C. E. (1994). Pulsed growth of fungal hyphal tips. *Proc. Natl. Acad. Sci. USA* 91:12228–12232.

Ma, H., Snook, L. A., Kaminskyj, S. G. W., and Dahms, T. E. S. (2005). Surface ultrastructure and elasticity in growing tips and mature regions of Aspergillus hyphae describe wall maturation. *Microbiology* 151:3679–3688.

Maheshwari, R. (2005). Nuclear behavior in fungal hyphae. *FEMS Microbiol. Letts.* 249:7–14.

Markham, P. and Collinge, A. J. (1987). Woronin bodies of filamentous fungi. *FEMS Microbiol. Rev.* 46:1–11.

Maruyama, J.-I. and Kitamoto, K. (2007). Differential distribution of the endoplasmic reticulum network in filamentous fungi. *FEMS Microbiology Let.*272: 1–7

Meyer, V., Arentshorst, M., van den Hondel, C. A., and Ram, A. F. (2008). The polarisome component SpaA localizes to hyphal tips of *Aspergillus niger* and is important for polar growth. *Fung. Genet. Biol.* 45:152–164.

Mishra, N. C. and Tatum, E. L. (1972). Effects of L-sorbose in polysaccharide synthetase of Neurospora. *Proc. Natl. Acad. Sci. USA* 69:313–317.

Momany, M., Richardson, E. A., Van Sickle, C., and Jedd, G. (2002). Mapping Woronin body position in *Aspergillus nidulans*. *Mycologia* 94:260–266.

Moore, D. (1995). Tissue formation. In: N. A. R. Gow and G. M. Gadd, eds., *The Growing Fungus*. Chapman and Hall, London, pp. 423–465.

Mouriño-Perez, R. R., Roberson, R. W., and Bartnicki-Garcia, S. (2006). Microtubule dynamics and organization during hyphal growth and branching in *Neurospora crassa*. *Fung. Genet. Biol.* 43:389–400.

Mouyna, I., Fontaine, T., Vai, M., Monod, H., Fonzi, W. A., Diaquin, M., Popolo, L., Hartland, R. P., Fonzi, W. A., and Latge, J.-P. (2000).Glycosylphosphatidylinositol-anchored glucanosyltransferases play an active role in the biosynthesis of the fungal cell wall. *J. Biol. Chem.* 276:14882–14889.

Pollack, J. K., Li, Z. J., and Marten, M. R. (2008). Fungal mycelia show lag time before regrowth on endogenous carbon. *Biotechnol. Bioeng.* 100:458–465.

Rayner, A. D. M. (1991). The challenge of the individualistic mycelium. *Mycologia* 83:48–71.

Reggiori, F. and Klionsky, D. J. (2002). Autophagy in the eukaryotic cell. *Euk. Cell* 1:11–21.

Riquelme, M., Reynaga-Peña, C. G., Gierz, G. and Bartnicki-Garcia, S. (1998). What determines growth direction in fungal hyphae? *Fung. Genet. Biol.* 24:101–109.

Roberson, R. W., Abril, M., Blackwell, M. S., *et al.* (2010). Hyphal structure. In: K. A. Borkovich and D. J. Ebbole, eds., *Cellular and Molecular Biology of Filamentous Fungi.* ASM Press, Washington, D.C.

Roberson, R. W. and Fuller, M. S. (1988). Ultrastructural aspects of hyphal tips of *Sclerotium rolfsii* preserved by freeze substitution. *Protoplasma* 164:143–149.

Sampson, K. and Heath, I. B. (2005). The dynamic behaviour of microtubules and their contributions to hyphal tip growth in *Aspergillus nidulans. Microbiology* 152: 1543–1555. Schoffelmeer, E., Klis, F. M., Sietsma, J. H., and Cornelissen, B. J. C. (1998). The cell wall of *Fusarium oxysporum. Fung. Genet. Biol.* 27:275–282.

Shatkin, A. J. and Tatum, E. L. (1959). Electron microscopy of *Neurospora crassa* mycelia. *J. Biophys. Biochem. Cytol.* 5:423–426 (plates 210 to 216).

Sietsma, J. H. and Wessels, J. G. H. (1979). Chemical analysis of the hyphal wall of *Schizophyllum commune. J. Gen. Microbiol.* 114:99–108.

Silverman-Gavrila, L. B. and Lew, R. R. (2000). Calcium and tip growth in *Neurospora crassa. Protoplasma* 213:203–217.

Silverman-Gavrila, L. B. and Lew, R. R. (2001). Regulation of the tip-high [Ca^{2+}] gradient in growing hyphae of the fungus *Neurospora crassa. Eur. J. Cell Biol.* 80:379–390.

Smith, M., Bruhn, J. N., and Anderson, J. B. (1992). The fungus *Armillaria bulbosa* is among the largest and oldest living organisms. *Nature* 356:428–431.

Srinivasan, S., Vargas, M. M., and Roberson, R. W. (1996). Functional, organizational, and biochemical analysis of actin in hyphal tip cells of *Allomyces macrogynus. Mycologia* 88:57–70.

Steinberg, G. (2007a). Hyphal growth: A tale of motors, lipids, and the Spitzenkörper. *Euk. Cell* 6:351–360.

Steinberg, G. (2007b). On the move: Endosomes in fungal growth and pathogenicity. *Nature Rev. Microbiol.* 5:309–316.

Talbot, N. J. (1997). Growing into the air. *Curr. Biol.* 7:78–81.

Trinci, A. P. J. and Righelato, R. C. (1970). Changes in constituents and ultrastructure of hyphal compartments during autolysis of glucose-starved *Penicillium chrysogenum. J. Gen. Microbiol.* 60:239–249.

Uetake, Y., Kojima, T., Ezawa, T., and Saito, M. (2002).Extensive tubular vacuole system in an arbuscular mycorrhizal fungus, *Gigaspora margarita. New Phytol.* 154:761–768.

van Peer, A. F., Wang, F., van Driel, K. G. A., de Jong, J. F., van Donselaar, E. G., Müller, W. H., Boekhout, T., Lugones, L. G., and Wösten, H. A. B. (2010). The septal pore cap is an organelle that functions in vegetative growth and mushroom formation of the wood-rot fungus *Schizophyllum commune. Environ. Microbiol.* 12:833–844.

Vera Meyer, V., Arentshorst, M., van den Hondel, C. A. M. J. J., and Ram, A. F. J. (2008). The polarisome component SpaA localises to hyphal tips of *Aspergillus niger* and is important for polar growth. *Fung. Genet. Biol.* 45:152–164.

Vinck, A., Terlou, M., Pestman, W. R., Martens, E. P., Ram, A. F., and van den Hondel, C. A. M. J. J. (2005). Hyphal differentiation in the exploring mycelium of *Aspergillus niger. Mol. Microbiol.* 58:693–699.

Wessels, J. G. H. (1996). Fungal hydrophobins: Proteins that function at an interface. *Trends Plant Sci.* 1:9–15.

Wittler, R., Baumgartl, H, Lübbers, D. W., and Schügerl, K. (1986). Investigations of oxygen transfer into *Penicillium chrysogenum* pellets by microprobe measurements. *Biotechnol. Bioeng.* 28:1024–1036.

Wolkow, T. D., Harris, S. D., and Hamer, J. E. (1996). Cytokinesis in *Aspergillus nidulans* is controlled by cell size, nuclear positioning and mitosis. *J. Cell Sci.* 109:2179–2188.

Wösten, H. A. B., Moukha, S. M., Sietsma, J. H., and Wessels, J. G. H. (1991). Localization of growth and secretion of proteins in *Aspergillus niger. J. Gen. Microbiol.* 137:2017–2023.

Wösten, H. A. B., Schuren, F. H. J., and Wessels, J. G. H. (1994). Interfacial self-assembly of a hydrophobin into an amphipathic membrane mediates fungal attachment to hydrophobic surfaces. *EMBO J.* 13:5848–5854.

Yin, Q. Y, de Groot, P. W. J, Dekker, H. L. de Jong, L, Klis, F. M., and de Koster, C. G. (2005). Comprehensive proteomic analysis of *Saccharomyces cerevisiae* cell walls: Identification of proteins covalently attached via glycosylphosphatidylinositol remnants or mild alkali-sensitive linkages. *J. Biol. Chem.* 280:20894–20901.

CHAPTER 2

The Multinuclear Condition

Our knowledge of cell biology of eukaryotes has been gained largely through studies of plant and animal cells that contain a single nucleus per cell. With the exception of unicellular yeast fungi, the majority of fungi composed of hypha contain several nuclei per hyphal compartment. A common method of visualizing nuclei in hyphae is the use of DNA-specific dyes (Figure 2.1). In the past few years the *Neurospora* chromosomal protein histone H1 gene was tagged with a gene from a jellyfish that encodes a green fluorescent protein (GFP) that glows bright green. Several filamentous fungi have been transformed by histone H1-GFP, and the tagged nuclei are visualized using a fluorescence microscope under blue-light excitation (488 mm) (see book cover). This enables the study of nuclear dynamics in living hyphae by video microscopy. Combined with the ease of isolating mutants in fungi and the availability of genome sequences of some species, this has stimulated studies of conditions that influence mitosis, the identification of nuclear genes involved in nuclear movement, their distribution in hypha, and morphogenesis. The feasibility of fusing fungal cells containing two different nuclear types into heterokaryon makes has made it possible to study cooperation or competition between nuclei when put in a common cytoplasm of fungal compartment.

2.1 NUCLEAR NUMBER AND HYPHAL GROWTH RATE

Unlike plant or animal tissues, nuclear division in fungi is not obligatorily coupled to cell division. As a consequence the hypha, even if produced from a uninucleate spore, becomes multinucleate. Some fungi form spores with varying nuclear numbers. For example, the macroconidia of *Neurospora crassa* contain one to four nuclei per cell. An interesting question is whether the nuclear number influences their rate of germination. Serna and Stadler (1978) measured the DNA content in mithramycin-stained germinating conidia by flow microfluorimetry as the individual cells passed in a single file past a laser beam. The plot of DNA per 10^6 cells (x axis) versus time (y axis) showed a steady rise over a period of several hours, contrary to the expected stepwise increase if the nuclear division was synchronous. Most likely the nuclei in conidia are arrested at various points in the nuclear division cycle. The rate of germination of conidia was not related to nuclear number. The significance of a multinuclear condition remains unknown.

2.2 CHROMOSOME NUMBERS

2.2.1 Classical Methods

The majority of the "true fungi" are haploid. The genome size of fungi varies from 15 to 45 megabases. This genome is smaller than that of the other eukaryotes and consequently the fungal chromosomes in the vegetative phase are at the limit of resolution of the light microscope.

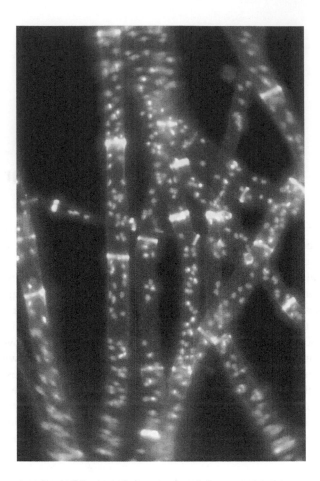

Figure 2.1   (See color insert.) The hyphae of *Neurospora crassa* stained with a DNA-binding fluorescent dye to stain nuclei and a chitin-binding dye to stain septa. The nuclei in living hypha can move in both directions through a central pore in the septum.

However, in the exceptionally large meiotic cells of certain species belonging to Ascomycotina (e.g., *Neurospora*), the chromosomes are duplicated and condensed. In such cells, the pachytene stage is often used for microscopic determination of the chromosome number after staining with aceto-orcein, iron-hematoxylin, Giemsa, or acriflavine. Some examples of chromosome numbers (where *n* is one chromosome complement) are: *Aspergillus nidulans* (8), *Alternaria* spp. (9–11), *Cochliobolus heterostrophus* (15), *Magnaporthe grisea* (7), *Nectria haematococca* (*Fusarium solani*) (10–14), *Neurospora crassa* (7), and *Podospora anserina* (7). The unicellular yeast *Saccharomyces cerevisiae* has 16 chromosomes. A highly successful technique used in *Botrytis cinerea* is to burst the hyphal tips of germlings by treatment with methanol-acetic acid on a glass slide, releasing the chromosomes, and staining them with Giemsa for counting their numbers (Shirane *et al.*, 1988). The length of chromosomes ranged from 0.6 to 1.3 μm and width from 0.2 to 0.3 μm; total length of the component was 13.1 μm.

2.2.2 New Methods

Intact, tangled chromosomes for chromosome numbers are prepared by bursting conidial germ tubes on a microscope slide, followed by staining. In another method the DNA molecules are separated by pulsed field gel electrophoresis (PFGE). In this technique the orientation of the applied electrical field on agarose gel is periodically changed, thereby reorienting and separating chromosomes. After staining with ethidium bromide, the chromosomes are resolved as bands and counted under UV light, and the chromosome numbers of fungi may be determined. For a PFGE showing separation of chromosomes of wheat pathogen *Mycosphaerella graminicola*, see Mehrabi *et al.* (2007). Germ-tube burst method (GTBM) analyses revealed the mitotic metaphase chromosomes, enabling chromosome quantification, which was congruent with the PFGE analyses. The fungus *Nectria haematococca* (Ascomycotina, asexual name *Fusarium solani*) causes disease in 100 genera of plants and opportunistic infections in humans. The species has a diverse range of habitats and unique genes on supernumerary chromosomes (14, 15, or 17). It has 17 chromosomes and is one of the largest fungal genomes, comprising 15,707 genes (Coleman *et al.*, 2009).

Another method of karyotyping is based on telomere sequence. Telomeres are specialized structures at the ends of eukaryotic chromosomes that ensure chromosome stability during replication cycles and allow the cell to separate itself intact from broken chromosomes. Telomere fingerprinting is reported to substitute for inadequately resolved chromosomes in electrophoresis karyotyping in estimation of chromosome numbers. An oligonucleotide constituted by telomere repeat sequence—(5k-TTAGGG-3k)8—is end-labeled with ^{32}P to serve as the probe. The number of bands in each lane is counted and halved to arrive at the chromosome number of that isolate. When the number of bands is odd, it is rounded up to the next even number. Using this method the chromosome number in the different isolates of insect pathogen *Beauveria bassiana* was found to vary from 5 to 10 by Padmavathi *et al.* (2003) (Figure 2.2).

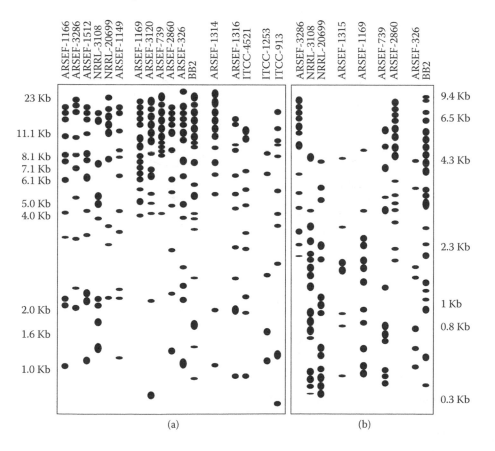

Figure 2.2 Estimation of nuclear numbers by telomere fingerprinting and electrophoresis in an entomogenous fungus *Beauveria bassiana*. (Tracing from Padmavathi, J. et al. 2003.)

2.3 MULTIPLE GENOMES IN INDIVIDUAL NUCLEI

Staining spores of arbuscular mycorrhiza (AM) fungi (*Gigaspora marginata* and *Scutellospora erythropa*) with 4,6-diamino-2-phenylindole (DAPI) showed that these contain approximately 20,000 nuclei per spore (Burggraff and Beringer, 1989). This number matched well with the amount of DNA extractable from crushed spores and by quantifying the DNA on the basis that average DNA content per nucleus is 0.4 picogram (Viera and Glenn, 1990). DNA from spores of *Scutellospora castanea* was analyzed by crushing one single spore and diluting the nuclear suspension to obtain one nucleus per tube (Figure 2.3). The internal transcribed spacer (ITS) sequence in ribosomal DNA was analyzed. The ITS sequences are located between the 18S and 5.8S rRNA-coding regions (ITS1) and between 5.8S and 25S rRNA-coding regions (ITS2). Since the ITS region is flanked by highly conserved coding regions, polymerase chain reaction (PCR) amplification of ITS1 and ITS2 was done using universal primers (Hijri *et al.*, 1999; Hosny *et al.*, 1999). The amplified products were digested with restriction enzymes and fractionated on a gel. Fragments of different lengths (RFLP, for restriction fragment length polymorphism) showed several ITS sequences among the nuclei, demonstrating that the spores are heterokaryotic. ITS fragments could be grouped into six types that were cloned and sequenced (Hijri *et al.*, 1999). Most of the sequences were very similar to those of *Scutellospora castanea*, and a few sequences matched to different Glomales genera, *Scutellospora* and *Glomus* (Hosny *et al.*, 1999). Subsequently fluorescent *in situ* DNA-DNA hybridization (FISH) was used to visually demonstrate coexistence of different nuclei in individual mycorrhizal fungus (Kuhn *et al.*, 2001). Using hybridization probes that specifically recognize divergent sequences, T2 and T4, in nuclei of *Scutellospora castanea*, it was found that 40% of nuclei contained

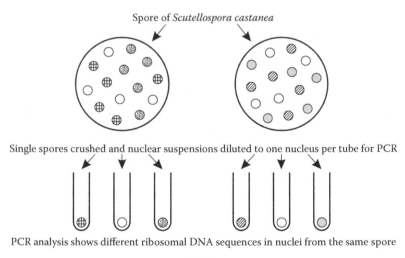

Spore of *Scutellospora castanea*

Single spores crushed and nuclear suspensions diluted to one nucleus per tube for PCR

PCR analysis shows different ribosomal DNA sequences in nuclei from the same spore

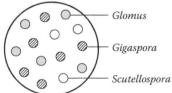

Glomus

Gigaspora

Scutellospora

DNA sequences in *Scutellospora castanea* match genera in *Gigasporaceae* and *Glomaceae*

Figure 2.3 Multiple genomes within individual nuclei of a mycorrhizal fungus. (Adapted from Sanders,1999, with permission.)

only the T2, 6–9% contained only T4, whereas 8–9% nuclei contained both T2 and T4, supporting that arbuscular mycorrhizal fungus spores contain a population of genetically different nuclei. This observation has raised the puzzling question: How do such divergent nuclei come together in an individual that lacks sexual reproduction? No evidence for sexual reproduction, hence recombination, has been found, suggesting that they reproduce clonally for their entire association with plants. A plausible explanation is hyphal anastomoses between mycelia of two genera, followed by exchange of nuclei.

2.4 NUCLEAR DIVISION CYCLE

The cell cycle is divided into four phases: G1 phase (between mitosis and the beginning of DNA synthesis), S phase (the period of DNA synthesis), G2 phase (the interval following the S phase and the beginning of mitosis), and M phase (separation of daughter nuclei). The G1, S, and G2 phases are typically longer parts of the cell cycle than the M phase (Figure 2.4). Whereas in animal and plant cells the nuclear membrane breaks down during cell division, fungi—including yeasts—have a closed mitosis: Spindle microtubules are inside the nucleus and the spindle pole bodies are formed within an intact nuclear membrane. For a transmission electron micrograph of closed mitosis in fungi, see Alexopoulos *et al.* (1996). Since nuclear division in fungi is generally not followed by cytokinesis, some prefer to use the term "nuclear division cycle" in place of cell cycle for the sequence of events by which the nucleus duplicates into two. However, because of the common use of the term "cell cycle," it is used here, often interchangeably.

Aspergillus nidulans is an attractive fungus for cell cycle studies because its conidia are uninucleate. Therefore, the rounds of division that a nucleus has undergone at different times can be determined by counting the number of stained nuclei. The first three divisions of nuclei are synchronous and result in the formation of eight nuclei in a conidium. In this fungus, the three nuclear divisions in a conidium are essential for emergence of the germ tube. The genes that are required for nuclear division as well as nuclear movement into the germ tube are identified by isolating temperature-sensitive (conditional) mutants in *A. nidulans*.

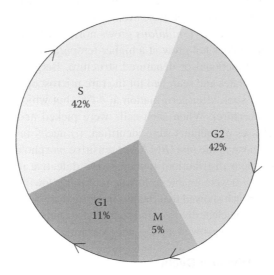

Figure 2.4 *Aspergillus nidulans* cell cycle. The duration of different phases is shown as percentage of total cell cycle.

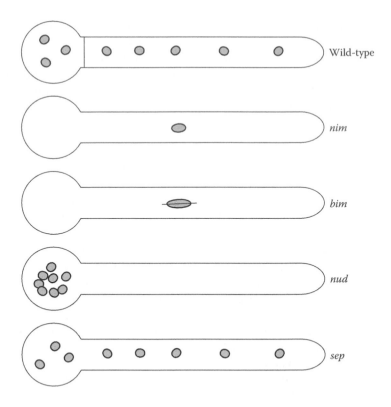

Figure 2.5 Temperature-sensitive mutants of *Aspergillus nidulans* defective in nuclear division and nuclear movement. (Redrawn from Plamann, M., *J. Genet*. 75, 351–360, 1996.)

2.4.1 Temperature-Sensitive Mutants

Nuclear division, nuclear movement, and septation are indispensable for hyphal growth. Therefore, to identify the genes that control these events, the mutants in which growth fails only at a higher temperature must be isolated. *A. nidulans* grows normally between 15 and 44°C, whereas a temperature-sensitive mutant does not grow at a higher temperature (44°C) as the mutated gene-encoded protein takes a nonfunctional or denatured structure. Bergen and Morris (1983) spread mutagenized conidia on agar plates and searched for the rare microscopic colonies in which growth was arrested either before or soon after germination at 44°C, but which grew normally if shifted to 32°C (a permissive temperature). When such cells were picked and grown at the permissive temperature, four major classes of mutants were identified: (i) nuclei in the *nim* (*n*ever *i*n *m*itosis) mutants did not divide; (ii) those in the *bim* (*B*MAA *i*nsensitive *m*orphology) mutants were blocked in mitosis; (iii) the *nud* (*nu*clear *d*istribution) mutants were defective in nuclear distribution; and (iv) the *sep* mutants (from *sep*ta) were unable to form a septum following the third nuclear division (Figure 2.5). The mutant approach showed that there is no obligatory coupling between nuclear division and cytokinesis. However, nuclear division and septation are required for hypha development as the *nud* and the *sep* mutants fail to grow at 44°C—a temperature at which the wild type can grow.

2.4.2 Stages in Nuclear Division Cycle

Using the *nim* mutant of *A. nidulans* blocked in the G_2 phase, the nuclear division cycle was deduced from the number of nuclei in the cell (Bergen and Morris, 1983). The first step was to synchronize the multinuclear germlings at the beginning of the S phase. Dormant conidia were germinated for 6 h in the presence of hydroxyurea (a DNA synthesis inhibitor) at the restrictive temperature. The conidia underwent one nuclear division, indicated by the appearance of binucleate germlings, and this synchronized the germlings at the beginning of second cycle, S, phase. The binucleate germlings were shifted to hydroxyurea-free medium at the permissive temperature and samples were removed at intervals and retreated with hydroxyurea before they became sensitive to it, i.e., complete S phase (become tetranucleate). The time required to pass the hydroxyurea-sensitive phase (S phase) was around 40 min. The rate of nuclear doubling was determined from a plot of nuclear division per cell vs. time, using the equation

$$\text{Division per cell} = \log\ (N/C) \times \log 2$$

where N is the number of nuclei, and C is the number of cells analyzed. The nuclear doubling time (generation time or G_t), calculated from the slope of the line, was 95 min. The length of the M phase, calculated from mitotic index (percentage mitotic figures in hyphal tips) of culture was 5 min. The G_2 phase, found by subtracting the aggregate length of M and S from the generation time, was 40 min. The G_1 phase, determined as G_t (S + G_2 + M), was 10 min.

In *N. crassa*, morphological differences in nuclear shape in different phases were used to determine the nuclear division cycle (Martegani *et al.*, 1980). The G_1 nuclei are compact and globular; in S and G_2 they are ring-shaped, and in M phase they are double ring or horseshoe shaped. Treatment with picolinic acid blocked nuclei in G_1. Release from picolinic acid inhibition was followed by a wave of synchronous DNA replication. From the frequencies of phase-specific nuclear shapes, the durations of the G_1 phase and S + G_2 phase were estimated. In conidia germinated in sucrose medium, the duplication (generation) time was 100 min (G_1 = 20 min, S = 30 min, G_2 = 40 min, and M = 10 min). The extended G_2, relative to G_1, is characteristic of fungi. These data on *A. nidulans* and *N. crassa* show that the cell cycle in fungi is completed faster than in animals or in plants. For example, a human cell growing in culture has G_1= 9 h, S = 10 h, G_2 = 4.5 h, and M = 30 to 45 min, with the complete cycle taking about 24 h.

2.5 ASYNCHRONOUS NUCLEAR DIVISIONS

Nuclear transplantation in frog eggs and cell fusion experiments with cultured animal cells had revealed that nuclear division (mitosis) is induced by diffusible factors present in cytoplasm. For example, when a cell in G1 phase was fused with a cell in M phase, the G1 nucleus in the fused cell prematurely entered into nuclear division (Rao and Johnson, 1970). This observation suggests that the division of fungal nuclei that are in a common cytoplasm is synchronous. Assuming complete synchrony, the uninucleate *A. nidulans* conidia (asexual spores) will yield a hypha containing 2^n nuclei, where n is the number of nuclear divisions after germination. A deviation from this value indicates lack of synchrony. Rosenberger and Kessel (1967) found that synchrony was lost after four to six divisions in individual hypha. In multinucleate conidia or in the wall-less slime mutant of *N. crassa*, synchrony was not observed even when the nuclei were present in close proximity in the same cytoplasm (Raju, 1984).

From work done with animal cells, a model of cell cycle regulation emerged in which heterodimers of cyclin proteins complexed with different cyclin dependent kinases (CDKs) driving mitosis, with destruction of cyclin occurring after each mitosis. The filamentous fungus *Ashbya gossypii*, which contains many nuclei in linear arrays in compartments, was chosen to follow nuclear pedigrees by time-lapse video-microscopy in growing mycelium in which nuclei were visualized by tagging histone protein with GFP (Gladfelter *et al.*, 2006). Cyclins were present at all stages; its degradation did not correlate with mitosis exit, leading the authors to speculate that multinucleated cells may have evolved an oscillator based on CDK inhibitors rather than control by cyclin protein.

Thus nuclei, although residing in a common cytoplasm, control their division individually through control of CDK activity.

2.6 NUCLEAR MIGRATION

In fungi, the nuclei divide in the hyphal tip (King and Alexander, 1969) and migrate through septal pores into hyphal compartments. The fungal hypha is therefore an excellent material to study the rates and mechanisms of long-distance nuclear movement. The nuclei may be quite variable in shape. Light and electron micrographs show dumbbell-shaped nuclei squeezing through septa or becoming thread-like while entering into a branch. A new technique to study migration of nuclei is to stain with DNA-binding fluorescent dyes or tag with the green fluorescent protein (GFP) and monitor movement by video-enhanced fluorescent microscopy. In *Aspergillus nidulans*, velocities from 0.1 to 40 μm min^{-1} have been observed (Suelmann *et al.*, 1997). Microscopy of live cells often shows nuclei to be stationary in a moving cytoplasm, dispelling the notion that cytoplasmic streaming carries nuclei.

The analysis of fungal mutants has provided evidence that nuclear movement requires a motor to move the nucleus through the cytoplasm, a track for the nucleus to move on, and a coupling mechanism to link the motor to the nucleus (Morris *et al.*, 1995). The *nud* (*nuclear distribution*) mutants of *A. nidulans* and the *ro* (*ropy*) mutants of *N. crassa* have clustered nuclei rather than evenly distributed as in the wild type. The wild-type *nud* and *ro* genes were isolated by transforming mutants at nonpermissive temperature with plasmids containing wild-type DNA and isolating the rare cells that grew into colonies (functional complementation). Plasmid DNA isolated from transformed colonies carried the wild-type *nud⁺* or *ro⁺* gene. Sequencing of *nud⁺* and *ro⁺* genes revealed that they encode components of the multiprotein complex called dynein, a mechanochemical enzyme that provides the motive force for movement of nuclei along the filamentous tracks of actin and microfilaments in the cytoplasm (Osmani *et al.*, 1990; Morris *et al.*, 1995; Minke *et al.*, 1999). Given the complexity of molecular motors, it can be expected that several genes will be found that control nuclear migration and positioning. The main motor is a large protein called dynein, the track is a long cylindrical protein called a microtubule, and the coupler is the protein dynactin (Figure 2.6). Presumably, different motor proteins move nuclei at different velocities. The use of benomyl—a drug that causes disintegration of microtubules—revealed that these proteins are important in nuclear migration. Nuclei move in opposite directions in the hyphal compartment to reach an initial branch point, suggesting individual regulation of nuclear mobility (Suelmann *et al.*, 1997). The observation suggested that branch initiation is independent of nuclear distribution, although germ tube growth does not occur until a nucleus has entered into it.

2.7 NUCLEAR POSITIONING AND GENE REGULATION

In the Basidiomycotina, the processes of hyphal fusion, nuclear migration, and subsequently the selective association of nuclei convert a monokaryotic hypha into a dikaryotic hypha. The dikaryotic hyphae produce the mushroom fruit bodies. A small, backwardly projecting outgrowth (hook) occurs from the end cell of hypha into which one of the two daughter nuclei passes (Figure 2.7). Septa form and the hook cell then fuses with the penultimate cell, forming clamp connections at the septa. The process is repeated. The simultaneous division of nuclei, formation of hook cell, and its fusion with the penultimate cell ensures that each cell of the mycelium contains two genetically distinct nuclei. Recent work suggests that whether nuclei are juxtaposed or separated is important in gene regulation (Schuurs *et al.*, 1998). The fruiting body of *Schizophyllum commune* (Basidiomycotina) is rich in proteins called hydrophobins (Chapter 1). The type of hydrophobins secreted was determined by

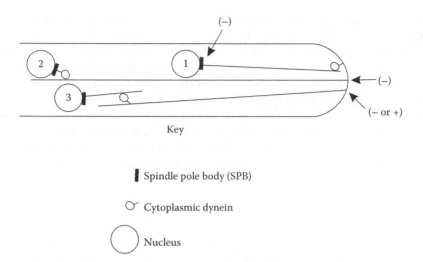

Key

Spindle pole body (SPB)

Cytoplasmic dynein

Nucleus

Figure 2.6 Possible mechanisms of nuclear migration. Nucleus #1 is migrating toward the hyphal tip. The minus end of the microtubule (MT) is postulated to be at the spindle pole body. Nucleus #2 has dynein attached to the spindle pole body (SPB) and is migrating toward the fungal tip. This model requires that the minus end of the MTs be at the tip. Nucleus #3 is migrating along an astral MT toward the minus end (at the SPB) and is anchored not at the tip but on an MT that is attached to the tip. (From Morris, N.R., Xiang, X., and Beckwith, S.M., *Trends Cell Biol.* 5, 278–282, 1995, with permission from Elsevier.)

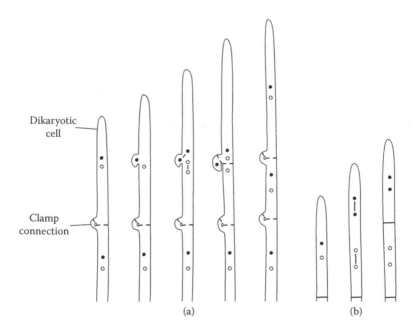

Dikaryotic cell

Clamp connection

(a)   (b)

Figure 2.7 (a) Clamp formation. Each cell of the dikaryotic hypha of a mushroom fungus (Basidiomycotina) has two nuclei (shown as white and black) of both types. Diagram showing clamp formation. (b) Nuclear types had clamps not formed. (From Ingold, C.T. and Hudson, J.H., *The Biology of Fungi,* Chapman and Hall, London, 1993. With kind permission of Springer Science and Business Media.)

immunochemical staining methods. The monokaryotic hyphae secreted hydrophobin SC3, but not hydrophobins SC4 and SC7. On the other hand, the type of hydrophobin secreted by the dikaryotic hyphae was determined by internuclear distance—which could be manipulated by growth on hydrophobic or hydrophilic surface. On a hydrophilic surface in which the nuclei were adjacent (< 2 μm), the hypha secreted SC4 hydrophobin (which coats air channels within the fruit body) and SC7 but not SC3 (which coats aerial hyphae and hyphae at the surface of fruit bodies); whereas on a hydrophobic surface the nuclei were separated (13–16 μm), and the hyphae secreted hydrophobin SC3 but not hydrophobins SC4 and SC7. The regulation of gene activity by modulating the internuclear distance may be a unique mechanism in fungal hyphae. It questions distribution of nuclei by bulk cytoplasmic flow. Rather specific recognition of nuclei and their positioning appears to be involved.

2.8 HETEROKARYOSIS

A consequence of multinuclear condition is heterokaryosis. This term refers to the coexistence of two or more genetically different nuclei inside the same hyphal cell. If single spores of the fungus *Botrytis cinerea* taken from a natural substratum are grown on an agar medium, a circular colony grows that forms sectors, noticeable by poor or heavy sporulation (Figure 2.8). Single spores from each sector produced nonsectoring colonies, whereas a spore from the main colony produced a culture that sectored like the parent one. Two distinct types of nuclei coexisted in the hyphae and differing types of these nuclei were incorporated in spores. The heterokaryotic condition can arise

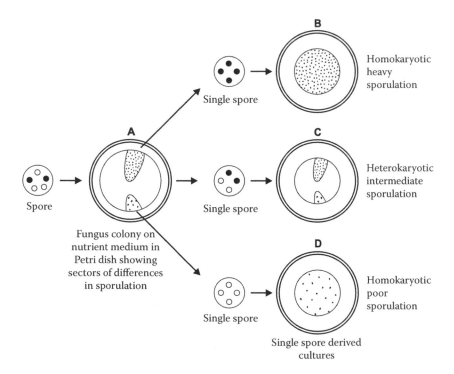

Figure 2.8 Discovery of heterokaryosis. (A) A colony of *Botrytis cinerea* derived from a single spore, producing sectors of poor and heavy spores. A single spore from each sector gave non-sectoring colonies (B, D) whereas a spore from the main colony gave a culture sectoring like the parent one (C). This condition is due to different types of nuclei represented by black and white. (Redrawn from Ingold, C.T. and Hudson, H.J., *The Biology of Fungi*, Chapman and Hall, London, 1993, with kind permission of Springer Science and Business Media.)

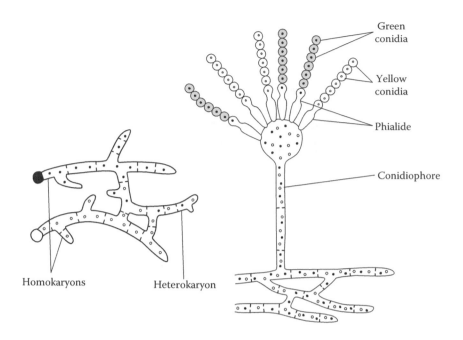

Figure 2.9 Formation of heterokaryon in *Aspergillus*. The two genetically different types of nuclei are shown as open and filled circles. The nuclear types are segregated during budding of uninucleate spores from phialide cell. The uninucleate spores (conidia) germinate to produce homokary- otic hyphae which may fuse to give a heterokaryon. (From Ingold, C.T. and Hudson, H.J., *The Biology of Fungi*, Chapman and Hall, London, 1993, with kind permission of Springer Science and Business Media.)

either by spontaneous mutation in some nuclei within an originally genetically homogeneous myce- lium, or by the fusion of genetically distinct hyphae followed by nuclear mixing. Because of their multinuclear condition, heterokaryosis is a unique feature of the fungi. It has been assumed to be an important mechanism in the adaptation of fungi to fluctuating environment through alteration in nuclear ratios (the proportion of nuclear types).

In some fungi, co-spotting a mixture of two mutant conidia of same mating type on agar medium results in the conidial germ tubes fusing among each other, followed by mixing of nuclei, allowing heterokaryotic hypha to be formed (Figure 2.9). The formation of heterokaryotic mycelium is used to distinguish if two strains of the same phenotype have mutation in the same gene or in two differ- ent genes in a biochemical pathway (test of allelism). If nuclei with nonallelic mutations coexist in a heterokaryon, the phenotype of heterokaryon is normal since what is lacking in one is present in the other, and all functions can be performed. Allelic mutations fail to complement because neither nucleus can perform the vital function (Figure 2.10).

2.8.1 Sheltering of Lethal Mutation

Although the mutation rate is estimated to be in the order of one in a million nuclei, a multi- nuclear mycelium can over time accumulate lethal mutations. The mutant genes would be sheltered by their normal alleles in other nuclei in the cytoplasm and the lethal mutation may be undetected in multinucleate hypha. Nuclei can, however, be extracted from mycelium in the form of uninucle- ate spores, and these can be grown into homokaryotic culture and examined for mutant phenotype (Maheshwari, 2000). A genetic proof has been given that heterokaryons form in nature (Pandit and Maheshwari, 1996). An example sheltering of recessive lethal mutation in coenocytic heterokary- otic hypha is given in Chapter 16.

a. Non-allelic mutations (*gene-1⁺* and *gene-2⁻*)

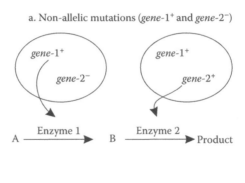

b. Allelic mutations (*gene–2ᵃ* and *gene-2ᵇ*)

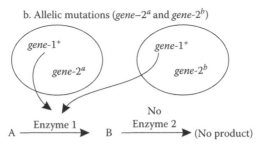

Figure 2.10   The heterokaryon test to determine whether two mutations are in the same gene. (a) Nuclei with non-allelic mutations complement each other, (b) nuclei with allelic mutations fail to complement. (From Davis, R.H., *The Microbial Models of Molecular Biology, Neurospora*, Oxford University Press, New York, 2003, chap. 3, with permission of the publisher.)

2.8.2 Change in Nuclear Ratio

Asynchronous nuclear division can lead to variation of the proportion of nuclear types in a mycelium. It has been thought that a factor accounting for the high adaptability of fungi is the change in the ratios of nuclear types due to asynchronous nuclear division. The first experiment to test this was done by Ryan and Lederberg (1946) who constructed a heterokaryon of *N. crassa* containing mutant leucineless (*leu*) nuclei by tagging the nuclei with genetic color markers (*al*bino, a mutant gene affecting the carotenoid biosynthesis in cultures). The prototrophic (*leu⁺*) nuclei contained wild-type allele (orange color). When this heterokaryon was grown with leucine supplementation, the mutant nuclei were selected over the prototrophic *leu⁺* nuclei. The degree of selection was so extreme that the appearance of the culture changed from orange to colorless due to selective reduction in wild-type *al⁺* nuclei. Furthermore the change occurred reversibly depending on whether leucine was provided or not.

Thus, it may be that a population of nuclei of different genotypes in multinuclear mycelium could be a means of rapid biochemical adjustment to environmental change through the selection of particular nuclear types. To demonstrate this by an actual experiment, the nuclei in different strains of *N. crassa* were tagged by color-marker genes and the strains used to construct a heterokaryon (Pitchaimani and Maheshwari, 2003) (Figure 2.11). The gene *al-1* has a white phenotype; the gene *al-2* has a rose-white phenotype. A heterokaryon was constructed between prototrophic *his-3⁺ al-1* and a histidine auxotrophic strain *his-3 al-2* and subcultured four times on minimal medium with or without histidine supplement. It turned a rose-white color on histidine medium—a change in growth environment—indicating a reduction in the number of *his-3⁺* nuclei marked with the *al-1* gene. When conidia from the histidine-grown heterokaryon were propagated in medium lacking histidine, the orange color was restored, indicating an increase in the proportion of transformed

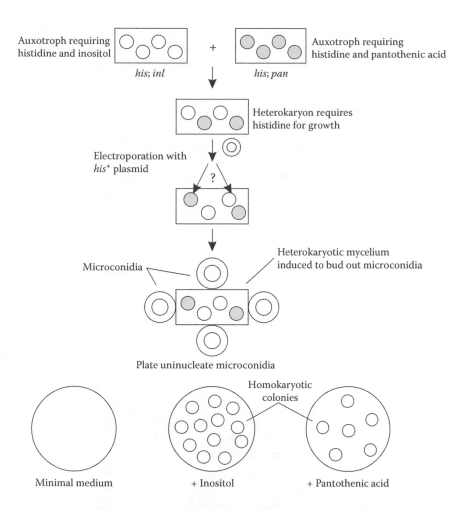

Figure 2.11 Nuclear competence. An experiment with *Neurospora crassa* to test whether two different nuclear types but each carrying same mutant *his(tidine)* allele, and hence requiring histidine supplementation for growth, are equally competent to accept and stably incorporate exogenous *his-3*⁺ DNA when the present in a heterokaryon. The genotype of nuclei in which transforming *his-3*⁺ DNA had entered was determined by extracting nuclei in the form of uninucleate spores (microconidia) and plating these cells on differential media.

nuclei and providing visual proof that the proportion of nuclear types changes in response to change in chemical environment. The results of this experiment—easily done with fungi—demonstrate that a diverse population of nuclei in mycelium can permit adaptation to fluctuating environmental conditions through changes in nuclear ratio in mycelium. "A fungus which combines heterokaryosis with sexual reproduction provides both for the present and the future" (Burnett, 1976).

Disproportionate ratios of wild-type and mutant nuclei in heterokaryon were used to determine if the enzyme activity encoded by the wild-type allele is related to the dosage of wild-type nuclei. A [*his-3* + *his-3(EC)*] heterokaryon of *N. crassa* was generated by transformation of a histidine auxotrophic strain (*his-3*) with a plasmid-containing gene for histidinol dehydrogenase (Pitchaimani and Maheshwari, 2003). Note that in genetic terminology, the genotype of strains combined in a heterokaryon is enclosed in parenthesis: A *Neurospora* gene that has been integrated ectopically by transformation is designated by appending "*(EC)*" to the gene symbol. Extreme nuclear disproportion occurred when the transformant was grown in the presence of histidine but not when grown without histidine. Despite the drastic change in nuclear ratio, visually monitored by incorporating a genetic color marker, the activity of histidinol dehydrogenase in mycelia grown in the two conditions was similar. The authors hypothesized that not all nuclei in hyphae may be active at any given time, with the rare active nuclei being sufficient to confer wild-type phenotype through biosynthesis of the enzyme.

2.8.3 Nuclear Competence

Do different nuclei, when combined in the same cytoplasm, retain their individual nuclear cycles (Pandit and Russo, 1992; Grotelueschen and Metzenberg, 1995)? Put differently, will nuclei of different genotypes but occurring in the same cytoplasm be equally competent to accept foreign DNA? As a hypothetical example, consider mutant strain 1, which is auxotrophic for markers X and Z, while strain 2 is auxotrophic for markers Y and Z. A heterokaryon formed by fusion of strains 1 and 2 is auxotrophic only for the Z marker because of complementation of X and Y gene products. When a heterokaryon that contains two nuclear types (Figure 2.12) is transformed with a plasmid carrying an exogenous copy of the Z gene, the heterokaryon will become prototrophic for the Z gene product and could be selected on a medium that allows growth only of the heterokaryotic cell.

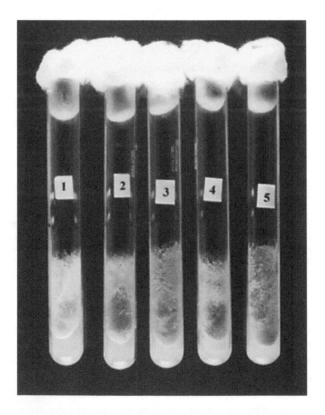

Figure 2.12 **(See color insert.)** An experiment using color mutants of Neurospora crassa to demonstrate that the proportion of number of nuclei in multinuclear mycelium can change in response to alteration of chemical composition of growth medium. A three-component heterokaryon [his-3 al-1 + his-3$^+$ (EC) al-1+ his-3 al-2] was constructed between albino mutants impaired in histidine biosynthesis and a histidine transformed strain containing al-2 nuclei. Tube # 1, his-3$^+$ (EC) al-1(white). Tube # 2, rosy phenotype of his-3 al-2. Tube #3, orange phenotype of his-3$^+$ (EC) al-1 + his-3 al-2 heterokaryon grown in minimal medium; Tube # 4, rosy phenotype of his-3$^+$ (EC) al-1 + his-3 al-2grown in histidine medium. Tube # 5, orange phenotype of his-3$^+$ (EC) al-1 + his-3 al-2 heterokaryon following transfer from histidine medium to histidine drop-out medium. Abbreviation: EC (ectopic).

In *N. crassa*, the individual nuclei from heterokaryotic cells (conidia) can be extracted in the form of uninucleate microconidia, germinated, and the phenotype of individual homokaryotic colonies analyzed (Dev and Maheshwari, 2002). This offered a means of analyzing the fate of exogenous DNA in a genetically forced heterokaryon, using the electroporation method for transformation. Results showed that in any particular event of transformation, the plasmid DNA enters into one type of nucleus. Genomic Southern analysis showed both homologous and ectopic integrations. The histidinol dehydrogenase activities varied in different transformants and were less than those of wild-type strains, indicating that gene expression is affected when the linear order of genes is disrupted by ectopic integration of foreign DNA.

2.9 PARASEXUAL RECOMBINATION

A consequence of heterokaryosis is the parasexual cycle, discovered in *Aspergillus nidulans* by G. Pontecorvo (see Davis, 2003). The parasexual cycle involves fusion of two genetically unlike nuclei to form a transient diploid nucleus. During subsequent multiplication of the diploid nucleus, a rare event of mitotic crossing-over occurs and members of each homologous pair of chromosomes assort independently of other pairs. The conidia produced during the cycle are genetically different from the original mycelium as a result of crossover that has occurred. The parasexual cycle is not a common phenomenon.

2.10 ARE ALL NUCLEI ACTIVE SIMULTANEOUSLY?

At the beginning of this chapter it was said that, with the exception of the unicellular yeast, fungi are multinuclear. This raises the question: Do all nuclei contribute to the phenotype of the organism? Or, at any particular time, are the majority of the nuclei silent (inactive, not transcribed)—i.e., the genes not expressed? If so, what determines the active and the inactive states of DNA, and how is this controlled during development of the fungus from the time a spore germinates to form a mycelium that produces spores?

Approximately half of the total genomic DNA in eukaryotes is assembled into highly condensed chromatin domains, called heterochromatin or silent chromatin. Heterochromatin is interspersed with relatively decondensed euchromatin regions. In mammals an entire X chromosome is kept in an inactive state, visually distinguished based on their staining reaction to DNA-specific fluorescence dyes; the dye binds well to the former, rendering it bright under a fluorescence microscope, whereas the latter will be stained poorly. Thus long tracts of DNA can be held in an inactive state. In eukaryotes the active and the inactive states are known as euchromatin and heterochromatin. In fungi the nuclei in young hyphae stain brightly whereas those in older cells do not.

Chemically, the euchromatin and heterochromatin are related to the degree of methylation status of DNA. Could the large population of nuclei in fungal mycelium be maintained in an inactive state by the addition of a methyl group to the 5 position of the cytosine or the number 6 nitrogen of the adenine ring? If so, what cellular machinery is involved in controlling the functional state of nuclei? How are the activities of hundreds of nuclei per hyphal cell and of thousands of nuclei in the interconnected compartment comprising the mycelium coordinated? In *Neurospora*, DNA demethylation, either with the drug 5-azacytidine or using a *defective in methylation* (*dim*) mutation, can lead to activation of transcription, suggesting that DNA methylation is the critical modification that makes chromatin refractory to transcription. There is therefore a need to understand how methylation affects condensing of DNA by associated histone into heterochromatin and its control in replication rate and the number of nuclei in growth and conidiation. The process by which chromatin is converted into inactive heterochromatin could provide clues to how nuclei in multinuclear cells are inactivated or marked for autophagy for recycling C, N, and P from DNA in supernumerary nuclei. There has been no *in situ* demonstration of transcription of nuclei in hypha. However, a change is believed to be due to the modification of bases in DNA after its replication and distribution to daughter cells, that is, by epigenetic modifications. Chemically, the euchromatin and heterochromatin are related to the degree of methylation status of DNA: The cytosine in DNA can be methylated by the addition of a methyl group to the 5 position of cytosine or the number 6 nitrogen of the adenine.

2.11 CONCLUDING REMARKS

A persistent multinuclear condition is a unique feature of fungi. There is as yet no evidence that the multinuclear condition confers an observable advantage of rapid growth rate. Considering that synthesis of a DNA molecule requires chemical energy, the question arises as to why fungi have so many nuclei. Indeed, this question has not received any serious thought or experiment to find an explanation. A pool of genetically different nuclei coexisting in a cell would allow a suitable nuclear genotype to be selected for rapid activity in a soil environment where physical and chemical conditions vary, conferring unusual tolerance and stability, and accounting for their ubiquity. Another possibility is that during conditions of nutrient availability fungi convert nitrogen and phosphorus in a protected organic form (DNA) that in supernumerary nuclei is recycled by regulated autophagy and translocated to the hyphal tips, giving the hyphal tips the capability of extension and colonizing substrata.

Among the intriguing questions raised by the multinuclear condition: Are all the nuclei active at any given time? What are the mechanisms involved in migration and positioning of nuclei? How is the interaction of genetically dissimilar nuclei that exist in cytoplasmic continuity affected? What is the influence of environment in altering nuclear ratios in heterokaryotic mycelium? What is the quantitative effect of nuclei on the phenotype? What are the mechanisms in recognition of self and nonself nuclei?

REFERENCES

Alexopoulos, C. J., Mims, C. W., and Blackwell, M. (1996). *Introductory Mycology*. Wiley, New York, chap. 2.

Bergen, I. G. and Morris, N. R. (1983). Kinetics of nuclear division cycle of *Aspergillus nidulans*. *J. Bact.* 156:155–160.

Burggraaff, A. J. P. and Beringer, J. E. (1989). Absence of nuclear DNA synthesis in vesicular–arbuscular mycorrhizal fungi during *in vitro* development. *New Phytol.* 111: 25–33.

Burnett, J. H. (1976). *Fundamentals of Mycology*. Edward Arnold, London, chap. 16.

Coleman, J. J., Rounsley, S. D., Rodriguez-Carres, M., Kuo, A., Wasmann, C. C., *et al.* (2009). The genome of *Nectria haematococca*: Contribution of supernumerary chromosomes to gene expansion. *PLoS Genet* 5(8):e1000618. doi:10.1371/journal.pgen.1000618.

Davis, R. H. (2003). *The Microbial Models of Molecular Biology, Neurospora*. Oxford University Press, New York, chap. 3.

Dev, K. and Maheshwari, R. (2002). Transformation in heterokaryons of *Neurospora crassa* is nuclear rather than cellular phenomenon. *Curr. Microbiol.* 44:309–313.

Gladfelter, A. S., Hungerbuehler, A. K., and Philippsen, P. (2006). Asynchronous nuclear division cycles in multinucleated cells. *J. Cell Biol.* 172:347–362.

Grotelueschen, J. and Metzenberg, R. J. (1995). Some property of the nucleus determines the competence of *Neurospora crassa* for transformation. *Genetics* 139:1545–1551.

Hosny, M., Gianinazzi-Pearson, V., and Dulieu, H. (1998). Nuclear DNA content of eleven fungal species in Glomales. *Genome* 41: 422–428.

Hijri, M., Hosny, M., van Tuinen, D., and Dulieu, H. (1999). Intraspecific ITS polymorphism in *Scutellospora castanea* (Glomales, Zygomycotina) is structured within multinucleate spores. *Fung. Genet. Biol.* 26: 141–151.

Ingold, C. T. and Hudson, H. J. (1993). *The Biology of Fungi*. Chapman and Hall, London.

King, S. B. and Alexander, L. J. (1969). Nuclear behaviour, septation, and hyphal growth of *Alternaria solani*. *Am. J. Bot.* 56:249–253.

Kuhn, G., Hijri, M., and Sanders, I. R. (2001). Evidence for the evolution of multiple genomes in arbuscular mycorrhizal fungi. *Nature* 414:745–748.

Maheshwari, R. (1999). Microconidia of *Neurospora crassa*. *Fung. Genet. Biol.* 26:1–18.

Maheshwari, R. (2005). Nuclear behavior in fungal hyphae. *FEMS Microbiology Letters* 249:7–14.

Martegani, E., Tome, F., and Trezzi, F. (1981).Timing of nuclear division cycle in *Neurospora crassa*. *J. Cell Sci.* 48:127–136.

Martegani, E., Levi, M., Trezzi, F., and Alberghina, L. (1980). Nuclear division cycle in *Neurospora crassa* hyphae under different growth conditions. *J. Bact.* 142:268–275.

Mehrabi, R., Taga, M., and Kema, G. H. J. (2007). Electrophoretic and cytological karyotyping of the foliar wheat pathogen *Mycosphaerella graminicola* reveals many chromosomes with a large size range. *Mycologia* 99:868–876.

Minke, P. F., Lee, I. H., Tinsley, J. H., Bruno, K. S., and Plamann, M. (1999). *Neurospora crassa ro-10* and *ro-11* genes encode novel proteins required for nuclear distribution. *Mol. Microbiol.* 32:1065–1076.

Morris, N. R. (1976). Mitotic mutants of *Aspergillus nidulans*. *Genet. Res.* 26:237–254.

Morris, N. R., Xiang, X. and Beckwith, S. M. (1995). Nuclear migration advances in fungi. *Trends Cell Biol.* 5:278–282.

Osmani, A. H., Osmani. S. A., and Morris, N. R. (1990). The molecular cloning and identification of a gene product specifically required for nuclear movement in *Aspergillus nidulans*. *J. Cell Biol.* 111:543–551.

Padmavathi, J., Uma Devi, K., Uma Maheswara Rao, C., and Nageswara Rao Reddy, N. (2003). Telomere fingerprinting for assessing chromosome number, isolate typing and recombination in the entomopathogen, *Beauveria bassiana* (Balsamo) Vuillemin. *Mycol. Res.* 107: 572–580.

Pandit, A. and Maheshwari, R. (1996) A demonstration of the role of *het* genes in heterokaryon formation in simulated field conditions. *Fung. Genet. Biol.* 20:99–102.

Pandit, N. N. and Russo, E. A. (1992). Reversible inactivation of a foreign gene, *hph*, during the asexual cycle in *N. crassa* transformants. *Mol. Gen. Genet.* 234:412–422.

Pitchaimani, K. and Maheshwari, R. (2003). Extreme nuclear disproportion and constancy of enzyme activity in heterokaryon of *Neurospora crassa*. *J. Genet.* 82:1–6.

Plamann, M. (1996). Nuclear division, nuclear distribution and cytokinesis in filamentous fungi. *J. Genet.* 75:351–360.

Raju, N. B. (1984). Use of enlarged cells and nuclei for studying mitosis in *Neurospora*. *Protoplasma* 121:87–98.

Rao, P. N. and Johnson, R. T. (1970). Mammalian cell fusion: Studies on the regulation of DNA synthesis and mitosis. *Nature* 225:159–164.

Rosenberger, R. F. and Kessel, M. (1967). Synchrony of nuclear replication in individual hyphae of *Aspergillus nidulans*. *J. Bact.* 94:1464–1469.

Ryan, F. J. and Lederberg, J. (1946). Reverse mutation and adaptation in leucineless *Neurospora*. *Proc. Natl. Acad. Sci., USA* 32:163–173.

Schuurs, T. A., Dalstra, H. J. P., Scheer, J. M. J., and Wessels, J. G. H. (1998). Positioning of nuclei in the secondary mycelium of *Schizophyllum commune* in relation to differential gene expression. *Fung. Genet. Biol.* 23:150–161.

Serna, L. and Stadler, D. (1978). Nuclear division cycle in germinating conidia of *Neurospora crassa*. *J. Bact.* 136:341–351.

Shirane, N., Masuko, M., and Hayashi, Y. (1989). Nuclear behaviour and division in germinating conidia of *Botrytris cinerea*. *Phytopathology* 78:1627–1630.

Suelmann, R., Sievers, N., and Fischer, R. (1997). Nuclear traffic in fungal hyphae: *In vivo* study of nuclear migration and positioning in *Aspergillus nidulans*. *Mol. Microbiol.* 25:759–769.

Tsuchiya, D. and Taga, M. (2001). Cytological karyotyping of three *Cochliobolus* spp. by the germ tube burst method. *Phytopathology* 91:354–360.

Viera, G. and Glenn, M. (1990) DNA content of vesicular-arbuscular mycorrhizal fungal spores. *Mycologia* 82: 263–267.

CHAPTER 3

Spores
Their Dormancy, Germination, and Uses

If most of life were destroyed by a holocaust of natural or man-made origin, a residuum in the form of spores, cysts, or seeds might remain to serve as the raw material of further evolution.

—P. Becquerel (1950) (in French, summary in Sussman and Halvorson, 1966)

Fungal biology is inextricably linked with spores. Experiments on fungi almost always begin with spores and end when the culture has begun to produce spores. In the middle of the 17th century when fascination with looking at the microscopic-sized objects was developing, a French farmer named Mathieu Tillet (1714–1791) collected a black dusty mass from diseased grains of wheat and applied them to a plot sown with healthy wheat seed. He showed that the powdery mass caused the bunt of wheat, establishing that the disease is seed borne. The Italian botanist Micheli (1679–1737) collected spores of fungi, sowed them on organic substrate (pieces of melon), and put forth the view that fungi arose from their own spores. He described germination of powdery wheat bunt spores for the first time, and this was confirmed by Prevost (1755–1819). The Tulasne brothers, Louis (1815–1885) and Charles (1817–1884), illustrated spores of several fungi. The German botanist Anton de Bary (1831–1888) traced the germination of spores, including those of the rust fungi (Figure 3.1), to mycelium inside the host plant tissue and its eventual external production of disseminative spores. It thus began to be understood that the vegetative mycelium of fungi is mostly hidden inside the substratum and that what is observed are only the externally produced colored spores. Most fungal spores are dark because of melanin pigment. Some fungi, however, produce colored spores. For example, *Penicillium* produces blue-green conidia, *Fusarium* puts forth pink conidia, and *Puccinia* forms pustules containing rust-colored urediospores, while the mushroom fungi discharge yellowish basidiospores.

3.1 MODELS FOR CELLULAR PROCESSES

Following Mendel's discovery of the laws of inheritance, the concept that spores of fungi are analogous to seeds of higher plants led to the use of progeny spores from genetic crosses as a convenient means of obtaining hereditary information. The American mycologist B. O. Dodge perceived the special advantage of the ascomycete fungi wherein the products of a single meiotic event are enclosed in an elongated sac, as for example in *Neurospora*. By dissecting out progeny ascospores in the order they are enclosed in a single ascus from a cross between two strains of mating types *A* and *a* and growing the individual ascospores, Dodge interpreted that the linearly arranged 2*A*:2*a*:2*A*:2*a* ascospores in the ascus implies that a crossover event occurred at the "four-strand stage."

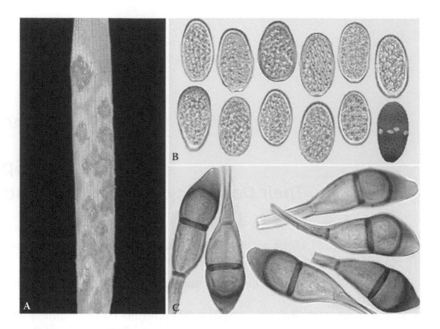

Figure 3.1   Spores of a parasitic rust fungus, *Puccinia graminis tritici.* (A) An infected wheat leaf showing ure-
dosori containing rust-colored urediospores. (B) Urediospores. Note germ pores in a urediospore at
bottom right. (C) Teliospores. (From Jacolyn Morrison, University of Minnesota, with permission.)

In *Coniochaeta tetraspora* only four of the original eight ascospores survive while four abort (Raju
and Perkins, 2000). One, and only one, of the two haploid nuclei entering the zygote must carry some
altered element that is segregated into two of the four meiotic products and is eliminated when ascospores
that contain it disintegrate. A mechanism of programmed cell death must exist that marks the nuclei for
destruction. Spores are thus excellent models for studying cell differentiation (Figure 3.2).

3.2 PLEOMORPHISM

The majority of fungi form more than one type of spore, either by asexual or sexual mode of
reproduction—a feature termed pleomorphism (Table 3.1). The most complex example of pleomorphism is seen in the parasitic rust fungi, which typically require two quite unrelated plant hosts
(wheat and barberry) to complete their life cycle (asexual and sexual stages), producing five different
types of spores: The urediospores, teliospores, basidiospores, pycniospores, and aeciospores. The
asexual spores, called conidia, are produced by successive mitotic divisions. These are typically
5–10 μm, thin walled, and are effective agents for long-range dispersal. The sexual spores are produced as a result of nuclear fusion, followed by meiosis. The sexual spores are about twice or more
as large as the asexual spores, have a multilayered cell wall (spore coat), and are capable of surviving for periods extending to years. A multilayered cell wall is also present in nondormant spores
such as the conidia of *Aspergillus nidulans*, raising doubt whether a thick spore wall is indeed a
factor in dormancy (Florance *et al.*, 1972).

The production of asexual and sexual spores varies in time and space. The asexual spores are
produced by the actively growing mycelium early in the life cycle and are liberated mostly externally
into air, whereas the sexual spores are formed late and may be submerged inside the substratum.
Not surprisingly, the asexual (anamorph) and sexual (teleomorph) stages of the same fungus may
be missed. Cases are known where the two stages of the same fungus were discovered separately
by different investigators and given different names. For example, *Septoria tritici* (anamorph), a

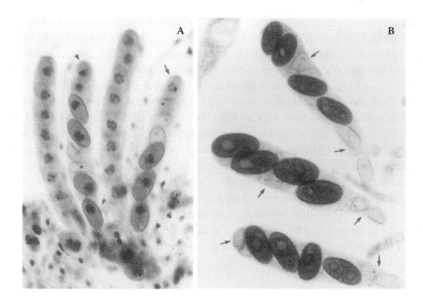

Figure 3.2 Programmed cell death in spores of *Coniochaeta tetraspora* visualized by Hematoxylin staining. (A) Four viable and four aborted ascospores in ascus (12 days) (B) Maturing asci, each with two pairs of large, normally developing, ascospores and two pairs of small, aborted ascospores (arrows; 15 days). (From Raju, N. B. and Perkins, D. D. 2000. *Fung. Genet. Biol.* 30:213–221. With permission of Elsevier.)

Table 3.1   Comparison of Three Different Types of Spores Produced by *Neurospora crassa*

	Macroconidia	Microconidia	Ascospores
Shape	Oblong to spherical	Ovoid	Spindle-shaped
Size	5–9 μm	2.5 × 3.5 μm	27–30 × 13–15 μm
Color	Pink-orange	Greenish-gray	Black
Pigments	Carotenoid	Melanin and carotenoid	Melanin
Development	Blastic (yeast-like budding)	Phialidic	By invagination of preformed ascospore membranes that enclose a nucleus and the surrounding cytoplasm
Nutritional conditions for development	High sugar and nitrogen	Low sugar and low nitrogen	Low sugar and low nitrogen
Production in nature	Liberated externally	In tissue pockets	Submerged, in tissue pockets
Function	Dispersal	Fertilization	Dispersal
Wall	Thin	Thin	Thick
Nuclear number	1–10	1	30–40
Function	Substrate conditioning for sexual reproduction	Fertilization	Genetic recombination and long-term survival
Endogenous reserves	Lipid	Lipid	Lipid and trehalose
Longevity	< 1 month at room temperature	< 1 week	> 1 year at room temperature
Germination	High	Low	High

plant pathogenic fungus, was first described in 1843 on living wheat plants, whereas *Magnaporthe graminicola* (teleomorph) was described 133 years later on dead host (wheat) stubble, undoubtedly on the proper balance of a number of factors. This example illustrates the fact that the conditions for production of asexual and sexual spores may be quite different. It is a challenging task to connect the anamorph and teleomorph states produced at different times or in different situations (see the Appendix). Even in the most intensively studied fungus, the roles of production of macroconidia and microconidia could be elucidated only after the life history of the fungus as found in nature was examined based on reconstruction experiments in the laboratory (Pandit and Maheshwari, 1996).

Considering the range of environmental conditions under which spores are formed, their morphology and dimensions are remarkably invariant and these parameters have been used in the classification schemes for fungi. As stated, some fungi exhibit pleomorphism; determining the respective roles of different spore types can be very challenging, involving single-spore isolation, months of continuous inoculation on possible substrates, and incubation. Even in the fungus *Neurospora*, the roles of relatively large-size macroconidia and small-size microconidia had remained undefined until recently. The term *spores* and *conidia* refer to any dispersive cell—uninucleate or multinucleate, produced asexually or sexually.

3.3 DISSEMINATION

Spores of certain fungi are unmistakably recognized, and the trajectory of their spread can be tracked by using adhesive-coated slides. In North America wheat is grown from south-central Mexico to the prairie provinces of Canada, a distance of about 2500 miles. In 1923, using spore traps at different places, E. C. Stakman and collaborators showed that wheat stem rust, *P. graminis tritici,* moved this distance in two months, allowing perpetuation of the obligate parasitic rust fungi. Currently, a race of wheat stem rust discovered in Uganda, designated Ug99, is a threat since the wheat varieties in its likely migration path (Figure 3.3) are susceptible to this race. Aided by Hurricane Ivan, the rust on soybean caused by *Phakopsora pachyrhizi* has reached North America from its center of origin in China.

3.4 LONGEVITY

Sussman and Halvorson (1966) collected data on the longevity of spores and vegetative cells and the conditions under which they survive. The water content of most air-dry spores is 10–15%. Spore types within a single organism can show great variations in longevity. For example, the conidia of *Neurospora crassa* from agar-grown cultures stored at 5°C remain viable for 2–3 years, whereas the ascospores remain viable for 18 years. The longevity of conidia of *Aspergillus oryzae*—a fungus of industrial importance is > 22 years, presumably under refrigerated conditions. One of the reasons for the use of *Neurospora*, *Aspergillus*, *Penicillium*, and *Fusarium* in research is the longevity of their conidia, allowing cultures to be maintained and revived as and when needed.

3.5 STRUCTURE

3.5.1 Topography

The surface properties of spores are being examined by atomic force, scanning (SEM) and transmission (TEM) electron microscopy (Figure 3.4). A common constituent of all air-dispersed spores is the presence on the surface of spore wall of a cystein-rich protein called hydrophobin, which is recognizable in SEM as nano-sized parallel groups of rodlets. For example, the dormant

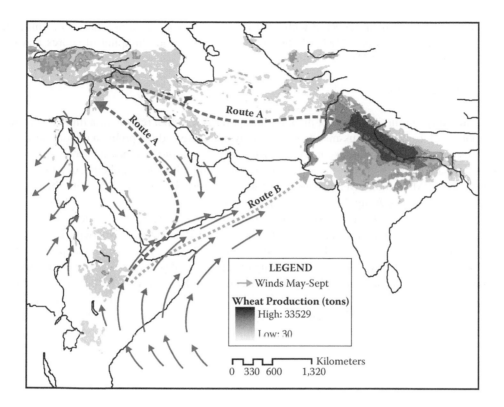

Figure 3.3 Urediospore dissemination by air currents. Theoretical dispersal of urediospores of *Puccinia graminis tritici* race Ug99 of wheat stem rust, which appeared in Uganda in 1998. (Reproduced from Singh, R.P., Hodson, D.P., Jin, Y., Huerta-Espino, J., Kinyua, M.G., Wanyera, R., Njau, P., and Ward, R.W., *CAB Reviews*, 1(54), 2006. With permission.)

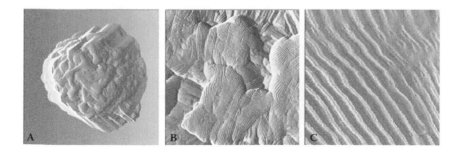

Figure 3.4 Topography of *Aspergillus nidulans* conidia captured using atomic force microscopy. (A) A single conidium, approximately 10 µm. (B) Distribution of fascicle hydrophobin rodlets on outer spore wall. Image width approximately 1 µm. (C) Enlarged image of hydrophobin rodlets. Image width approximately 0.1 µm. (From Mark R. Marten, with permission.)

conidia of *Aspergillus* sp. are covered with 10±1 nm rodlets grouped in parallel and forming fascicles in different orientations. Hydrophobin prevents wetting and clumping of spores and gives protection to spores against UV.

3.5.2 Wall Structure

The difficulty in thin sectioning of thick-walled spores is the chief reason for a limited knowledge of the spore wall. Lowry and Sussman (1968) partially circumvented the problem of poor penetration of fixatives and sectioning of ascospores of *Neurospora tetrasperma* by cracking spores in a solution of permanganate between glass, after which the spores were dehydrated and embedded in epon for sectioning for TEM. A multilayered structure of wall was revealed (Figure 3.5), with the outermost spore layer having ridges. A highly complex wall structure was observed in the spores of a mycorrhizal fungus, *Gigaspora margarita*, retrieved from soil by a sieving and decanting technique (Sward, 1981). Its spore wall is four-layered, 8–13 μm thick. The spores stained dark purple-blue with Toluidine blue, light blue with Coomassie brilliant blue, faint blue/black with Sudan black B, and light pink following the periodic acid Schiff (PAS) reaction, indicating the presence of polysaccharide, protein, and lipid. Following surface sterilization and plating of spores on agar, the germ-tube initially proceeded with the breakdown of the inner spore wall layers (III and IV) followed by the intrusion of cytoplasm from the germination plate region. Although dormancy is commonly ascribed to the presence of a thick wall, there is no conclusive evidence to support this. Heat-activated ascospores of *Neurospora* are reversibly deactivated by subsequent storage at 4°C for one or two days or incubated under anaerobic conditions. Transmission electron microscopy of germinating ascospore of *Neurospora tetrasperma*

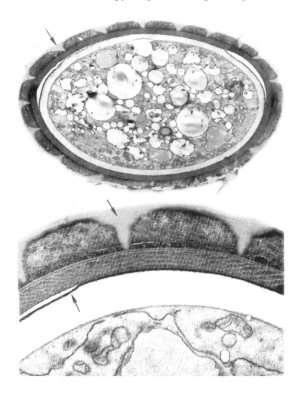

Figure 3.5 *Neurospora tetrasperma* ascospore image of a section by electron microscope. The outer arrow points to a thick outer layer of multi-layered spore wall. The inner arrow points to inner wall layer detached in sectioning. (From Lowry, R.J. and Sussman, A.S., *J. Gen. Microbiol.* 51, 403–409, 1968. With permission.)

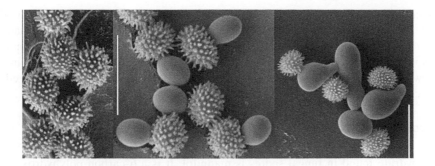

Figure 3. 6 Ascospores of *Talaromyces macrosporus*. Images by low-temperature scanning electron micros-copy. Left, spinulose outer wall. Middle, prosilition 120 min after heat treatment. Right, a hyphal tip development in prosilited cells after 350 min. Bar is 10 μm. (From Dijksterhuis, J., van Driel, K.G.A., Sanders, M.G., Molenaar, D., Houbraken, J.A.M.P., Samson, R.A., and Kets, E.P.W., *Arch. Microbiol.* 178:1–7, 2002. With permission.)

revealed that the germ tube is continuous with the innermost layer of spore wall and it emerges through a germ pore—an area where spore wall is thin (Sussman and Halvorson, 1966, p. 124). Indeed careful examination has revealed germ pores in thick-walled spores of several fungi. There is no conclusive evidence that spore wall thickness is related to dormancy.

3.6 WATER RELATIONS

The water content in spores, estimated as weight decrease upon heating, has varied from 6 to 25%, but no standardized temperature or time of drying was used in estimation of water content. For the conidia of powdery mildews, which germinate without exogenous water, it is 52–75% of their fresh weight. Van Laere *et al.* (1987) determined the water content of conidiospores of *Phycomyces blakesleeanus* using proton nuclear magnetic resonance (NMR) spectroscopy. These spores have about 70% water. Since up to 35% of spore dry wt is trehalose, this translates into an intracellular trehalose concentration greater than 1 M.

Ascospores of *Talaromyces macrosporus* (Figure 3.6) survive a heat treatment at 85°C for 100 min or high pressurization at 1000 MPa for 5 min and have withstood 17 years of storage (Dijksterhuis *et al.*, 2007). Their germination is triggered by a short heat treatment at 85°C. At the time of germination, the ornamented spore coat of ascospores splits in seconds and the complete inner spore is released (prosilition). Using a labeled spin that penetrates spores, the viscosity of spore cytoplasm has been correlated with dormancy by electron spin resonance spectroscopy, which provides information about the freedom of molecular movement within the region of the membrane, indicating phase change. The cytoplasmic viscosity is much higher in ascospores than in conidia of several *Penicillium* species. The cytoplasmic viscosity in the activated ascospores decreases upon swelling and this is accompanied by increase in respiratory activity. Another example of highly resistant ascospores is *Neosartorya fisheri*, which survive heating at 95°C for 60 min. Heat-resistant spores cause contamination problems in canning of food.

3.7 ENDOGENOUS SUBSTRATES

3.7.1 Lipids

After the ascospores of *Neurospora* have been heat activated by moist heat at 60°C for 20–30 min, they are fully capable of germination in distilled water. The respiratory quotient (RQ; CO_2

evolved/O_2 consumed) of dormant spores measured by the manometric method was 0.57. It rose dramatically to 1.04 upon activation, indicating that dormant spores utilize lipid and change over to lipid + sugar (trehalose). The spores of mycorrhizal fungi have higher lipid content (approximately 45%) than spores of nonmycorrhizal fungi (1 to 35%). Time-lapse photomicrography of mycorrhizal hyphae showed that lipid bodies move along the hyphae, indicating that lipid is utilized until the germ tubes connect to the roots of the surrounding plants.

3.7.2 Polyols

A carbohydrate found in fungal spores is mannitol. It is a 6-carbon polyhydroxy sugar that is synthesized by mycelium from fructose via fructose 6-phosphate and is reduced to mannitol 1-phosphate by NAD-dependent mannitol 1-phosphate dehydrogenase (MPDH) (Figure 3.7). Dephosphorylation of mannitol 1-phosphate by mannitol 1-phosphate phosphatase (MPP) yields mannitol. It comprises 15% of the dry weight of *A. niger* conidia. Its roles include as a reserve carbon source, as an antioxidant, for regenerating reducing power in NADPH (nicotinamide adenine dinucleotide phosphate hydrogen), and as a carbon compound that is transported through hyphae (Ruijter *et al.*, 2003). Complementation of *E. coli mtlD* mutant lacking mannitol phosphate dehydrogenase by transformation with an *A. niger* cDNA library and selection for transformants capable of growth on mannitol as the sole carbon source led to cloning and characterization of the *A. niger mpdA* gene. A deletion strain *A. niger* lacking mannitol dehydrogenase enzyme formed conidia. However, since mannitol supplied in the medium repaired this deficiency, mannitol appears to be essential for the protection of *A. niger* spores against high temperature rather than providing an energy source. It is widely assumed that polyols (mannitol, ribitol, arabitol)—compatible solutes—protect proteins from denaturation through direct interactions with enzymes and other proteins.

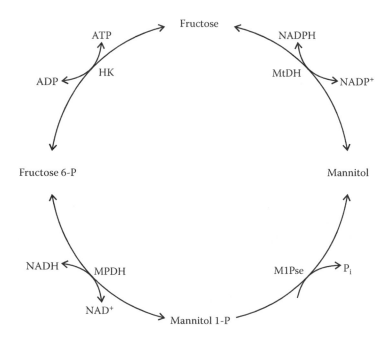

Figure 3.7 Mannitol cycle.

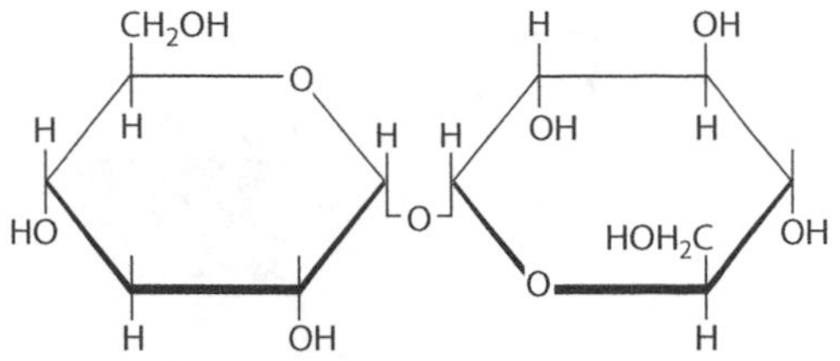

Figure 3.8 Chemical structure of trehalose.

3.7.3 Trehalose

Spores of fungi generally contain trehalose (Figure 3.8). The ascospores of *Neurospora* contain up to 10% dry wt of trehalose, whereas conidia of *Aspergillus* have about 4% of spore dry wt as trehalose. Interestingly, in spite of carbon reserves, many spores still require exogenous carbon for germination, raising a question on the significance of occurrence of trehalose in spores. Conidia of *Botrytis cinerea* conidia contain approximately 1.8% trehalose (Doehlemann *et al.*, 2006). A sequence encoding a putative trehalose-6-phosphate phosphatase was also identified in the as-yet not annotated *B. cinerea* genome, indicating that the trehalose synthesis pathway is identical to that in yeast and *A. nidulans*. It is now thought that trehalose has a regulatory role in the utilization of carbohydrate through the glycolysis pathway.

3.8 LINK BETWEEN SPORULATION AND SECONDARY METABOLISM

Various secondary metabolites have been isolated from sporulating cultures that induce precocious sporulation in cultures (Yu and Keller, 2005). For example, in cultures of *Aspergillus nidulans*, discs soaked in 0.1 mg linoleic acid induced sexual development, visualized by halo of yellow Hülle cells, whereas 0.5- and 1-mg treatments induced asexual development as visualized by the production of brown conidia. The precocious sporulation inducers are abbreviated as psi factor.

Coupling of secondary metabolism with morphological development of the fungus appears to be a universal constant in filamentous fungi. Mycotoxin contamination of seeds of peanut and other plants and feed by spores of *Aspergillus flavus* and other species presents a serious food safety issue on a global scale; of current interest is how the sporulation process is linked with production of the mycotoxins aflatoxin, trichothecin, and pigments; whether the genes controlling their production occur in clusters; the involvement of G protein–mediated signaling; and the regulatory elements.

3.9 SELF-RECONDITIONING OF SUBSTRATUM

In the late 19th century, based on experiments with pure cultures, Klebs formulated certain principles of sporulation in fungi. Importantly, the production of the sexual stage depends on conditions quite the opposite of those favoring asexual spores (conidia). Spores can be viewed as containing packaged food—carbon and nitrogen compounds—drawn heavily by mycelium from substratum. The continuous production and dissemination of astronomical numbers of conidia into the environment could be a sessile fungus's way of altering the internal chemical environment of the substratum for sexual reproduction, genetic recombination, and production of resistant propagule. The use of artificial media containing low carbon and nitrogen sources favoring sexual reproduction for genetic experiments attests to the fairly general rule that conditions for asexual and sexual

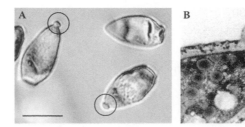

Figure 3.9   Attachment of spore attachment on surface. Left, dry conidia of *Venturia inaequalis* showing extra-cellular matrix material (circles) on their apex (bar represents 10 μm), light microscopy. Right, transmission electron microscopy. (From Schumacher, C.F.A., Steiner, U., Dehne, H.-W., and Oerke, E.-C., *Phytopathology* 98, 760–768, 2008. With permission. Original photograph from E.-C.Oerke, with permission.)

reproduction are quite different: The latter is favored by limiting carbohydrate and nitrogen. Thus, though seemingly wasteful, the mycelium penetrates into the substratum and produces countless asexual conidia externally to create a microhabitat in the growth substrate for female sexual structures to be formed and to be "pollinated" by male cells through the agency of vector microfauna.

3.10 GERMINATION TRIGGERS

3.10.1 Adhesion

Spores do not germinate in hanging drops. Rather, attachment to a hard surface primes the spore for germination. In plant parasitic fungi, firm attachment of the germ tube is necessary for differentiation of specialized infection structures for forcing its entry into the host through stomata. Attachment can be measured by measuring the force required to detach conidia allowed to settle down on various surfaces inside a flow-chamber (Chaky *et al.*, 2001). Attachment of nongerminated spores of *Venturia inaequalis*, which causes scab of apples, is through preformed glue that is extruded immediately after contact with a surface (Figure 3.9) (Schumacher *et al.*, 2008). Based on positive reactions with protein and carbohydrate dyes and the lectin Concanavalin, the glue is a glycoprotein that is extruded from the spore tip. Conidia of *Phyllosticta* have evolved a requirement for attachment to glass and polymeric plastics to trigger germination (Shaw *et al.*, 2006). Adhesion of conidia by spore-secreted mucilage could protect the spore against desiccation. It also may be important in determining the site of the germ tube.

To define the topographic features to quantify the minimum germ tube contact, Apoga *et al.* (2004) microfabricated arrays of pillars (1.4–20 μm) on the surface and found that the germ tubes required approximately 4.5 μm of continuous contact for appressorium development. The *Colletotrichum* spores are surrounded by a fibrillar spore coat comprised of several major glycoproteins. Treatment with protease or a monoclonal antibody that recognized the glycoprotein(s) was able to inhibit adhesion, suggesting that an intact spore coat is required for adhesion to a hydrophobic surface (Rawlings *et al.*, 2007) and for appressorium formation by germ tube. Physical contact of spores with the surface apparently is important in recognition of surface texture in the early phase of their interaction with the host. Since an antagonist compound completely inhibited conidial germination and appressorium formation in *Colletotrichum* conidia, this suggested that CaM is involved in this process. A putative CaM kinase (CaMK) cDNA was cloned with transcripts from hard-surface-treated conidia. A selective inhibitor KN93 inhibited adhesion and germination. Adhesion of conidia signals germination and appears to select the site of emergence of germ tube.

3.10.2 Heat

The observations of fungal blooms on burnt trees or burnt-over soils following a forest fire led to the view that dormant spores are activated by heat. The preferential occurrence of species of *Ascobolus* on herbivore dung is explained on the basis that the violently discharged ascospores stick on plant material and are activated as the grazed material passes through the alimentary canal (37°C). Based on this observation B. O. Dodge developed the heat-shock procedure of activating dormant ascospores of *Neurospora*. In this case heat activation is one of the advantages in genetic studies. Tetrad analysis requires that each meiotically formed spore be free from contaminating mitotically produced asexual conidia. A heat shock (30 min at 60°C) kills contaminating asexual conidia while it activates the ascospore. Surprisingly, the mechanism of heat activation of ascospores remains unknown. In crosses with wild type and the *per-1* mutant of *N. crassa*, ascospores germinate without heat shock and are killed by a heat treatment normally used to activate ascospores. This mutant may be useful in investigating the heat-activation phenomenon.

Because of long-distance aerial dissemination, consider the scenario of urediospores of the wheat stem rust *Puccinia graminis* f. sp. *tritici* blown over the Arctic and landing on the moist surfaces of wheat plants. Will these spores retain a capability of infecting plants? Following exposure of urediospores to temperatures below freezing, germination of urediospores is markedly reduced, even after prolonged thawing at room temperature (Maheshwari and Sussman, 1971) but is restored if the spores were briefly warmed to 40°C. This phenomenon was called "cold dormancy." However, cold dormancy is not reversed if the cold-exposed urediospores contact liquid water, as they undoubtedly would encounter on a moist wheat-leaf surface for germination. If cold-dormant urediospores showing no external symptoms of damage are suspended in liquid without prior heat shock, there is an immediate leakage of metabolites. Such leakage was two to three times greater than from untreated or heat-shocked cold-dormant spores and accounts for up to 70% of the soluble pool of metabolites normally present in germinating urediospores. Cold exposure transformed the air-dry spores into a state of hypersensitivity to liquid water, resulting in irreversible permeability damage.

3.10.3 Light

A common fungus in soil that derives nutrition through decomposition of chitin and cellulose is *Trichoderma atroviride*. Since the fungus can also parasitize species of root-infecting pathogenic fungi, it has potential in biocontrol of root-borne pathogenic fungi; hence large amounts of conidia will be required for its use in the field. It is therefore of practical interest to determine the environmental factors that control sporulation. In *T. atroviride*, a pulse of blue/UVA light induces the synchronous production of conidia situated at the colony perimeter where the pulse was received. The early physiological responses in *Trichoderma viride* induced by light include changes in protein phosphorylation patterns and membrane potential (Gresik *et al.*, 1988, 1989), an increase in the activity of adenyl cyclase (Kolarova *et al.*, 1992), as well as an increase in the levels of cyclic adenosine monophosphate (cAMP) and adenosine-5'-triphosphate (ATP) (Gresik *et al.*, 1988).

3.10.4 Chemicals

Germination of spores of several species of fungi is specifically stimulated by a variety of chemicals (Sussman and Halvorson, 1966). Apparently the chemical stimulants were discovered by chance. While a synthetic minimal medium was being used in which D-xylose was the carbon source, Mary Emerson observed a consistently high percentage of spontaneous germination of ascospores of *Neurospora crassa* that was traced to the production of furfural. Furfural can be produced from xylose upon heating. Xylan, which is a polymer of xylose, is a common constituent of plant cell walls. Pandit and Maheshwari (1997) gave evidence that furfural is released from burnt vegetation in soil and is a natural activator of ascospores active at 10^{-4} M to 10^{-5} M. Many volatile chemicals carry hydroxyl (–OH), carbonyl (–C=O), aldehyde (–CH=O), carboxyl (–COOH), amine (–NH2), sulfhydryl (–SH), or other functional groups. In addition, they may be linear, branched, or cyclic, and saturated or unsaturated, possessing ethylene (–CH=CH–) or acetylene (–C–C–) bonds to stimulate spore germination. French (1992) illustrates *in situ* germination of urediospores of several species of *Puccinia* on infected wheat plants kept in dew chambers. The mycorrhizal fungi are almost indispensable for plant growth. Following a finding that sesquiterpene lactones are seed-germination stimulants for the parasitic weeds *Striga* and *Orobanche*, strigolactone exuded from host roots has been identified as an inducer of germination and germ-tube branching in arbuscular mycorrhiza (AM) fungi (see Figure 3.8). A synthetic analogue of strigonolactone induced hyphal branching at picogram to nanogram levels in the assay.

3.11 MATERIAL FOR PROBING FUNGAL PHYSIOLOGY

3.11.1 Prerequisites for Development

An early recognizable event in germination is about a 1.5–2-fold increase in the diameter of spores (Gottlieb and Tripathi, 1968). That the process of swelling is not a simple passive phenomenon due to uptake of water was demonstrated by trapping $^{14}CO_2$ during spore swelling under anaerobic conditions. Conidia of *Penicillium atroventum* required exogenous glucose, phosphate, and oxygen and the presence of a nitrogen source in the medium for swelling. A nitrogen source in the medium was required for germ-tube formation. The optimum pH for both swelling and germ-tube formation was between 4.0 and 8.0. Spores harvested from cultures more than four weeks old resulted in reduced percentages of swollen spores. Swelling was dependent on respiratory and synthetic processes, and carbon dioxide was produced during the entire swelling period. The inhibition of energy formation by azide and 2,4-dinitrophenol as well as fluoride and malonate prevented swelling and carbon dioxide production. During swelling, RNA, DNA, protein, lipids, and cell wall components were synthesized from glucose. Inhibitors of protein synthesis prevented swelling. During swelling ^{14}C uracil was incorporated into RNA, and ^3H-labeled thymidine into DNA. Using ^{14}C glucose, it was demonstrated that RNA, DNA, and protein synthesis of RNA occurred during swelling. Since swelling was stopped by azide, it was dependent on an energy source. Yanagita (1969) found that swelling of *Aspergillus niger* did not occur in CO_2-free air; the maximum number of spores swelled in 0.1% CO_2 (v/v). Swelling involves entry of water inside the spores and increased respiratory activity. An analysis of *A. fumigatus* showed the presence of prepackaged mRNA transcripts (Lamarre *et al.*, 2008) in dormant conidia. During incubation at 37°C in a nutritive medium, 19% of the total number of genes were up-regulated, whereas 22% of the genes were down-regulated, with most modifications occurring during the first 30 min of germination.

3.11.2 Release of Self-Inhibitors and Efflux of Carbon Compounds

Allen (1972) found that floating spores *en masse* for a few minutes in water leaches the natural inhibitors present in the spores. From the spore efflux, methyl ferulate and methyl 3,4-dimethoxy-cinnamate was purified. These compounds inhibited urediospore germination at about 7 and 140 nM, respectively. Another self-inhibitor identified in rust urediospores is methyl-*cis*-3,4-dimethoxy-cinnamate. In an experiment, carbon-labeled urediospores collected from infected wheat plants that had been allowed to photosynthesize $^{14}CO_2$ were suspended in a buffer and the released radioactivity was monitored in the presence and absence of a semi-purified preparation of self-inhibitor. The

Table 3.2   Cold-Induced Dormancy in Urediospores of *P. graminis* f. sp. *tritici* and Its Reversal by Heat Shock

Treatment	Radioactivity (cpm)	% Germination in medium × 10⁻²
Control	464	55
CS	1393	0
CS-HS	510	52
CS-HS-CS	1439	0
CS-HS-CS-HS	597	48
CS-HS-CS-HS-CS	1439	0
CS-HS-CS-HS-CS-HS	624	61

Note: Radiolabeled urediospores of this obligate plant pathogenic fungus were produced by infecting wheat plants and allowing the plants to photosynthesize $^{14}CO_2$ in a sealed chamber. CS, cold shock of 30 min at −196°C; HS, heat shock of dry spores sealed in an ampoule for 10 min in a water bath at 40°C prior to incubation in liquid.
Source: Data from Maheshwari, R. and Sussman, A.S., *Plant Physiol.* 47, 289–295, 1971.

self-inhibitor of *Rhizopus oligosporus* is nonanoic acid. Self-inhibition has the obvious advantage of preventing rapid germination of all spores at the same time and place, which ensures survival in fluctuating environmental conditions and stimulates dissemination in nature. The mode of action of fungal self-inhibitors has not been elucidated. The presence of innate inhibitory compounds may be a timing device—it ensures that germination does not occur unless the spore is removed from the parent and is spread to ensure survival of the fungus.

3.11.3 Cold Dormancy

In Section 3.3 we noted that air currents disseminate urediospores over long distances. Do these spores remain viable and capable of infecting after long-distance dissemination through freezing conditions? Spores were cold shocked (CS) for 30 min in liquid nitrogen at −196°C and subsequently briefly heat shocked. Air-dry ^{14}C-labeled urediospores were sealed in seven tubes and treated as shown in Table 3.2 to temperatures below freezing. Even after prolonged thawing at room temperature, urediospores did not germinate after several cycles of alternate cold shock and heat shock, a phenomenon that was called cold dormancy. Germination of urediospores was fully restored by a brief heat shock or by hydration through the vapor phase but not if these were placed in water. Suspension of cold-dormant urediospores in liquid without a prior heat shock enhanced loss of metabolites. Therefore, sensitivity to environmental factors, rather than resistance, characterizes cold dormancy in urediospores. In view of the generalized nature of leakage from cold-shocked spores, it is thought that temperature causes physical changes in the lipoprotein of cytoplasmic membranes. Restoration of germination of cold-dormant urediospores by vapor phase hydration is analogous to the situation in dry pollen or seeds that leak solutes. In these systems the mechanism of leakage is explained on the basis of thermotropic phase transition of the phospholipid cell membrane from the gel to liquid crystalline phase (Hoekstra *et al.*, 1998). Phase transition due to dehydration or chilling is thought to underlie the leakage from chilled urediospores, as has been observed from dry seeds and pollen grains. It is significant that either a heat shock or prehumidification through vapor phase prevented leakage and preserved viability. Good germination of urediospores (Figure 3.10) always coincided with efflux over the course of germination, showing that growing systems exchange material with the environment.

3.11.4 Genome Activation

Almost all fungal spores examined show an increase in protein synthesis prior to RNA synthesis. A notable exception reported were the urediospores of the obligatory parasitic rust fungi.

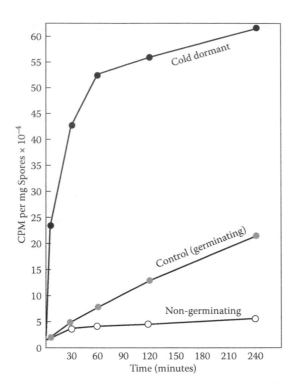

Figure 3.10   Kinetics of efflux of carbon compounds from ¹⁴C-labeled urediospores of *Puccinia graminis tritici*.

The urediospores germinate in pure water, but still must increase their DNA, RNA, and protein content as evidenced by nuclear division and cellular differentiation of the germ tube upon contacting a hydrophobic membrane (described in Chapter 6). Here we shall merely note that dependency of spore germination and early events on RNA and protein synthesis could be inferred simply by studying their germination in the presence of metabolic inhibitors (Figure 3.11).

3.11.5 Dark Fixation of CO_2

Yanagita (1957) reported that in flasks equipped with a center well, such as the Warburg flasks, containing KOH or when CO_2-free air was bubbled into the medium, spores of *Aspergillus niger* neither swelled nor germinated, implying the requirement of CO_2 for germination. Conidia of *Penicillium atroventum* required exogenous glucose, phosphate, and a nitrogen source, indicating that swelling is dependent on respiratory and synthetic processes. The inhibition of energy formation by azide and 2,4-dinitrophenol as well as fluoride and malonate prevented swelling and carbon dioxide production. That swelling and early germination are dependent on CO_2 suggests that a critical event in the exit from dormancy is anaplerotic reactions for providing the tricarboxylic acid (TCA) cycle intermediates and ATP for biosynthesis of RNA, DNA, protein, lipids, and cell wall components. This view is supported by demonstration of activities of anaplerotic enzymes: Pyruvate carboxylase and phosphoenolpyruvate carboxykinase in spores.

3.11.6 Respiratory Increase

The respiration of spores markedly increases upon germination. One exception to this rule was reported for urediospores of wheat stem rust *Puccinia graminis tritici*. An analysis of the reasons

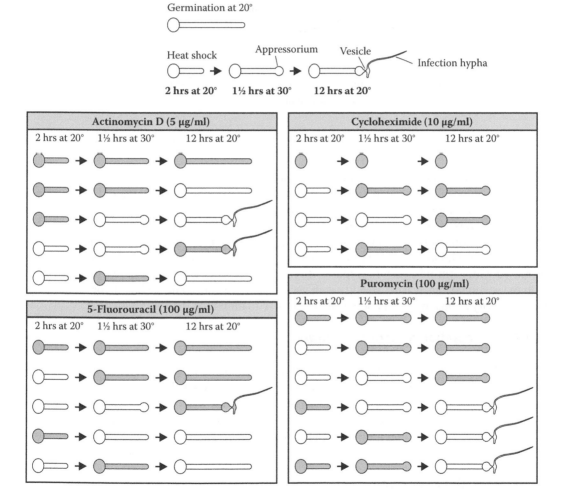

Figure 3.11 Use of inhibitors to demonstrate macromolecule synthesis in urediospores floated on aqueous medium.

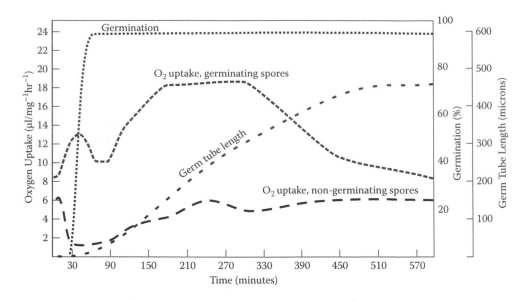

Figure 3.12 Development of respiratory activity during germination of urediospores *Puccinia graminis tritici*. (Data from Maheshwari, R. and Sussman, A.S., *Phytopathology* 60, 1357–1364, 1970.)

for the reported lack of increased respiration revealed that the results could be faulted because measurements were essentially of spores that, although put in an aqueous medium, did not germinate because of self-inhibition. Hence respiration was measured by two methods: Warburg's manometric method and polarographically using an oxygen electrode. When urediospores were germinated in suspension using a buffer and a non-ionic detergent in the presence of 10^{-4} M nonyl alcohol to counteract self-inhibition, and the respiration measured using an oxygen electrode, their respiration increased with germ-tube growth (Figure 3.12). Of significance is the biphasic respiratory activity, with oxygen uptake increasing steadily with germ-tube elongation. The self-inhibitor affects respiratory activity although germination is more sensitive to inhibition than respiration.

3.11.7 Regeneration of Reducing Power

The near ubiquitous occurrence of mannitol in fungi and its disappearance from germinating spores suggests that it plays a role in the storage of carbon or energy. The enzymes of the mannitol cycle are present in many fungi and it has been proposed to have a role in NADPH production in the fungus *Alternaria alternata*. Although there is yet no evidence that mannitol turnover is required for NADPH—an electron donor production in reductive biosynthesis—the presence of mannitol is easy to understand.

3.11.8 Synthetic Pathways

Since the majority of carbon stored in AM fungal spores is as lipid and not as carbohydrate, one would expect the conversion of lipids to hexose to be important at this stage of the fungal life cycle. Stimulation of this catabolic pathway would provide a higher amount of acetyl-CoA and accelerate the citric acid and glyoxylate cycles, leading to higher NADPH and ATP synthesis required for more active anabolism. Using ^{13}C-labeled substrates and nuclear magnetic resonance spectroscopy to study carbon fluxes during spore germination of *Glomus intraradices*, Bago *et al.* (1999) found labeling patterns consistent with significant carbon fluxes through gluconeogenesis, the glyoxylate cycle, the tricarboxylic acid cycle, glycolysis, nonphotosynthetic one-carbon metabolism, the pentose phosphate pathway, and most or all of the urea cycle. They concluded that (a) sugars are made from stored lipids; (b) trehalose (but not lipid) is synthesized as well as degraded; (c) glucose and fructose, but not mannitol, can be taken up and utilized; (d) dark fixation of CO_2 is substantial; and (e) arginine and other amino acids are synthesized. A betaine compound was detected during germination.

3.11.9 Sensing, Signaling, and Transcriptional Changes

A variety of environmental signals in different fungi—availability of water, contact with a hard surface, availability of a carbon source, and in some cases some ions such as Ca^{2+}—signal a change in the spore from isotropic growth (swelling) to polarized growth (germ-tube emergence). In almost all cases the signal is relayed to the nucleus for DNA synthesis to commence and nuclei to replicate before the germ tube emerges from the thick wall. It is therefore believed that, as in other eukaryotes, the key event is sensing of the stimulus and signaling to the genome (Figure 3.13). The sequencing of genomes of filamentous fungi has made it possible to undertake a comprehensive analysis of the conidial mRNAs since this may reveal which proteins need to be translated at the very beginning of germination and control conidium dormancy. It has identified the heterotrimeric G protein first identified in mammalian cells and conserved in all eukaryotes: G_α, G_β, and G_γ subunits. The molecular details of germination conidia of *Aspergillus nidulans* were identified using spore germination-deficient mutants that are blocked in this process at the restrictive temperature. These mutants defined eight genes that, based on a homology search, were as involved in translation, protein folding, and degree of homology to malonyl CoA synthetase. Conidia contained the *ras* homologue in *A. nidulans*, which germinated in the absence of an inducing carbon source,

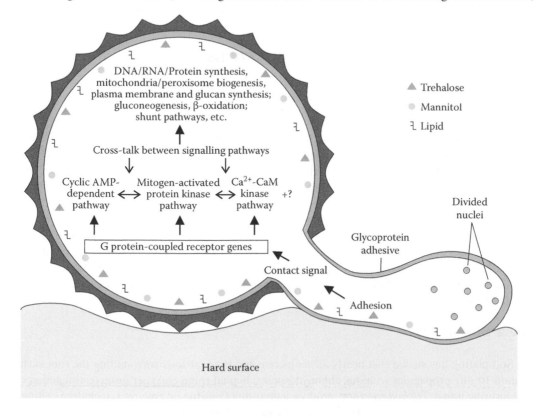

Figure 3.13 Model of cell-signaling events in germinating spore.

suggesting an important role of an as-yet-to-be-determined signaling cascade to a carbon-source-sensing apparatus.

The conidia of *Aspergillus fumigatus* inhaled into the lung cause aspergillosis; hence there is much interest in learning more about this fungus. The genome of *A. fumigatus* was sequenced in 2005. A transciptome analysis with PCR fragments identifies more than 3000 genes (Lamarre *et al.*, 2008). In the first 30 min of incubation of one-week-old conidia, a total of 787 genes (25%) were differentially expressed, in which up- and down-regulated genes accounted respectively for 598 (19%) and 189 (6%) genes. The major results of this analysis are the following: (i) Conidia stored prepackaged mRNA transcripts (27% of genes have transcripts in the resting conidia); (ii) incubation at 37°C in a nutritive medium induced up- and down-regulation of genes, where 19% of the total number of genes deposited on the array were up-regulated, and 22% of the genes with prepackaged mRNA in the resting conidia were down-regulated; (iii) most modifications were seen during the first 30 min of germination, whereas very little modification of gene expression occurred during the following hour; and (iv) one-year-old conidia and one-week-old conidia behaved similarly at the transcriptional level. The functional categories of these genes belong mainly to fermentative metabolism and oxidoreductase activity, whereas up-regulated genes were overrepresented within the RNA and phosphorus metabolism, amino acid and protein biosynthesis, and protein complex assembly. The exit from dormancy is associated with a shift from a fermentative metabolism to a respiratory metabolism as well as a trend toward immediate protein synthesis. A finding is the putative role of transposable elements in the activation of the conidial metabolism during exit from dormancy. In humans, 45% of the genome is composed of transposable elements. Gene expression data were used to predict the cellular and physiological state of each developmental stage for known processes. Predictions were confirmed for several previously unreported developmental events, such as occurrence of peroxisomes and nuclear division within macroconidia prior to spore germination.

3.12 APPLICATIONS

3.12.1 Biocatalysis

The ability of spores of *Penicillium roqueforti*—which can form methyl ketones from fatty acids, giving flavor to blue cheese—had led to research on the use of mold spores in pharmaceutical preparation of regio- and stereoselective hydroxylation. In 1968 it was reported that intact conidia of *Aspergillus ochraceus* converted progesterone into 11α hydroxyprogesterone. Conidia of imperfect fungi and ascomycetes have been used for a variety of conversions: Hydroxylation in 11α, 11β, dehydrogenation in position 1–2; oxidation in 11 and 17; reduction in 20; and cleavage of the side chain with 17 keto formation and complete cleavage. By choosing suitable concentrations of conidia, progesterone, and incubation time, 80–90% of the progesterone in a buffer could be converted to 11α hydroxyprogesterone (Chapter 7). Glucose (0.2–0.4%) was required for complete conversion (Vezina *et al.*, 1963). Fungal steroid hydroxylase has been identified as cytochrome P-450-dependent monooxygenases. Apparently spores have a NADPH-generating system for NADPH-dependent steroid hydroxylation.

3.12.2 Biocontrol

Soil plating has shown that nearly all temperate and tropical soils surrounding the root surface contain 10^1–10^3 propagules (conidia, chlamydospores, hyphal fragments) per gram (estimated as colony-forming units). *Trichoderma* spp. produce a mixture of antifungal enzymes, including chitinases and β-1,3 glucanases. Species of *Trichoderma* are mycoparasitic; that is, the hyphae attach to mycelia of other fungi and upon contact the *Trichoderma* hyphal coils form an attachment organ called an appressorium, whereupon the hyphae secrete fungal cell-wall-degrading enzymes (β-1,3 glucanases), effecting destruction of target fungi. Hence *Trichoderma* strains have emerged for biocontrol of certain root-infecting fungi of crop plants (Harman, 2006). This requires application of spores into the soil in sufficient numbers. In order to monitor growth dynamics and survival of *Trichoderma* spores, studies are currently being made to transform strains incorporating green fluorescent protein to obtain information on important abiotic and biotic factors affecting biocontrol efficacy.

3.13 CONCLUDING REMARKS

Several spores, especially the ascospores of Ascomycotina, are activated upon brief exposure to heat although the mechanism of heat activation is unknown. Spore dormancy is analogous to a state of anhydrobiosis that is reversed by heat shock or slow rehydration through vapor phase. It is likely that dormancy involves alteration in the molecular architecture of the membrane. Some basic questions that may be asked are: What is the chemistry of the spore wall? What is the mechanism of heat activation of spores? What are the functions of different types of spores produced by the same fungus? How does the membrane of dry spores differ from that of hydrated spores? Do dormant spores contain stable mRNA molecules for synthesis of vegetative enzymes? What is the significance of increased heterotrophy of spores compared to mycelium? How does the cytoplasmic microviscosity in short-lived cells (conidia) compare with that in dormant cells (ascospores) of the same species? How do the properties of enzymes in spores differ from those of the same enzymes in the vegetative cells? Why do some spores containing endogenous reserves of carbon still require an exogenous source of carbon compound for initiating germination? What is the mechanism by which multinucleated spores with multiple genomes per nucleus are formed (mycorrhizal fungi)? Additionally, spores continue to be exploited for understanding mechanisms in genetic recombination and in survival under stressful environmental conditions, particularly anhydrobiosis.

REFERENCES

Akiyama, K. and Hayashi, H. (2006). Strigolactones: Chemical signals for fungal symbionts and parasitic weeds in plant roots. *Ann. Bot.* 97:925–931.

Allen, P. J. (1972). Specificity of the cis-isomers of inhibitors of uredospores germination in the rust fungi. *Proc. Natl. Acad. Sci. USA* 69:3497–3500.

Apoga, D., Barnard, J., Craighead, H. G., and Hoch, H. C. (2004). Quantification of substratum contact required for initiation of *Colletotrichum graminicola appressoria*. *Fung. Genet. Biol.* 41:1–12.

Bae, Y.-S. and Knudsen, G. R. (2000). Cotransformation of *Trichoderma harzianum* with β-glucuronidase and green fluorescent protein genes provides a useful tool for monitoring fungal growth and activity in natural soils. *Appl. Env. Microbiol.* 66:810–815.

Bago, B., Pfeffer, P. E., Douds, D. D., Jr., Brouillette, J., Bécard, G., and Shachar-Hill, Y. (1999). Carbon metabolism in spores of the arbuscular mycorrhizal fungus *Glomus intraradices* as revealed by nuclear magnetic resonance spectroscopy. *Plant Physiol.* 121:263–271.

Brodhagen, M. and Keller, N. P. (2006). Signaling pathways connecting mycotoxin production and sporulation. *Mol. Plant Pathol.* 7:285–301.

Brunner, K., Zeilinger, S., Ciliento, R., Woo, S. L., Lorito, M., Kubicek, C. P., and Mach, R. L. (2005). Improvement of the fungal biocontrol agent *Trichoderma atroviride* to enhance both antagonism and induction of plant systemic disease resistance. *Appl. Environ. Microbiol.* 71:3959–3965.

Chaky, J., Anderson, K., Moss, M., and Vaillancourt, L. (2001). Surface hydrophobicity and surface rigidity induce spore germination in *Colletotrichum graminicola*. *Phytopathology* 91:558–564.

Corina, D. L. and Munday, K. A. (1971). Studies on polyol function in *Aspergillus clavatus*: A role for mannitol and ribitol. *J. Gen. Microbiol.* 69:221–227.

Dijksterhuis, J., Nijsse, J., Hoekstra, F. A., and Golovina, E. A. (2007). High viscosity and anisotropy character-ize cytoplasm of fungal dormant stress-resistant spores. *Euk. Cell* 16:157–170.

Dijksterhuis, J., van Driel, K. G. A., Sanders, M. G., Molenaar, D., Houbraken, J. A. M. P., Samson, R. A., and Kets, E. P. W. (2002). Trehalose degradation and glucose efflux precede cell ejection during germination of heat-resistant ascospores of *Talaromyces macrosporus*. *Arch. Microbiol.* 178:1–7.

Doehlemann, G., Berndt, P., and Hahn, M. (2006). Different signaling pathways involving a Gα protein, cAMP and a MAP kinase control germination of *Botrytis cinerea* conidia. *Mol. Microbiol.* 59:821–835.

Dunkle, L. D., Maheshwari, R., and Allen, P. J. (1969). Infection structures from rust urediospores: Effect of RNA and protein synthesis inhibitors. *Science* 158:481–482.

Dute, R. R., Weete, J. D., and Rushing, A. E. (1989). Ultrastructure of dormant and germinating conidia of *Aspergillus ochraceus*. *Mycologia* 81:772–782.

Florance, E. R., Denison, W. C., and Allen, T. C., Jr. (1972). Ultrastructure of dormant and germinating conidia of *Aspergillus nidulans*. *Mycologia* 64:115–123.

French, R. C. (1992). Volatile chemical germination stimulators of rust and other fungal spores. *Mycologia* 84:277–288.

Golovina, E.A., Hoekstra, A., Hemminga, M. A. (1998). Drying increases intracellular partitioning of amphiphilic substances into the lipid phase. Impact on membrane permeability and significance for des-iccation tolerance. *Plant Physiol.* 118:975–986.

Gottlieb, D. and Tripathi, R. K. (1968). The physiology of swelling phase of spore germination in *Penicillium atrovenetum*. *Mycologia* 60:571–590.

Gresik, M., Kolarova, N., and Farkas, V. (1988). Membrane potential, ATP, and cyclic AMP changes induced by light in *Trichoderma viride*. *Exp. Mycol.* 12:295–301.

Harman, G. E. (2006). Overview of mechanisms and uses of *Trichoderma* spp. *Phytopathology* 96:190–194.

Harman, G. E., Howell, C. R., Viterbo, A., Chet, I., and Lorito, M. (2004). *Trichoderma* species—Opportunis-tic, avirulent plant symbionts. *Nature Rev. Microbiol.* 2:43–56.

Hess, W. M., Sassen M. M. A., and Remsen, C. C. (1968). Surface characteristics of *Penicillium* conidia. *Mycologia* 60:290–303.

Hoekstra, F. A., Golovina, E. A., Van Aelst, A. C., and Hemmings, M. A. (1999). Imbibitional leakage from anhydrobiotes revisited. *Plant Cell Environ.* 23:1121–1131.

Hughes, H. B., Carzaniga, R. S., Rawlings, S. L., Green, J. R., and O'Connell, R. J. (1999). Spore surface gly-coproteins of *Colletotrichum lindenuthianum* are recognized by a monoclonal antibody which inhibits adhesion to polystyrene. *Microbiology* 145:1927–1936.

Kasuga, T., Townsend, J. P., Tian, C., Gilbert, L. B., Mannhaupt, G., Taylor, J. W., and Glass, N. L. (2005). Long-oligomer microarray profiling in *Neurospora crassa* reveals the transcriptional pro-gram underlying biochemical and physiological events of conidial germination. *Nuc. Acids Res.* 33:6469–6485.

Kim, Y. K., Li, D., and Kolattukudy, P. E. (1998). Induction of Ca21-Calmodulin signaling by hard-surface con-tact primes *Colletotrichum gloeosporioides* conidia to germinate and form appressorium. *J. Bacteriol.* 180:5144–5150.

Knights, J. K, Day, M. R, and Lucas, J. A. (1982). The ultrastructure of dormant, germinating, and photo-inhibited urediospores of the rust fungus *Puccinia graminis* f. sp. *tritici*. *Protoplasma* 113:57–68.

Kolarova, N., Haplova, J., and Grešík, M. (1992). Light-activated adenyl cyclase from *Trichoderma viride*. *FEMS Microbiology Lett.* 93:75–78.

Kwon, Y. H. and Epstein, L.(1997). Involvement of the 90 kDa glycoprotein in adhesion of *Nectria haemato-cocca* macroconidia. *Physiol. Mol. Plant Pathol.* 51:287–303.

Lamarre, C., Sokol, S., Debeaupuis, J.-P., Henry, C., Lacroix, C., Glaser, P., Coppée J.-Y., François, J.-M., and Latgé, J.-P. (2008). Transcriptomic analysis of the exit from dormancy of *Aspergillus fumigatus* conidia. *BMC Genomics* 9:417. doi:10.1186/1471-2164-9-417.

Liebmann, B., Muller, M., Braun, A., and Brakhage, A. A. (2004). The cyclic AMP-dependent protein kinase a network regulates development and virulence in *Aspergillus fumigatus*. *Infect. Immun.* 72:5193–5203.

Loo, M. (1976). Some required events in conidial germination of *Neurospora crassa*. *Dev. Biol.* 54:201–213.

Lowry, R. J. and Sussman, A. S. (1958). Wall structure of ascospores of *Neurospora tetrasperma*. *Am. J. Bot.* 45:397–403.

Lowry, R. J. and Sussman, A. S. (1968). Ultrastructural changes during germination of ascospores of *Neurospora tetrasperma*. *J. Gen. Microbiol.* 51:403–409.

Maheshwari, R. and Sussman, A. S. (1970). Respiratory changes during germination of urediospores of *Puccinia graminis* f. sp. *tritici*. *Phytopathology* 60:1357–1364.

Maheshwari, R. and Sussman, A. S. (1971). The nature of cold induced dormancy in urediospores of *Puccinia graminis tritici*. *Plant Physiol.* 47:289–295.

Mims, C. W., Richardson, E. A., Clay, R. P., and Nicholson, R. L. (1995). Ultrastructure of conidia and the conidium aging process in the plant pathogenic fungus *Colletotrichum graminicola*. *Int. J. Plant Sci.* 156:9–18.

Osherov, N. and May, G. S. (2001). The molecular mechanisms of conidial germination. *FEMS Microbiol. Lett.* 199:153–160.

Pandit, A. and Maheshwari, R. (1996). Life-history of *Neurospora intermedia* in a sugar cane field. *J. Biosci.* 21:57–79.

Raju, N. B. and Perkins, D. D. (2000). Programmed ascospore death in the homothallic ascomycete *Coniochaeta tetraspora*. *Fung. Genet. Biol.* 30:213–221.

Rawlings, S. L., O'Connell, R. J., and Green, J. R. (2007). The spore coat of the bean anthracnose fungus *Colletotrichum lindemuthianum* is required for adhesion, appressorium development and pathogenicity. *Physiol. Mol. Plant Pathol.* 70:110–119.

Ricci, M., Krappmann, D., and Russo, V. E. A. (1991). Nitrogen and carbon starvation regulate conidia and protoperithecia formation of *Neurospora crassa* grown on solid media. *Fung. Genet. Newsl.* 38:87–88.

Roberts, A. N., Berlin, V., Hager, K. M., and Yanofsky, C. (1988). Molecular analysis of a *Neurospora crassa* gene expressed during conidiation. *Mol. Cell. Biol.* 8:2411–2418.

Roberts, A. N. and Yanofsky, C. (1989). Genes expressed during conidiation in *Neurospora crassa*: Characterization of *con-8*. *Nuc. Acid Res.* 17:197–214.

Ruijter, G. J. G., Bax, M., Patel, H., Flitter, S. J., van de Vondervoort, P. J. I., de Vries, R. P., van Kuyk, P. A., and Visser, J. (2003). Mannitol is required for stress tolerance in *Aspergillus niger* conidiospores. *Euk. Cell* 2:690–698.

Sachs, M. S. and Yanofsky, C. (1991). Developmental expression of genes involved in conidiation and amino acid biosynthesis in *Neurospora crassa*. *Dev. Biol.* 148:117–128.

Salcedo-Hernandez, R., Escamilla, E., and Ruiz-Herrera, J. (1994). Organization and regulation of the mito-chondrial oxidative pathway in *Mucor rouxii*. *Microbiology* 140:399–407.

Schumacher, C. F. A., Steiner, U., Dehne, H.-W., and Oerke, E.-C. (2008). Localized adhesion of nongermi-nated *Venturia inaequalis* conidia to leaves and artificial surfaces. *Phytopathology* 98:760–768.

Seon, K-Y, Zhao, X., Xu, J. R., Güldener, U., and Kistler, H. C. (2008). Conidial germination in the filamentous fungus *Fusarium graminearum*. *Fung. Genet. Biol.* 45:389–399.

Shaw, B. D., Carroll, G. C., and Hoch, H. C. (2006). Generality of the prerequisite of conidium attachment to hydrophobic substratum as a signal for germination among *Phyllosticta* species. *Mycologia* 98:186–194.

Singh, R. P., Hodson, D. P., Jin, Y., Huerta-Espino, J., Kinyua, M. G., Wanyera, R., Njau, P., and Ward, R. W. (2006). Current status, likely migration and strategies to mitigate the threat to wheat production from race Ug99 (TTKS) of stem rust pathogen. *CAB Reviews* 1(54).

Sussman, A. S. (1992). Procedure for mass-production of *Neurospora tetrasperma* ascospores. *Fung. Genet. Newsl.* 39:84–85.

Sussman, A. S. and Halvorson, H. O. (1966). *Spores: Their Dormancy and Germination*. Harper & Row, New York, p. 25.

Sward, R. J. (1981). The structure of the spores of *Gigaspora margarita*. I. The dormant spore. *New Phytol.* 87:761–768

Thines, E., Weber, R. W., and Talbot, N. J. (2000). MAP kinase and protein kinase A-dependent mobilization of triacylglycerol and glycogen during appressorium turgor generation by *Magnaporthe grisea*. *Plant Cell* 12:1703–1718.

Thomas, M. B. and Read, A. F. (2007). Fungal bioinsecticide with a sting. *Nature Biotechnol.* 23:1367–1368.

Tsitsigiannis, D. I. and Keller, N. P. (2007). Oxylipins as developmental and host–fungal communication sig-nals. *Trends Microbiol.* 15:109–118.

van der Aa, B. C., Asther, M., and Dufrene, Y. F. (2002). Surface properties of *Aspergillus oryzae* spores inves-tigated by atomic force microscopy. *Colloids and Surfaces. B: Biointerfaces* 24:277–284.

van Laere, A., Francois, A., Overloop, K., Verbeke, MC., and van Gerven, L. (1987). Relation between ger-mination, trehalose and the status of water in *Phycomyces blakesleeanus* spores as measured by proton-NMR. *J. Gen Microbiol.* 133:239–245.

Vézina, C., Sehgal, S. N., and Singh, K. (1963). Transformation of steroids by spores of microorganisms. I. Hydroxylation of progesterone by conidia of *Aspergillus ochraceus*. *Appl. Environ. Microbiol.* 11:50–57.

Wolken, W. A. M, Tramper, J., and van der Werf, M. J. (2003). What can spores do for us? *Trends Biotechnol.* 21:338–345.

Xue, T., Nguyen, C. K., Romans, A., and May, G. S. (2004). A mitogen-activated protein kinase that senses nitrogen regulates conidial germination and growth in *Aspergillus fumigatus*. *Euk. Cell* 3:557–560.

Yanagita, T. (1969). Biochemical aspects on the germination of conidiospores of *Aspergillus niger*. *Arch. Microbiol.* 26:329–344.

Zahiri, A. R., Babu, M. R., and Saville, B. J. (2005). Differential gene expression during teliospore germination in *Ustilago maydis*. *Mol. Genet. Genomics* 273:394–403.

PART II

Fungi in Biosphere and Human Health

CHAPTER 4

Fungi as Scavengers

Lignin is the second most abundant constituent of the cell wall of vascular plants, where it protects cellulose towards hydrolytic attack by saprophytic and pathogenic microbes. Its removal represents a key step for carbon recycling in land ecosystems, as well as a central issue for industrial utilization of plant biomass.

—**Francisco J. Ruiz-Dueñas and Ángel T. Martínez (2009)**

Vast quantities of litter comprised mostly of dead plant cell walls are continuously decomposed. The functioning of the ecosystem depends on this crucial process in which communities of microorganisms, with fungi being the principal player, break litter down into smaller molecules. During this process the mineral ions required by living organisms are dissociated from the organic substances with which they were complexed and released into the soil. With their hyphal tips shaped as spears, the fungi penetrate through the pit apertures into the cell lumen, secreting enzymes that break pectin in the middle lamella and effecting separation of cells in dead plant material. Eventually, the cell wall is broken down into myriad small molecular substances for further decomposition into carbon dioxide and water. The fungi eke out a living by absorbing the released nutrients as a source of carbon and energy. Cellulose and hemicelluloses are the chief polysaccharide constituents of litter and wood. However, these polymeric substances are encrusted with lignin, a highly refractory insoluble compound resistant to microbial attack. Hence the key process in the recycling of carbon in nature is lignin biodegradation. The most striking lignin-degrading fungi are those that form fruiting bodies that project out from the woody trunk of trees (Figure 4.1). The isolation and identification of fungi from decomposing litter, their growth in pure cultures, and the characterization of the enzymes and of compounds have provided some insights into this process. The scavenging activity occurs through a number of extracellularly produced glycoside hydrolases, extracellular hydrogen-peroxide-producing enzymes, and low-molecular-weight oxidants, particularly hydroxyl radicals.

4.1 DECAY OF WOOD AND LITTER

Strictly speaking, communities other than fungi comprised of ants, termites, bacteria, and actinomycetes also participate in this process; however, it is the fungi that are considered to play the key role. The occurrence of a fungus where cellulose or lignin is abundant is not enough to infer that the fungus is responsible for decomposing these polymers, since a noncellulolytic fungus may live commensally on products formed by a cellulolytic fungus. The removal of certain plant cell-wall polysaccharides by one species may improve the accessibility of another species to cellulose (polysaccharide composed of glucose molecules joined in β-1,4 linkage) or hemicelluloses (noncellulosic

Figure 4.1   **(See color insert.)** Fruit bodies of basidiomycete fungus on a dead fallen tree trunk in a rain forest. Basidiomycete fungi are specialized for the decomposition of lignin, which is a major structural
component of cell wall in wood. (From Jean Paul Ferrero/Ardea.)

polysaccharide, composed of β-linked pentose with side chains). Direct microscopic observation of
sporulation structures and culture techniques demonstrate that decomposition of biomass involves
activities of a mixed microflora comprising fungi, actinomycetes, and bacteria. The most important
decay-causing organisms are the basidiomycete fungi (Basidiomycotina). Their leathery or woody
fruit bodies (basidiocarps) project out from tree trunks, signaling that the fungal mycelium inside
the plant tissue has been slowly attacking lignin and polysaccharide constituents. Since fruit bodies are frequently seen on living standing trees, a possibility is that the mycelium producing the
fruit bodies started out as mycorrhiza, worked its way up as endophytic mycelium, and produced
basidiocarps (Figure 4.2). The early stages of decay of standing or fallen trees may be due to several
species, but soon competition sets in and only a single individual mycelium may extend to several
meters (Boddy and Rayner, 1982).

4.2 CLUES FROM MICROSCOPY

Light and scanning electron microscopy have revealed many aspects of the interactions
between fungi and wood structure. Figure 4.3 shows hyphae of the soft-rot fungus *Trametes versicolor* inside the lumen of fibers of birch wood, indicating loss of lignin from the lumen toward

Figure 4.2 Fruit bodies of a wood-rot fungus on a living tree. The photo suggests that many species of wood-rot fungi may be endophytic starting as mycorrhizal fungi. (Courtesy of T. Suryanarayana.)

the middle lamella. *Phialophora melinii* extends into the middle lamella of secondary cell walls in birch wood, causing diamond-shaped cavities (Figure 4.4). Wood fibers separate as a consequence of advanced delignification.

The decay of timber is monitored by a chemical extraction procedure: Lignin is quantitatively extracted by chlorite-acetic acid. Hemicelluloses are extracted as alkali-soluble material, whereas cellulose is insoluble in alkali. Extractions and estimations of these materials show that the various wood-decay fungi differ in their abilities to digest lignin, cellulose, and xylan. The hyphae grow adhering to the cell wall, enzymatically fragmenting it and transforming the decayed plant material into humus—a dark substance with a content of phenolic compounds. Lignin is not used as the sole carbon source as demonstrated by the inability of the wood decay fungi *Phanerochaete chrysosporium* and *Trametes versicolor* to evolve $^{14}CO_2$ from ^{14}C-labeled lignin in culture conditions. The presence of a co-substrate from which hydrogen peroxide can be generated is essential for the growth of fungus and for lignin to be depolymerized. This suggests that lignin degraders simultaneously break down polysaccharide constituents of litter. New equipment for monitoring decay in standing trees include the Arborsonic Decay Detector, Fractometer, Digital MicroProbe, increment borers, Picus Sonic Tomograph, portable compression meter, Resistograph, Shigometer, Stress Wave Timer, and Elastometer.

4.3 WHITE- AND BROWN-ROT FUNGI

The decay of lignocellulose is brought about by three major types of fungi. The *white-rot* fungi (Basidiomycotina) preferentially degrade the brown-colored lignin, leaving the white cellulose and giving the wood a bleached or pale appearance and transforming it into a fibrous mass that crumbles with a blow. These fungi produce polyphenol oxidase and laccase, which oxidize phenol compounds, mineralizing lignin all the way to CO_2 and H_2O. Electron microscopy of wood decayed by white-rot fungi using immunogold labeling of lignin peroxidase showed the highest concentrations of gold particles in the middle lamella and secondary wall (Blanchette *et al.*, 1989; Daniel *et al.*, 1989). Once lignin is removed, the middle lamella cells separate.

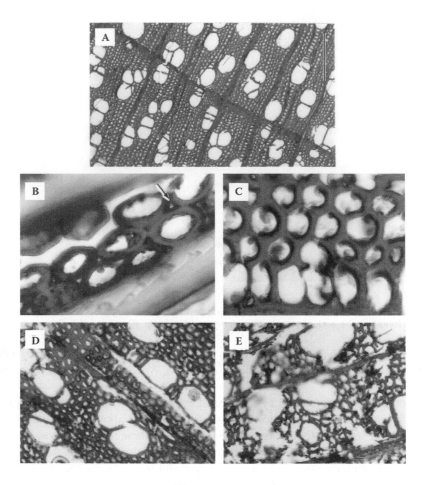

Figure 4.3 Microscopy of birch wood delignification by white-rot fungi. (A) Cryostat section of uninoculated wood. (B) Delignification of ray cells 5 weeks after inoculation of wood with *Trametes* (*Coriolous*) *subvermispora*. (C) 10 weeks after and 25% weight loss. (D) White rot caused by *Trametes versicolor* after 6 weeks and 30% weight loss. (From Srebotnik, E. and Messner, K., *Appl. Environ. Microbiol.* 60, 1383–1386, 1994; with permission. Original photos courtesy of Ewald Srebotnik.)

The *brown-rot* fungi (Basidiomycotina), although they cannot grow on lignin alone, cause demethylation of aromatic methoxyl groups. These selectively depolymerized hemicelluloses and celluloses in plant cell wall. The wood breaks into pieces that crumble into a brown powder. The brown-rot fungi depolymerize the cellulose component, leaving a modified lignin residue. To learn the functioning of this group of fungi, the genome sequence of *Postia placenta* revealed that the fungus grown on cellulose as a carbon source lacks conventional cellulase enzymes. Rather, the fungal hyphae cause cellulose depolymerization from a distance of the advancing hyphae. Two different mechanisms of decay are recognized: One that involves hydrolytic enzymes and another that involves oxidative reactions.

4.4 LITTER DECOMPOSERS

The third type of scavenger is identified by plating tissue macerates on suitable nutrient agar or by direct observation of the decaying tissue incubated in a damp atmosphere and identifying these on the basis of microscopic morphology. The soil-inhabiting fungi—*Penicillium*

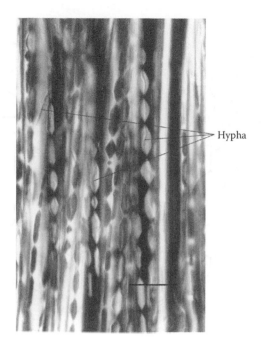

Figure 4.4 Ultrastructure of soft rot showing disruption of middle lamella. (From Susan E. Anagnost, with permission.)

(Fungi Anamorphici), *Humicola* (Fungi Anamorphici), *Trichoderma* (Fungi Anamorphici), *Mucor* (Straminipila), *Collybia* (Basidiomycotina), *Hydnum* (Basidiomycotina), *Marasmius* (Basidiomycotina), *Mycena* (Basidiomycotina,) and others—are typical litter-decomposing fungi. Scanning electron microscopy of tissue allows identification of fungi that are actually growing on the plant litter. There is a tendency to underestimate the role of small soil animals such as earthworms, slugs, millipedes, and mites since fragmentation of substrate by their activity would increase the surface area for microbial attack. In temperate regions the agarics are among the most active agents of decomposition. They are active producers of laccase or polyphenol oxidase, which detoxify litter phenolics. Litter decomposition in tropical forests where moisture and warm conditions prevail has been little studied.

4.5 DEGRADATION OF CELL WALL POLYMERS

4.5.1 Lignin

Lignin is a polymeric substance formed by random cross-linking of phenylpropanoid units: *p*-hydroxycinnamyl (coumaryl) alcohol, 4-hydroxy-3-methoxycinnamyl (coniferyl) alcohol, and 3,5-dimethoxy-4-hydroxycinnamyl (sinapyl) alcohol with several different carbon–carbon and carbon–oxygen linkages (Figure 4.5). Lignin encrusts the cellulose microfibrils within the plant cell walls, giving the vascular plant rigidity, and protects the plant from weather, insects, and pathogenic organisms. The procedure generally used to screen ligninolytic fungi is to inoculate wood blocks with the mycelium isolated from fungal fruiting bodies and, after a 10-week incubation period, to estimate lignin loss by phloroglucinol staining or by weight loss. In Norway spruce wood, the highest lignin losses after 10 weeks were seen for *Ceriporiopsis subvermispora* (44%) and *Physisporinus rivulosus* (39%) (Hakala *et al.*, 2004). In other studies lignin-degrading

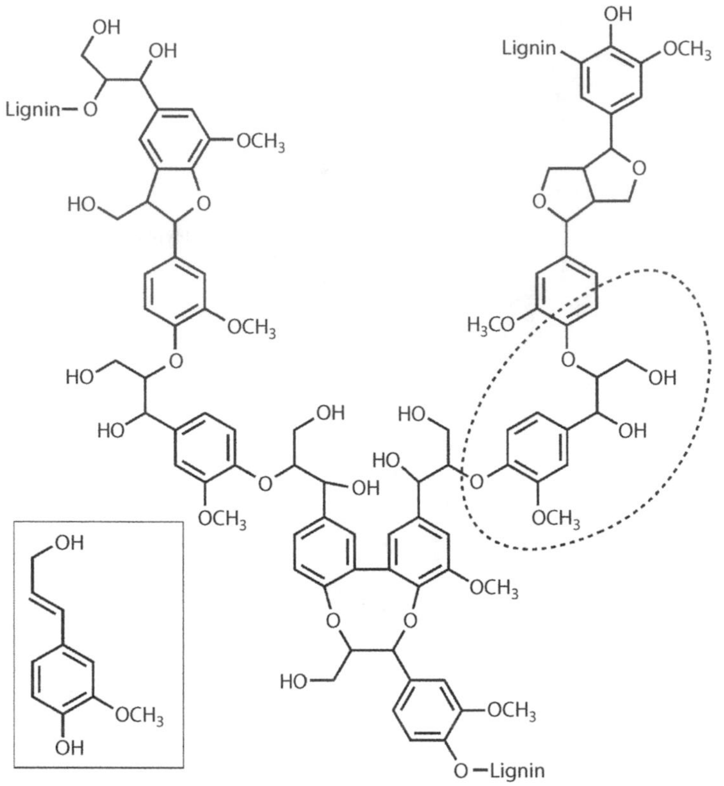

Figure 4.5 Schematic structure of lignin showing phenylpropane units connected by different C–C and C–O–C linkages. Inset shows coniferyl alcohol, the phenylpropanoid building block. The major arylglycerol-ß-aryl ether structure is circled. Soft-rot-causing fungi cleave the carbon–carbon bond; cleavage of propyl side chains, loss of methoxyls, oxidation of benzene hydroxyls, and ring opening are mechanically consistent with a free radical mechanism. (From Kenneth Hammel, with permission.)

activity was measured as evolution of $^{14}CO_2$ from ^{14}C-labeled synthetic lignin prepared by polymerizing ^{14}C-labeled p-hydroxycinnamyl alcohols with horseradish peroxidase, or by the oxidation of a lignin model compound, veratryl (3,4-dimethoxybenzyl) alcohol to veratryl aldehyde in the presence of H_2O_2, that appeared after primary growth of the P. chrysosporium had ceased (idiophase) due to nitrogen limitation (Jeffries et al., 1981). Although this observation raised uncertainty whether lignin degradation is a growth-associated process, a different picture emerged when growth and secretion of enzymes by the hyphae were monitored simultaneously. The fungus was grown sandwiched between two perforated membranes placed over an agar medium (Moukha et al., 1993a). This permitted autoradiographic detection of growth by the application of radioactively labeled acetylglucosamine (a chitin precursor) and the simultaneous detection of secreted enzymes using lignin peroxidase antibodies. Though the radial growth of the fungal colony had stopped, new short branches were initiated at the colony center that secreted Mn^{2+}-dependent lignin peroxidase, suggesting that ligninase is produced by a specialized type of hyphae that develop after much of the assimilable carbon source has been consumed, supported by recycling of intracellular metabolites. In surface-grown cultures, this period coincided with accumulation of RNA transcripts and secretion of ligninase. Therefore lignin degradation, as cellulose degradation, is a growth-associated process, although carried out by a specialized new hyphal growth from parent mycelium (see Section 3.2.4).

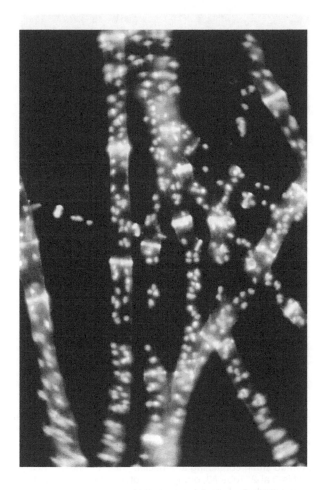

Figure 2.1 The hyphae of *Neurospora crassa* stained with a DNA-binding fluorescent dye to stain nuclei and a chitin-binding dye to stain septa. The nuclei in living hypha can move in both directions through a central pore in the septum.

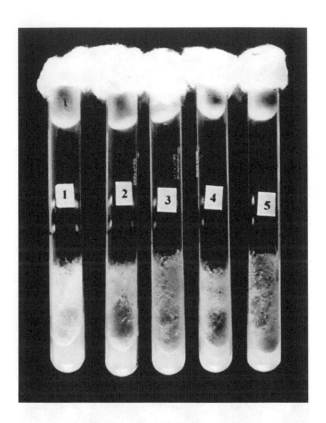

Figure 2.12 An experiment using color mutants of *Neurospora crassa* to demonstrate that the proportion of number of nuclei in multinuclear mycelium can change in response to alteration of chemical composition of growth medium. A three-component heterokaryon [his-3 al-1 + his-3⁺ (EC) al-1+ his-3 al-2] was constructed between albino mutants impaired in histidine biosynthesis and a histidine transformed strain containing al-2 nuclei. Tube # 1, his-3⁺ (EC) al-1(white). Tube # 2, rosy phenotype of his-3 al-2. Tube #3, orange phenotype of his-3⁺ (EC) al-1 + his-3 al-2 heterokaryon grown in minimal medium; Tube # 4, rosy phenotype of his-3⁺ (EC) al-1 + his-3 al-2grown in histidine medium. Tube # 5, orange phenotype of his-3⁺ (EC) al-1 + his-3 al-2 heterokaryon following transfer from histidine medium to histidine drop-out medium. Abbreviation: EC (ectopic).

Figure 4.1 Fruit bodies of basidiomycete fungus on a dead fallen tree trunk in a rain forest. Basidiomycete fungi are specialized for the decomposition of lignin, which is a major structural component of cell wall in wood. (From Jean Paul Ferrero/Ardea, with permission.)

Figure 5.2 Ectomycorrhiza. Some postage stamps featuring fruiting body (sporocarp) of basidiomycete fungi. Presumably all species are ectomycorrhizal (From left to right) *Lactarius deliciosus, Amanita muscaria, Boletus edulis, Krombholzia rufesc, Amanita caesarea*. Some species are edible, others are poisonous.

Figure 8.8 *Neurospora crassa* growing on a sugarcane factory waste dump. The genes encoding pink-orange color have been used as a visual reporter system in gene-silencing experiments. (From P. Maruthi Mohan, with permission.)

Electron microscopy of wood decayed by white-rot fungi revealed that lignin is degraded at some distance from the hyphae, suggesting that the hyphae produce a highly reactive oxygen species that diffuses out and depolymerizes lignin. An extracellularly produced lignin-degrading enzyme resembling peroxidase (haem protein) in spectral properties was isolated from *Phanerochaete chrysosporium* (Basidiomycotina). Hydrogen peroxide is a powerful oxidant and may be produced by oxidases that oxidize sugars to sugar lactones (Hammel, 1997). Immunochemical localization using gold-labeled antiserum in sections of white-rotted wood showed that lignin peroxidase is present in cell walls undergoing delignification. Two types of ligninase have been found: Those that require manganese for catalytic activity (manganese peroxidase or MnP) and those that do not require a metal ion for activity (lignin peroxidase or LiP). The enzymes can oxidize aromatic compounds containing free phenolic groups by removal of one electron from the aryl rings to form an aromatic cation radical. (A *radical* is a molecular fragment having one or more unpaired electrons. It pairs up with other electrons to make new chemical bonds, making radicals highly reactive.) The radicals break down the lignin polymer and can explain how lignin is degraded at sites some distance away from the hyphae. In white-rot fungi, the extracellular H_2O_2 required for the activity of lignin peroxidases is produced from oxidation of glyoxal and methylglyoxal metabolites secreted by the fungus (Figure 4.6). Wood decay by white-rot fungi requires the cooperative action of lignin peroxidases and glyoxal oxidase. White-rot fungi synthesize and secrete veratryl alcohol, a substrate for its own peroxidase enzymes.

In liquid-grown cultures of *P. chrysosporium*, ligninolytic activity appeared after primary growth had ceased as a consequence of nutrient limitation (Jeffries *et al.*, 1981). However, a different picture emerged from the use of a technique that allowed simultaneous monitoring of growth and enzyme secretion by hyphae. Moukha *et al.* (1993b) grew the fungus on an agarose medium that was sandwiched between two perforated membranes that allowed free uptake of nutrients and autoradiographic detection of growth by the application of radioactively labeled acetylglucosamine (a chitin precursor) and the detection of secreted enzyme using lignin peroxidase antibodies. Though net growth of fungal colony had stopped, short hyphal branches were initiated at the colony center. The mycelium secreted Mn^{2+}-dependent-lignin peroxidase after carbohydrate was depleted apparently by recycling of the intracellular carbon compounds. In surface-grown cultures, this period coincided with accumulation of RNA transcripts and secretion of ligninase, ruling out secretion of ligninase from the autolysing hyphae. Electron microscopy, cytochemistry, and immunogold labeling of *P. chrysosporium* growing in wood has shown that lignin-degrading activity (LiP and MnP) occurs in high amounts in the periplasm, that is, between the cell wall and cell membrane (Daniel

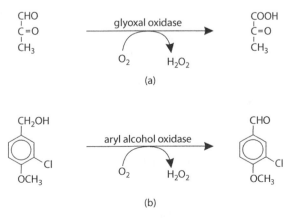

(a)

(b)

Figure 4.6 Mechanism in production of extracellular hydrogen peroxide by wood-rotting fungi.

et al., 1989) or associated with the mucilage (glucan) sheath at the apical regions of the hyphal cell wall surface (Ruel and Joseleau, 1991), suggesting that direct contact between lignin-degrading enzymes and lignin may be necessary during wood degradation by this fungus. A different mechanism of lignocellulose breakdown appears to operate in the case of brown-rot fungi, which use a Fenton system (vanden Wymelenberg *et al.*, 2010). In a lecture given in 1985 at Falun, Sweden, T. Kent Kirk gave an account of lignin degradation:

> We in Madison have taken two complementary approaches to learning how lignin is degraded by *P. chrysosporium* and other white-rot fungi: A chemical approach, and a biochemical one. The two approaches, pursued simultaneously, have now merged into one. During the last 15 years both approaches have contributed importantly to our progress.
>
> The chemical approach was quite straightforward, but not by any means simple. We partially decayed extractive-free wood with *P. chrysosporium* or other white-rot fungi, then isolated and identified or characterized the degradation intermediates originating from lignin (Figure 4.7). Based on the structural information gained, we were able to deduce the general chemical pathways and transformations that the fungi had used to degrade the major lignin substructures. That in turn gave us insight into the nature of the enzymes that had degraded the polymer. We isolated two kinds of lignin degradation products from the decayed wood: High molecular weight modified pieces, and low molecular weight fragments [Figure 3]. We compared the high molecular weight degraded lignin with undegraded (sound) lignin, using virtually all of the chemical and spectroscopic methods that lignin chemists had devised through the years; approximately 30 different types of analytical determinations were made. The low molecular weight fragments, most present in only trace amounts, were identified, primarily by gas chromatography/high resolution mass spectrometry, enabling structures to be assigned to over 100 compounds. From all of this work, we were able to formulate degradation schemes for the major substructures of lignin, as I have mentioned. An example is given in Figure 4.8. We could see that a few types of reactions are of major importance in polymer degradation: Demeth(ox)ylation of aromatic methoxyl groups, hydroxylation of aromatic nuclei, cleavage of aromatic nuclei, and cleavage between

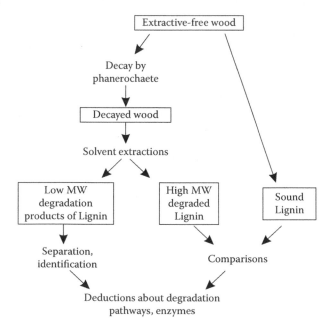

Figure 4.7   The chemical approach to studying lignin biodegradation involved extracting the lignin degradation products from wood that had been partially decayed by *P. chrysosporium* (or other white-rot fungi), and characterizing the high- and low molecular weight components. (After Kirk, T. K., The Marcus Wallenberg Foundation, September 12, 1985, 1985S-791 80, Falun, Sweden. With permission.)

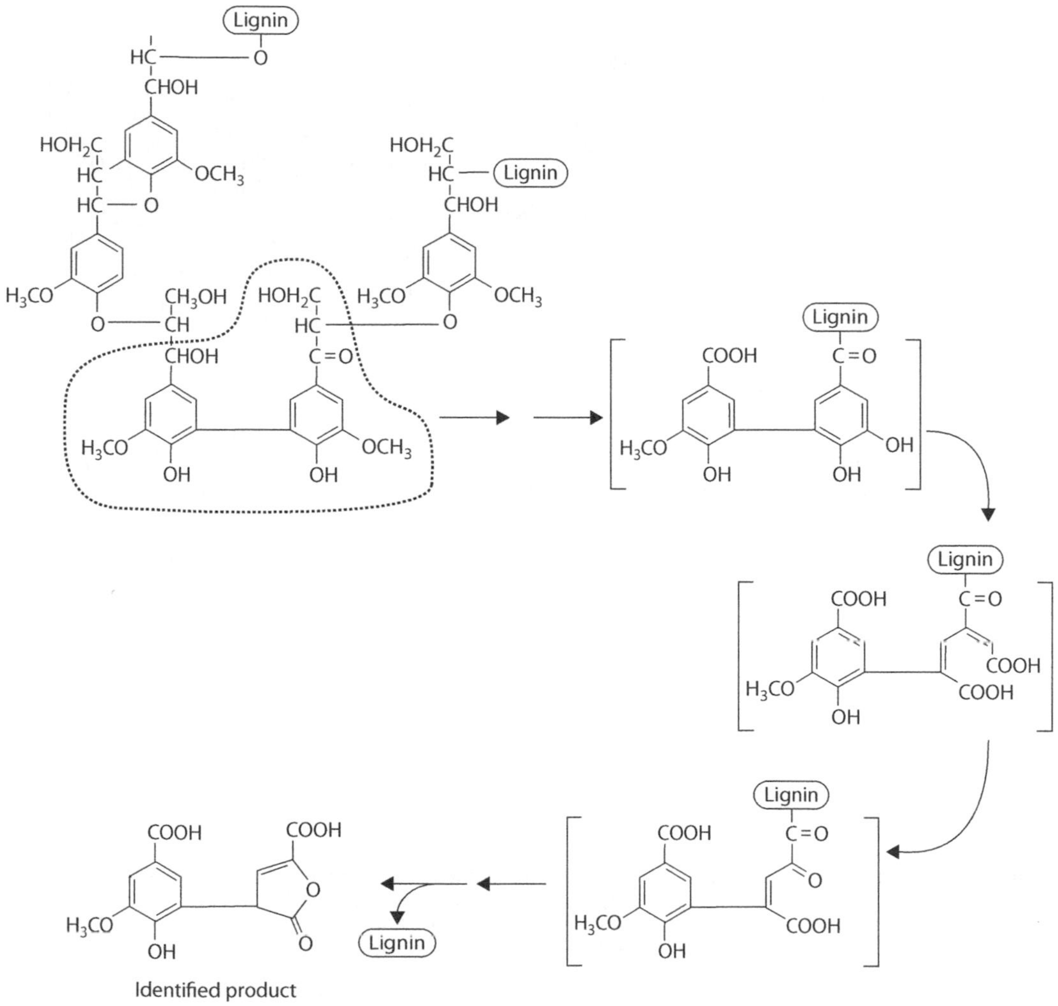

Figure 4.8 Hypothetical pathway leading to the formation of the product at lower left from biphenyl structures (enclosed by dashed line). The pathway is based not only on the identification of the product shown but on extensive studies of the partially degraded lignin polymer. (After Kirk, T. K., The Marcus Wallenberg Foundation, September 12, 1985, 1985S-791 80, Falun, Sweden. With permission.)

the first and second carbon atoms of the propyl side chains (cleavage). The last-named reaction, cleavage, later proved to be of central importance in the biochemical research, as described below. We were able to reach three general conclusions bearing on the biochemistry: 1) The degradative process involves primarily oxidative enzymes; 2) these enzymes are primarily extracellular; and 3) the enzymes are probably quite non-specific. These conclusions, already apparent ten years ago, have been found in the biochemical approach to be correct.

In the biochemical approach, our objective is to identify the enzymes responsible for individual reactions of lignin degradation. Two requirements were recognized at the outset: First, we had to identify specific degradative reactions; and second, we had to be able to make the fungus produce its lignin-degrading enzymes in a controlled and reproducible way. Both of these requirements proved difficult and time consuming to meet. We thought at first that the chemical approach described above would reveal specific reactions suitable for the biochemical investigations. In fact, it only revealed types of reactions. For example, we could deduce that cleavage in the side chains, mentioned above, is an important reaction, but we could not deduce the specific structures that were cleaved, or the products that

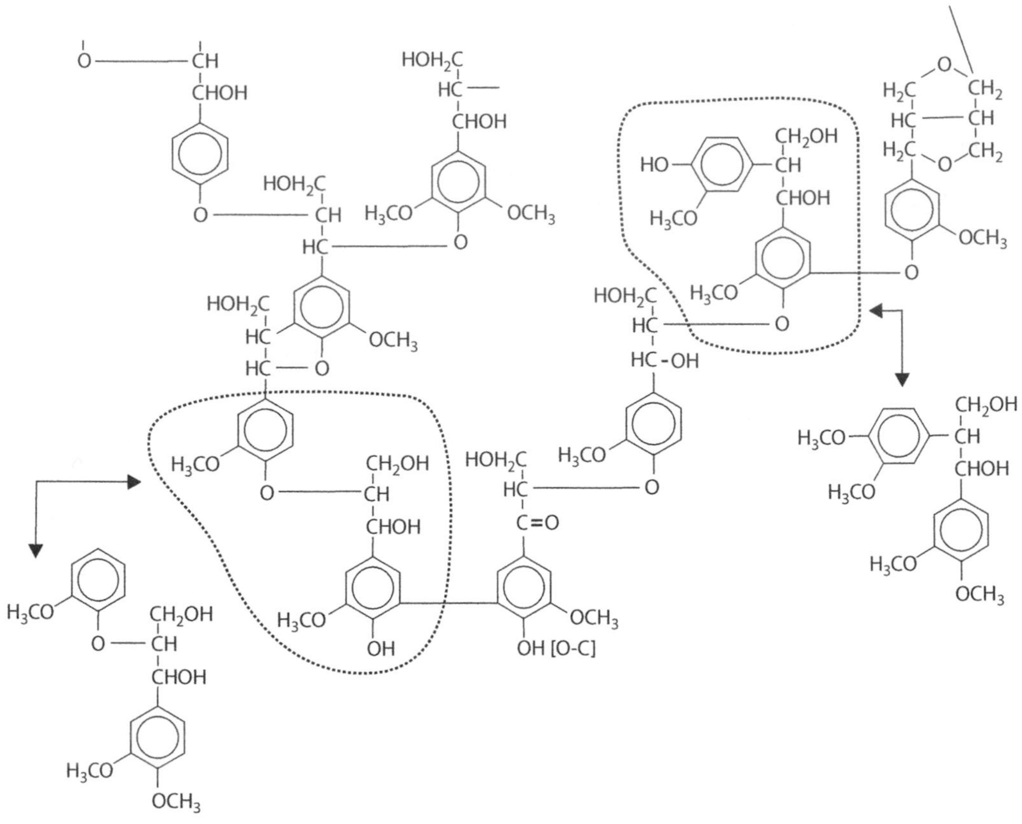

Figure 4.9 Synthetic model compounds such as the two shown here represent substructures (within dashed line enclosures) of the lignin polymer. These and other models have been highly valuable in elucidating the mechanism of lignin biodegradation. (After Kirk, T. K., The Marcus Wallenberg Foundation, September 12, 1985, 1985S-791 80, Falun, Sweden. With permission.)

were first formed, or even the type of chemical reaction involved—whether hydrolytic, eliminative, or oxidative. Being unable to make accurate deductions about specific reactions in research on lignin is a common problem in lignin chemistry, and it is solved by substituting synthetic model compounds for the complex polymer (Figure 4.9). Their use greatly simplifies the research. We proceeded with model compounds, thinking that we would rapidly identify specific reactions.

In the early 1970s we cultivated several white-rot fungi under favorable conditions for growth, and added model compounds to the cultures. To our surprise, the fungi completely ignored them. This meant one of two things: Either the models were not representative enough, or the culture conditions did not allow expression of the lignin-degrading enzymes. We decided the cultures were the problem, and set about to optimize them for lignin degradation—i.e., to maximize their production of lignin-degrading enzymes. This, too, sounded easy until we found, in contrast to older reports (which were based on inadequate methodology) that fungi will not grow on lignin as a food source; they must have cellulose or other carbohydrate as a co-substrate. We realized, too, that even when we supplied an alternate food source together with lignin we did not know how to determine whether the lignin was being degraded by the cultures. Methods for measuring lignin were inadequate for our purposes. Lignin is simply too heterogeneous and complex to measure accurately by spectroscopic 34 means, or by calorimetric or other chemical procedures. And for the culture optimization work, model compounds, even though easily measured, could not be trusted: They can get inside the fungal cells where they might be metabolized by enzymes that have nothing to do with lignin polymer degradation, which has to occur outside the cells. It did not take us long to decide that about the only *way* to measure lignin degradation rapidly and unequivocally was to synthesize carbon-14-labeled lignin and assay its

Table 4.1   Culture Parameters That Influence Lignin Degradation by
Phanerochaete chrysosporium

Alternate carbon/energy source is required

Cultures must be in a secondary metabolic state (brought on by limitation
  for nutrient carbon, nitrogen, or sulfur)

pH control and buffer choice are important

Agitation suppresses lignin degradation

Trace elements and their relative proportions are important

Good aeration is critical

Strain choice is important

Source: Based on Kirk 1985.

oxidation by the fungi to $^{14}CO_2$; $^{14}CO_2$ can easily be trapped and quantified. During 1973 and 1974 we synthesized ^{14}C-lignins in the laboratory using ^{14}C-labeled precursors, which we made and polymerized using procedures developed in Austria, Germany, and Sweden.

It took us another two years to learn how *P. chrysosporium* wants to be treated so it will degrade the synthetic lignins to $^{14}CO_2$. When we first put the synthetic lignins into the cultures, they—just like the model compounds—were not degraded. Through rather tedious empirical studies, however, we were able gradually to work out the required culture parameters (Table 4.1), some of which were quite unexpected. For example, we would never have expected that the organism needs to be nutrient limited and in a "secondary" phase before it will degrade lignin, or that culture agitation prevents degradation. And we were never able to demonstrate lignin degradation in the absence of an alternate food source, even in nutrient-limited cultures. In any event, we had finally learned how to make *P. chrysosporium* produce its lignin-degrading enzymes in amounts, under conditions, and at rates that we could study.

The finding that lignin degradation by *P. chrysosporium* is starvation-induced intrigued us, and we spent two to three years looking into it. One of the interesting things to come out of those studies is that the fungus synthesizes *de novo* an aromatic compound, veratryl alcohol. Veratryl alcohol biosynthesis, like lignin biodegradation, is observed only when the fungus is starved (Figure 4.10). The structural relationship of veratryl alcohol to lignin units intrigued us. As is described below, further investigation has provided a possible explanation for the apparent connection between biosynthesis of the alcohol and biodegradation of lignin.

As soon as we had determined how to grow the fungus so that it will degrade ^{14}C-lignin to $^{14}CO_2$, we not only looked at the intriguing physiological questions, we also returned to the study of model compound degradation. We soon found that our new ligninolytic cultures rapidly and completely degraded model compounds. Our focus thus became to identify single specific reactions that would facilitate our search for enzymes. After going down several blind alleys—i.e., using models that gave confusing and complex reactions—we eventually uncovered the desired single reaction. The reaction is the one I have mentioned before, cleavage (Figure 4.11). With the optimized cultures and suitable models labeled with isotopes, we are able to learn several things about that reaction, including the fact that it involves simultaneous cleavage and Cβ-hydroxylation, that the oxygen atom of the new hydroxyl group is derived from molecular oxygen, that hydrogen atoms and Cβ are not lost during cleavage, and that cleavage exhibits a lack of stereoselectivity (Figure 4.11). We were now ready, we thought, to find the first lignin-degrading enzyme, the one that catalyzes the cleavage. We did not realize, however, that *one* other fact was still

Table 4.2   *Phanerochaete chrysosporium*
as an Experimental Organism

Grows rapidly

Degrades lignin rapidly

Forms asexual spores in abundance

Completes the sexual cycle readily

Has a relatively high temperature optimum

Produces no lactase

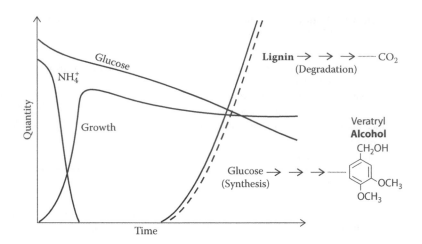

Figure 4.10 Nutrient-limited cultures of *P. chrysosporium* not only degrade lignin to CO_2 while growing on glucose, they also synthesize the compound veratryl alcohol from glucose at the same time. Here the cultures are limited for nutrient nitrogen (supplied as ammonium NH_4, salts), but they have an excess of other nutrients, including carbon (supplied as glucose). As a result of nitrogen depletion, growth stops, but the fungus continues to thrive in a "secondary" phase. Only in the secondary phase does it synthesize veratryl alcohol or degrade lignin. (After Kirk, T. K., The Marcus Wallenberg Foundation, September 12, 1985, 1985S-791 80, Falun, Sweden. With permission.)

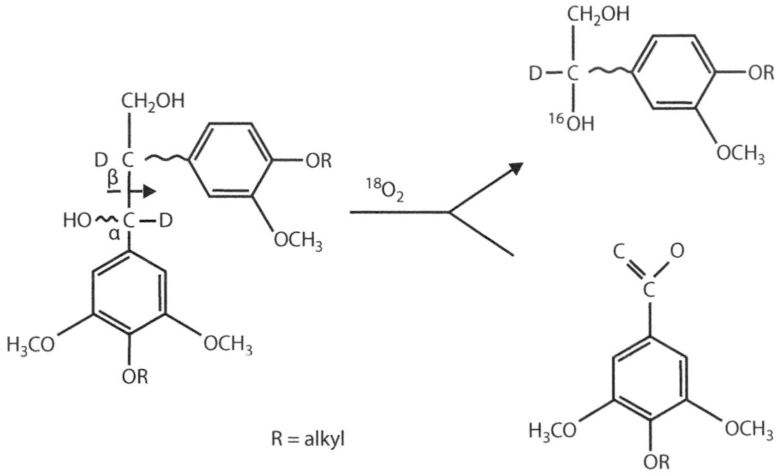

Figure 4.11 Model compounds are readily degraded by ligninolytic cultures of *P. chrysosporium*. Models of the type shown are degraded by an initial cleavage between carbon atoms and ß in the propyl side chains as indicated, forming an aromatic aldehyde product from the C_α moiety, and phenylglycol product from the C_β moiety. Studies with several different models (summarized in this scheme) provided insight into the cleavage mechanism, as shown. Illustrated are the facts that the new hydroxyl group in the phenylglycol product is derived from molecular oxygen, that the hydrogens (deuteriums) on C and C_β are retained during cleavage, and that the cleavage exhibits no stereoselectivity. (After Kirk, T. K., The Marcus Wallenberg Foundation, September 12, 1985, 1985S-791 80, Falun, Sweden. With permission.)

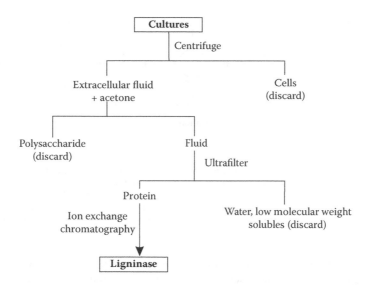

Figure 4.12 Schematic showing the procedure used to purify ligninase from cultures of *P. chrysosporium*. Centrifugation removed the fungal cell, ultrafiltration concentrated the proteins and removed low molecular weight materials, and ion exchange chromatography isolated the ligninase. (After Kirk, T. K., The Marcus Wallenberg Foundation, September 12, 1985, 1985S-791 80, Falun, Sweden. With permission.)

required before that discovery could be made. The missing information was that hydrogen peroxide (H_2O_2) is required for the cleavage.

The story behind the discovery of the H_2O_2 requirement is an involved one. In brief, it stemmed from investigations into the possibility that enzymes are not the agents of lignin oxidations by fungi, instead "diffusible activated oxygen species"—themselves produced by enzymes—are the agents. That hypothesis caused several laboratories, including ours, to examine the production of "activated" oxygen species by ligninolytic cultures of *P. chrysosporium*. The important facts coming out of those efforts are that extracellular H_2O_2 is produced by cultures and that it is required for lignin degradation. Its removal (by adding catalase, an enzyme that destroys it) stops the degradation of lignin. Because H_2O_2 itself is not a strong enough oxidant to affect lignin under the culture conditions, it was apparent that it is involved as a cofactor.

We discovered the first lignin-degrading enzyme in October 1982, even as the "activated oxygen" idea was being debated. The activity was detected as $C\alpha$-C_β cleavage in a model compound of the type shown in Figure 4.12; as expected, the activity was found in the cell-free culture fluid; and not surprisingly it took place only when a small amount of H_2O_2 was added. At first we did not know that the reaction was being catalyzed by an enzyme. For all we knew it could be an "activated oxygen species" produced from the peroxide. But it was only a matter of days before we knew that we had in fact discovered an enzyme. Not long thereafter we had purified it. We found that the enzyme partially depolymerizes lignin, and named it simply ligninase. It was the major extracellular enzyme in the cultures, and its purification was straightforward.

The mechanism of action of ligninase is rapidly being clarified. They, and some of the proteins with as yet unknown roles that are associated with them, are hemeproteins, just as is hemoglobin in our red blood cells. The ligninase enzymes are peroxidases whose catalytic cycle seems to be similar to that of the other peroxidases. But the ligninases are more powerful oxidants than previously studied peroxidases. The reaction (C_α–C_β cleavage) that allowed us first to detect them is only one of the many oxidative reactions that the ligninase brings about...if we treat the model compound with pure ligninase we get a variety of products that reflect several different reactions. This puzzled us for a while, until we discovered that a single common reaction underlies all of the others. The

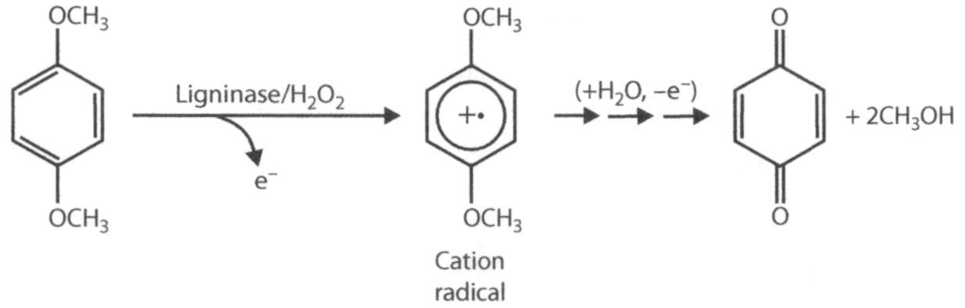

Figure 4.13 The underlying mechanism of ligninase as discovered through studies with simple methoxy-benzenes, which are oxidized by the enzyme to unstable species called cation radicals. They decompose spontaneously. The cation radical from the compound shown, 1,4-dimethoxyben-zene, decomposes by reacting with water to produce methanol and benzoquinone. (After Kirk, T. K., The Marcus Wallenberg Foundation, September 12, 1985, 1985S-791 80, Falun, Sweden. With permission.)

reaction is single electron oxidation of aromatic nuclei to produce unstable species called cation radicals. These radicals undergo a variety of further reactions, many of which do not involve the enzyme. A simple example is shown in Figure 4.13.

Lignin degradation is not complete without removal of side chains from the derived phenolic acids and the cleavage of the derived aromatic benzene ring into respirable compounds catechol and protocatechuic acid. Because of biological interest in the degradation of lignin, the genome of *P. chrysosporium* has been sequenced (Martinez *et al.*, 2004).

4.5.2 Cellulose

Cellulose is the most abundant organic compound in nature. It is a homopolymer of 10,000 or more D-glucose units in β-(1,4)-glycoside linkage (Figure 4.14). This linkage results in the aggregation of the flat glucan chains, side by side and above each other. Hydrogen bonds between the ring oxygen atom and the hydroxyl groups of glucose molecules result in the tight aggregation of the glucan chains in microfibrils. Their packing is so ordered that cellulose exists primarily in the crystalline state and is impermeable to water; it is therefore generally resistant to microbial attack (Figure 4.15).

A few species of fungi degrade paper cellulose completely under growth conditions (Bhat and Maheshwari, 1987). Commonly, three types of hydrolytic enzymes are found in culture filtrates of cellulolytic fungi: (i) Endoglucanase, or β-(1,4)-D-glucan glucanohydrolases (molecular weight 25 to 50 kDa), which cleaves β-linkages at random, commonly in the amorphous part of cellulose; (ii)

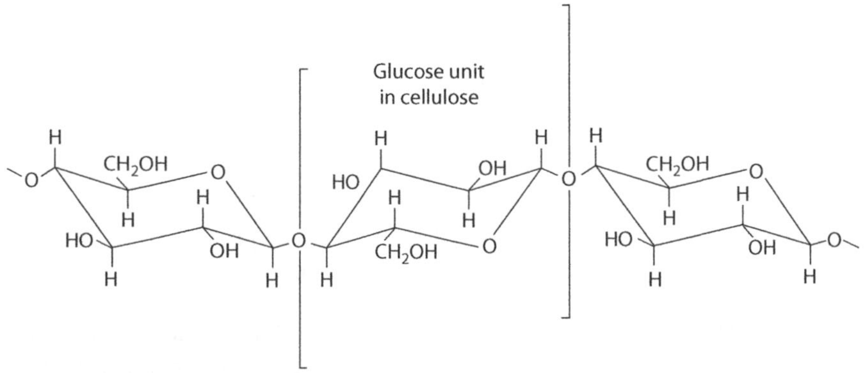

Figure 4.14 Chemical structure of cellulose.

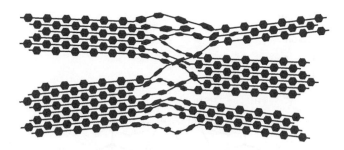

Figure 4.15   Schematic diagram showing crystalline and amorphous parts in cellulose microfibril due to tight or loose packing of glucose chains.

the exoglucanase or exo(1,4)β-glucanases (molecular weight 40 to 60 kDa), which releases cellobiose from the crystalline parts of cellulose; and (iii) the β-glucosidase (molecular weight 165 to 182 kDa), which releases cellobiose and short chain cellooligosaccharides. One of the most active fungi capable of utilizing cellulose rapidly and completely is a thermophilic fungus *Sporotrichum thermophile* (Figure 4.16). Its rate of cellulose utilization in shake-flask cultures is even faster than that of *Trichoderma viride*, even though its secreted levels of endoglucanase and exoglucanase enzymes are about one tenth of the latter fungus. This raised the question whether endo- and exocellulases and β-glucosidases are the primary enzymes required for extracellular solubilization and utilization of cellulose. In experiment the growth of *S. thermophile* in shake-flask cultures was interrupted by

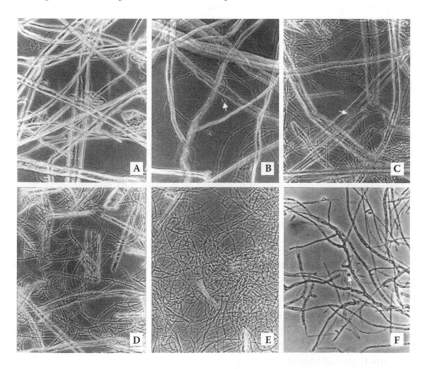

Figure 4.16   Utilization of cellulose by *Sporotrichum thermophile*. Light microscopy of samples from shake-flask cultures of *Sporotrichum thermophile*. (A) Initial appearance of fibers. (B) 16-h culture showing germination of conidia and no perceptible changes in structure of cellulose fibers. (C) 30-h culture showing extensive fragmentation of fibers. (D) Magnified view of fibers from 30-h samples showing fragmentation at "weak spots" (↑). (E) 48-h culture showing extensive fungal growth and nearly complete utilization of cellulose). (F) 72-h culture showing total utilization of cellulose fibers and beginning of sporulation (↑). (From M. K. Bhat, with permission.)

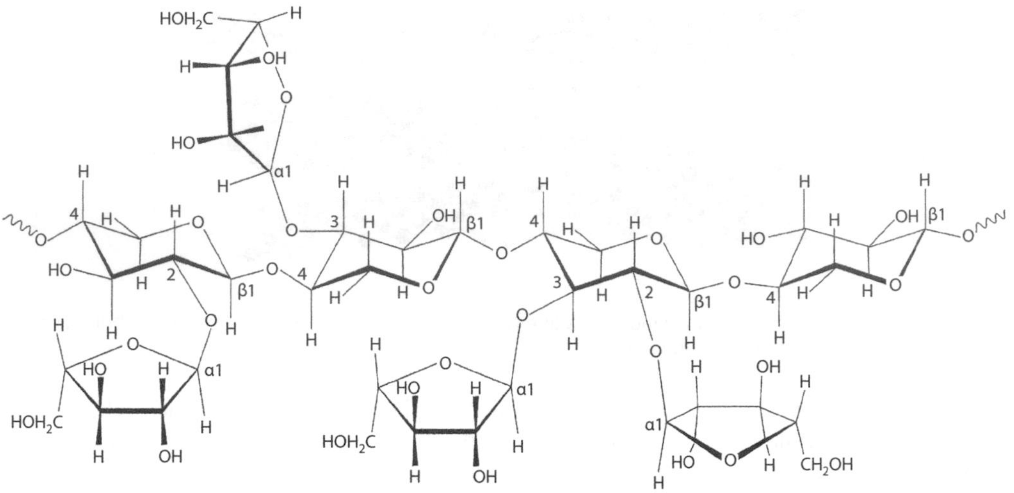

Figure 4.17   Xylan. C-1 of one five-carbon sugar ring is linked to C-4 of another sugar ring by a β-glycosidic bond.

the addition of cycloheximide, a protein synthesis inhibitor. Even though the endoglucanase, exoglucanase, and β-glucosidase that had been secreted prior to the inhibitor treatment were in the growth medium, the utilization (solubilization) of cellulose in the culture medium was interrupted. It was hypothesized that some crucial factors that are replenished by growing of the mycelium are involved in the utilization of cellulose, implying that cellulose degradation is a growth-associated process. The cellulose-grown culture filtrates of *S. thermophile* had limited action under the *in vitro* conditions.

4.5.3 Hemicellulose

Hemicelluloses (noncellulosic polysaccharides) are composed of β-(1,4)-linked pentoses with side chains consisting of sugars, sugar acids, and acetyl esters that prevent the aggregation of chains as in cellulose. Hemicelluloses are hydrogen bonded to cellulose and covalently bonded to lignin. Commonly occurring hemicelluloses are pectin, xylan, arabinan, and rhamnogalactouranan. Next to cellulose, xylan (Figure 4.17) is the most abundant structural polysaccharide in wood cell walls. There is indirect evidence to suggest that in cell wall xylan is external to cellulose. Therefore, to access cellulose, fungi must produce xylanase. Indeed, highly cellulolytic fungi elaborate xylanase but fungi capable of utilizing xylan may not be cellulolytic, as for example *Melanocarpus albomyces*, a thermophilic fungus belonging to Ascomycotina (Prabhu and Maheshwari, 1999). The complete hydrolysis of xylan requires cooperative action of the endoxylanase that randomly cleaves β-(1,4)-linked xylan backbone, the β-xylosidases that hydrolyze xylooligomers, and the different branch-splitting enzymes that remove the sugars attached to the backbone, e.g., glucuronic acid and arabinose. Xylanase of fungi show a multiplicity of forms with molecular mass ranging from 20 to 76 kDa. Analysis of the breakdown products of delignified cell walls using xylanase sequentially and simultaneously with cellulase showed that a mixture of these enzymes is more effective in degrading cell walls than individual enzymes.

4.6 REACTIVE OXYGEN SPECIES

The limited solubilization of lignocellulose under *in vitro* conditions is the revival of the idea that extracellular reactive oxygen species—the hydroxyl radical (·OH) produced by Fenton reaction. The cell-free culture filtrates of some fungi, such as *Trichoderma viride* (Fungi Anamorphici), *Humicola insolens* (Fungi Anamorphici), *Chaetomium thermophile* (Ascomycotina), and *Thermoascus aurantiacus* (Ascomycotina), solubilized cellulose much more rapidly under aerobic conditions than under anaerobic conditions, indicating that an oxidative reaction is involved in breakdown of cellulose. An oxidative enzyme, cellobiose dehydrogenase, is present in some fungi, which in the presence of Fe^{2+} can generate reactive hydroxyl radicals, $H_2O_2 + Fe^{2+} + H^+ \rightarrow H_2O + Fe^{3+} + \cdot HO$. Using a radical trap in the culture medium physiological amount of extracellular reactive species was detected in two species of brown-rot fungi grown on cellulose. In cooperation with cellulase, these radicals can depolymerize cellulose. It is thought that fungi produce H_2O_2, which in the presence of iron Fe^{2+} (Fenton reagent) loosens the plant cell-wall structure and allows the diffusion of enzymes (Koenigs, 1974a, 1974b; Kirk and Farrell, 1987).

4.7 UNSOLVED PROBLEMS

Studies with pure cultures show that the enzymes involved in biomass degradation are synthesized in significant amounts only when the substrates are present. How the insoluble polymeric substrates induce the synthesis of the enzymes remains poorly understood. Decay of biomass by fungi is an aerobic process: Oxygen is required not only for growth but also for the oxidation of lignin phenols. Moisture is vital not only for the synthesis of fungal protoplasm but also for the swelling of the substrate and the diffusion of the digestive enzymes into the substratum through pits and pores.

Apart from the complex nature of the substrate, other problems encountered in elucidation of lignocellulose breakdown are the type of mycelial morphology, employing different decay mechanisms and involving unknown numbers of microfauna and microflora in undetermined ways. A further difficulty in understanding the process of decomposition is due to the failure of growing mixed-microbial cultures in the laboratory and of monitoring the process. Microbiologists and biochemists have therefore used pure (axenic) cultures, in stationary liquid cultures, that facilitate periodic removal of samples for microscopic monitoring of the changes in morphology of the organism, its chemical role by elaboration of different enzymes, and the structural transformation in the substrate during the mineralization process. Liquid cultures are, however, unlike those in nature where the fungus grows on a solid and opaque substrate. Significantly, from fungi grown on translucent agar medium it now appears that fungi produce two types of mycelia: That growing on the surface of substratum and that penetrating below the surface (Vinck *et al.*, 2005).

Electron microscopy, cytochemistry, and immunogold labeling of lignin-degrading fungi growing in wood have shown that lignin-degrading peroxidase activities (LiP and MnP) are associated with extracellular mucilage sheath (Ruel and Joseleau, 1991; Barassa *et al.*, 1998) containing a β- (1,3)-linked backbone with glucose attached in β-(1,6) linkage. The sheath forms an intimate attachment of hyphae to wood, although in decaying wood the sheath dissociated from the hyphal wall and became concentrated with the secondary wall of the wood. The sheath could immobilize lignin- and polysaccharide-degrading enzymes by surface interaction and initiate attack at specific regions of the secondary wall of the plant.

4.8 CLUES FROM GENOME SEQUENCE

The genome of white-rot fungus *P. chrysosporium* (Martinez *et al.*, 2004) and of brown-rot fungus *Postia placenta* (Martinez *et al.*, 2009) have been sequenced. *P. chrysosporium* has an array of genes encoding secreted oxidase (alcohol oxidase), multiple genes for LiP and MnP and hydrolytic enzymes that cooperate in wood decay, and numerous polyketide synthases (enzymes that polymerize fatty acids into branched chain fatty acids) or large carbocyclic rings (macrolides) or aromatic rings fused into polyclic structures. It has genes for flavin adenine dinucleotide (FAD)-dependent oxidase for generating extracellular hydrogen peroxide, suggesting that this fungus can form highly reactive hydroxyl radicals. *P. placenta* has a larger number of genes of oxidase, reflecting its ability to cleave lignin-derived aromatic compounds. The genes for ligninolytic enzymes occur in clusters. The microarray analysis of genes up-regulated when grown on lignocellulose rather than on glucose revealed an array of genes encoding secreted oxidase, peroxidases, and hydrolytic enzymes that apparently cooperate in wood decay. The unexpected absence of conventional cellulase enzymes indicates fundamental differences in the physiology of white-rot and brown-rot fungi. Genome sequencing led to the surprising finding that the fungus *Trichoderma reesei* (syn., *Hypocrea jecorina*), which is a powerful degrader of agricultural crop residues, encodes fewer plant cell-wall polysaccharide-degrading cellulases and hemicellulases than any other sequenced fungus (Martinez *et al.*, 2008). The current view is that fungi secrete a variety of peroxidases that degrade carbon–carbon and carbon–oxygen linkages in lignin, and presumably in polysaccharides in litter.

4.9 PHYSIOLOGICAL PROCESSES

Among properties beneficial for scavenging are the sensing of nutrients in the environment by hyphae, with a hypha being able to sense nutrients and extend its growth toward the source, and the presence of systems to transport the nutrients inside the hypha against concentration gradients. A hyphal form is ideally suited for nutrient acquisition. Certain fungi switch between a unicellular yeast form and a multicellular invasive filamentous form, a phenomenon known as dimorphism. When starved of nitrogen, the budding yeast *Saccharomyces cerevisiae* yeast cells remain connected to produce pseudohyphal growth that allows them to forage fruits and wood (Gimeno *et al.*, 1992). This type of growth is also seen in the smut pathogen (Chapter 12) and in fungi parasitic on humans and animals. Yeast uses plasma-membrane localized sensors to obtain information from the extracellular environment, including the availability of amino acids, ammonium, and glucose. The G-protein-coupled receptor (Gpr1) functions as a dual sensor of both abundant C source and N starvation, and regulates pseudohyphal differentiation. Very low and limiting concentrations of external nutrients result in rapid and good mycorrhiza formation, suggesting that the sensing of environment is important and the nature of substrate influences morphology of the fungus.

4.10 CONCLUDING REMARKS

Though central to the carbon cycle and to exploitation of fungi for potential commercial processes, including production of biofuel, much remains to be learned about how fungi act as scavengers. As presently understood the process involves extracellular reactive oxygen species and oxidoreductase enzymes that act in conjunction with extracellular enzyme systems. New clues may come from the study of the morphology and physiology of a fungus when growing and degrading an insoluble polymer than when it is growing on their derived (soluble) products. More clues may come from investigation of the mucilage sheath containing ligninolytic peroxidase and laccase, as well as aryl-alcohol oxidase associated with fungal hyphae in decaying lignocellulose material (Ruel and Joseleau, 1991), and also from comparative investigations of biomass decomposition by pure cultures with those with mixed cultures. When the mechanism by which microbes convert waste plant material into energy is learned, the knowledge would be useful in development of a practical process for converting cellulose into biofuel ethanol.

REFERENCES

Barassa, J. M., Gutieŕrez, A., Escao, V., Guillán, F., Martinez, M. J., and Martinez, A. T. (1998). Electron and fluorescence microscopy of extracellular glucan and aryl-alcohol oxidase during wheat-straw degradation by *Pleurotus eryngii*. *Appl. Environ. Microbiol.* 64:325–332.

Bhat, K. M. and Maheshwari, R. (1987). *Sporotrichum thermophile* growth, cellulose degradation, and cellulase activity. *Appl. Environ. Microbiol.* 53:2175–2182.

Blanchette, R. A. Screening wood decayed by white-rot fungi for preferential lignin degradation. *Appl. Environ. Microbiol.* 48:647–653.

Blanchette, R. A., Abad, A. R., Farell, R. L., and Leathers, T. D. (1989). Detection of lignin peroxidase and xylanase by immunocytochemical labeling in wood decayed by basidiomycetes. *Appl. Environ. Microbiol.* 55:1457–1465.

Boddy, L. and Rayner, A. D. M. (1982). Population structure, intermycelial interactions and infection biology of *Stereum gausapatum*. *Trans. Brit. Mycol. Soc.* 78:337–351.

Cohen R., Jensen, K. A., Houtman, C. J., and Hammel, K. E. (2002). Significant levels of extracellular reactive species produced by brown rot basidiomycetes on cellulose. *FEBS Lets.* 531:483–488.

Daniel, G., Nilsson, T., and Petterson, B. (1989). Intra- and extracellular localization of lignin peroxidase during degradation of solid wood and wood fragments by *Phanerochaete chrysosporium* by using transmission electron microscopy and immuno-gold labeling. *Appl. Environ. Microbiol.* 55:871–881.

Dighton, J. (2003). *Fungi in Ecosystem Processes*. Marcel Dekker, New York.

Gaikwad, J. S. and Maheshwari, R. (1994). Localization and release of β-glucosidase in the thermophilic and cellulolytic fungus *Sporotrichum thermophile*. *Exp. Mycol.* 18:300–310.

Gimeno, C. J., Ljungdahl, P. O., and Fink, G. R. (1992). Unipolar cell divisions in the yeasts *S. cerevisiae* lead to a filamentous growth: Regulation by starvation and RAS. *Cell* 68:1077–1090.

Hakala, T. K., Maijala, P., Konn, J., and Hatakka, A. (2004). Evaluation of novel wood-rotting polypores and corticoid fungi for the decay and biopulping of Norway spruce (*Picea abies*) wood. *Enzyme Microb. Technol.* 34:255–263.

Hammel, K. E. (1997). Fungal degradation of lignin. In G. Ladisch and K. E. Giller, eds., *Plant Litter Quality and Decomposition*. CAB International, Wallingford, pp. 33–45.

Hammel, K. E., Kapich, A. N., Jensen, K. A., Jr., and Ryan, Z. C. (2002). Reactive oxygen species as agents of wood decay by fungi. *Enzyme Microbiol. Technol.* 30:445–453.

Harley, J. L. (1966). Mycorrhiza. In *Fungi*, Vol. 3, G. C. Ainsworth and A. S. Sussman, eds. Academic Press, New York.

Jeffries, T. W., Choi, S., and Kirk, T. K. (1981). Nutritional regulation of lignin degradation by *Phanerochaete chrysosporium*. *Appl. Environ. Microbiol.* 42:290–296.

Kirk, T. K. (1985). The discovery and promise of lignin-degrading enzymes. The Marcus Wallenberg Foundation, September 12, 1985S-791 80, Falun Sweden. ISSN 0282–4647.

Koenigs, J. W. (1974a). Hydrogen peroxide and iron: A proposed system for decomposition of wood by brown-rot basidiomycetes. *Wood Fiber* 6:66–79.

Koenigs, J. W. (1974b). Production of hydrogen peroxide by wood-decaying fungi in wood and its correlation with weight loss, depolymerization and pH changes. *Arch. Microbiol.* 99:129–145.

Martinez, D., Berka, R. M., Henrissat, B, Saloheimo, M., *et al.* (2008). Genome sequencing and analysis of the biomass-degrading fungus *Trichoderma reesei* (syn. *Hypocrea jecorina*). *Nature Biotech.* 26:553–560.

Martinez, D., Challacombe, J., Morgenstein, I., Hibbett, D., *et al.* (2009). Genome, transcriptome, and secretome analysis of wood decay fungus *Postia placenta* supports unique mechanisms of lignocellulose conversion. *Proc. Natl. Acad. Sci. USA* 106:1954–1999.

Martinez, D., Larrondo, L. F., Putnam, N., *et al.* (2004). Genome sequence of the lignocellulose degrading fungus *Phanerochaete chrysosporium* strain *RP78*. *Nature Biotech.* 22:695–700.

Moukha, S. M., Wösten, H. A. B., Asther, M., and Wessels, J. G. H. (1993a). *In situ* localization of the secretion of lignin peroxidases in colonies of *Phanerochaete chrysosporium* using a sandwiched mode of culture. *J. Gen. Microbiol.* 139:969–978.

Moukha, S. M., Wösten, H. A. B., Mylius, E.-J., Asther, M., and Wessels, J. G. H. (1993b). Spatial and temporal accumulation of mRNAs encoding two common lignin peroxidases in *Phanerochaete chrysosporium*. *J. Bacteriol.* 175:3672–3678.

Prabhu, K. A. and Maheshwari, R. (1999). Biochemical properties of xylanases from a thermophilic fungus, *Melanocarpus albomyces*, and their action on plant cell walls. *J. Biosci.* 24:461–470.

Quispel, A. (1951). Some theoretical aspects of symbiosis. *Antonie Van Leeuwenhoek.* 17:69–80.

Ruel, K. and Joseleau, J.-P. (1991). Involvement of an extracellular glucan sheath during degradation of *Populus* wood by *Phanerochaete chrysosporium. Appl. Environ. Microbiol.* 57:374–384.

Schwendener, S. (1872). Erörterung zur Gonidienfrage. *Flora,* N. R. 30:161–166.

Srebotnik, E. and Messner, K. (1994). A simple method that uses differential staining and light microscopy to assess the selectivity of wood delignification by white rot fungi. *Appl. Environ. Microbiol.* 60:1383–1386.

Tien, M. and Kirk, T. K. (1984). Lignin-degrading enzyme from *Phanerochaete chrysosporium:* Purification, characterization, and catalytic properties of a unique H_2O_2-requiring oxygenase. *Proc. Natl. Acad. Sci. USA* 81:2280–2284.

Tien, M. and Tu, C. P. (1987) Cloning and sequencing of cDNA for a ligninase from *Phanerochaete chrysosporium. Nature* 326:520–523.

Urzúa, U., Kersten, P. J., and Vicuña, R. (1998) Manganese peroxidase-dependent oxidation of glyoxylic and oxalic acids synthesized by *Ceriporiopsis subversmispora* produces extracellular hydrogen peroxide. *Appl. Environ. Microbiol.* 54:68–75.

Vanden Wymelenberg, A., Gaskell, J., Mozuch, M., Sabat, G., Ralph, J., Skyba, O., Mansfield, S. D., Blanchette, R. A., Martinez, D, Grigoriev, I., Kersten, P. J., and Cullen, D. (2010). Comparative transcriptome and secretome analysis of wood decay fungi *Postia placenta* and *Phanerochaete chrysosporium. Appl. Environ. Microbiol.* 76:3599–3610.

Vinck, A., Terlou, M., Pestman, W. R., Martens, E. P., Ram, A. F., van den Hondel, C. A. M. J. J., and Wösten, H. A. B. (2005). Hyphal differentiation in the exploring mycelium of *Aspergillus niger. Mol. Microbiol.* 58:893–899.

CHAPTER 5

Fungi as Symbiotic Partners

Two organisms don't enter into symbiosis to give something to the partner, but in order to take as much advantage of the partner as possible.

—A. Quispel

5.1 MYCORRHIZA

Fossil specimens of early land plants *Aglaophyton major* (originally *Rhynia major*) and *Nothia aphylla* discovered in Scotland had a rhizome and an approximately 18-cm-long dichotomously branched stem. Histological examination showed spores with attached hypha, suggesting that fungal association with plants is an ancient phenomenon. Indeed, it is believed that land plants never had an independence from fungi, for if they had, they could never have colonized land. Careful excavations of living root tips from soil followed by clearing in alkali and staining with a blue dye has shown that about 98% of all terrestrial plants have roots colonized by fungi. It is therefore thought that mycorrhizal (*myco*, fungus; *rhiza*, root) symbiosis evolved some 360–410 million years ago. The commonly seen fungi with aboveground fruiting bodies (basidiocarps) comprise edible or poisonous species are mycorrhiza forming (Figure 5.1). However, convincing evidence that plants are dependent on fungi has come only in recent years. The fungal partner entirely covers the root tip, with hyphae extending into the soil particles or leaf litter, tapping a larger volume of soil and increasing the plant's access to water and relatively immobile nutrients. The underground mycelium can interconnect plants. Many trees are so dependent on their mycorrhizal partners that they languish or die without them. The recognition of a mycorrhizal association between roots and fungi provided an explanation for the paradox of the luxuriance of rain forests growing in soil from which soluble minerals have been leached by torrential rains over millennia and are of extremely low fertility. Mycorrhizal fungi provide the primary mechanism for the uptake of nutrients by the forest trees and thus contribute to the green cover on Earth. In most natural habitats mycorrhizal plants compared with nonmycorrhizal plants show greater uptake of mineral ions such as nitrogen, potassium, and particularly of phosphorus.

5.1.1 Types of Mycorrhiza

Depending upon whether the bulk of the fungus is outside or inside the root, mycorrhiza is broadly divided into ectomycorrhiza (EM) or arbuscular mycorrhiza (AM), respectively. Most fruiting structures like mushrooms and puffballs are produced by EM fungi (Figure 5.2). These have a conspicuous mass of hyphae coating the root, between the cortical cells of the root, and outward, spreading considerably into surrounding soil and litter. In a cross section of root, the hyphae are seen penetrating the cortex and forming a thick sheath, with an intercellular network of hyphae that develop between root cells, called the Hartig net. Intracellular hyphal coils formed within the

Figure 5.1 Ectomycorrhiza. (A) *Boletus edulis,* one of the Basidiomycotina that forms mycorrhiza. (B) Seedling of Douglas fir (*Pseudotsuga manziesii*) colonized by *Lecinium* sp. The fungal mycelium has formed ectomycorrhiza on the root and has produced a basidiocarp above ground. (C) Short roots ensheathed by an ectomycorrhizal fungus. (D) Transverse section of a Eucalyptus/ Pisolithus ectomycorrhiza showing the external (EM) and internal (IM) mantles of hyphae; the fungal hyphae penetrating between epidermal cells of the root cortex (RC) to form Hartig net (HN). Extramatrical hyphae (EH) are exploring the medium. (From Martin, F., Duplessis, S., Ditengou, F., Lagrange, H., Voiblet, C., and Lepyrie, F., *New Phytol.* 151, 145–154, 2001. With permission of John Wiley-Blackwell.)

Figure 5.2 **(See color insert.)** Ectomycorrhiza. Some postage stamps featuring fruiting body (sporocarp) of basidiomycete fungi. Presumably all species are ectomycorrhizal (From left to right) *Lactarius deliciosus, Amanita muscaria, Boletus edulis, Krombholzia rufesc, Amanita caesarea.* Some species are edible, others are poisonous.

epidermal cells are called pelotons. The ectotrophic mycorrhiza can connect plants within communities and exchange resources through a common hyphal network. Orchids that have insufficient photosynthesis gained 6–14% of mass when linked to an autotrophic plant but lost 13% when not linked (Simard *et al.*, 1997). Most trees have arbuscular mycorrhiza in which the fungal hyphae ramify into tree-like structures called arbuscules within the root cells, invaginated by the host plasma membrane. The arbuscules provide a large-surface symbiotic interface for exchange of mineral nutrients from the fungus to the plant. Regardless of the mycorrhiza category, fungal hyphae breach cell walls but remain separated from the cell cytoplasm by a plant-derived membrane.

5.1.2 Identification

Mycorrhizal fungi are biotrophic—i.e., they can grow only on living plants. The root cultured *in vitro* is being used for their identification. Identification of individual fungi is based on the matching of gene sequences—for example, ribosomal 5.8S internal transcribed spacer (ITS) DNA regions—with the data of identified fungi in the fungal databases. The fungal ITS region in DNA extracts of mycorrhiza is polymerase chain reaction (PCR) amplified using universal primers, and the amplified DNA is digested separately with restriction enzymes, following which the fragments are separated by electrophoresis on agarose gels. The ITS regions are sequenced and compared with known fungal sequences in the GenBank database. Using this method, it has been found that the mushroom fungi, recognized by a simple basidiocarp consisting of stipe and cap, and the fungal species that produce leather-like fruit bodies on wood are mycorrhizal species. A single fungus can associate simultaneously with multiple plants while nonphotosynthetic plants such as *Monotropa* spp. have highly specific fungal associations (Bidartondo and Bruns, 2005).

5.1.3 Development

A cross-talk between the plant and a fungus in the form of chemical signals precedes establishment of a symbiotic relationship. Chemicals secreted by roots signal a mycorrhizal fungus spore to germinate and the germ tube to branch, thereby improving the fungus's chance to make contact with a plant root. The root epidermal cells transiently reorganize cellular structures for accommodating a fungal penetration structure without rejection. Akiyama *et al.* (2005) identified the chemical signals stimulating branching by concentrating water containing root exudates from aseptically grown plants and concentrating the factor by absorption on resin, followed by a sensitive bioassay in which branching activity was monitored using a paper disk placed on water agar containing spores in a diffusion assay. The purified factor active at picomolar concentration in bioassay was identified as strigolactone (5-deoxy-strigol) by mass spectroscopy. A variety of chemicals released from roots into the rhizosphere—auxins, alkaloids, cytokinins, phenylglycoside, and terpenes—could be involved in signaling and alteration of the hyphal shape and metabolism, guidance of the hyphae toward the root surface, and initiation and sustenance of *in planta* growth.

Root cultures (Section 5.1.2) infected with spores produced in root cultures in which both cytoskeleton and endoplasmic reticulum had been tagged by green fluorescent protein labeling have shown coordinated intracellular changes (Genre *et al.*, 2005). An animated movie shows that following contact but before actual penetration, the epidermal cell of the nucleus is positioned directly below the hypha contact by means of a swollen structure called the appressorium (Figure 5.3) and a cytoskeleton/endoplasmic reticulum structure is assembled within the epidermal cell in response to appressorium contact. The ramifying intercellular fungal hypha develops highly branched tree-like fungal structures called arbuscules inside the root cortex cells. Arbuscules occupy a major portion of plant cell volume but are separated from the host protoplast by a membrane that has greatly increased surface area and is the site of nutrient and signal exchange. Arbuscule development is accompanied by plastid proliferation, pointing to a highly regulated exchange of compounds and/

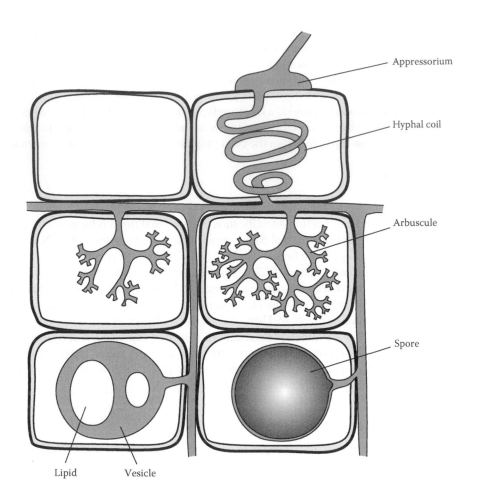

Figure 5.3 A composite diagram of different structures of mycorrhizal fungi in host roots. (Redrawn from Ingold, C. T. and Hudson, H. J., *The Biology of Fungi*, Chapman and Hall, 1993. With permission.)

or signals between the two partners (Hans *et al.*, 2004). Arbuscules degenerate in a few days and the fungus develops ovoid or spherical vesicles that become thick walled and contain fat globules. These asexually formed spores (chlamydospores) are 20–1000 μm or more and persist in soil for long periods. The identification of the vesicular arbuscular mycorrhizal fungi is based on the size, color, number of wall layers, and surface features of the spores.

5.1.4 Carbohydrate Transfer

To study the metabolism of carbohydrate received from a host (Figure 5.4), ^{14}C-labeled sucrose was supplied to mycorrhiza via beech root tissue (Harley, 1965). The fungus converted sugars taken up in the root compartment into trehalose (a typical fungal sugar not found in the higher plants) and into mannitol and lipids. This conversion could be a strategy of creating a sink in the roots for plant-made carbohydrate for its absorption by mycorrhiza mycelium. The results have been confirmed using cultured carrot roots colonized by *Glomus intraradices*. ^{13}C$_1$ glucose was provided either to the roots or the extraradical hyphae that grew out of root, and analyzed by nuclear magnetic resonance spectroscopy. Labeling patterns indicated that ^{13}C-labeled glucose and fructose were taken up by the fungus within the root and converted into trehalose, mannitol, and lipids (Pfeffer *et al.*, 1999, 2004). Triacylglycerol is the main form of carbon stored by the mycobiont at all stages of its life

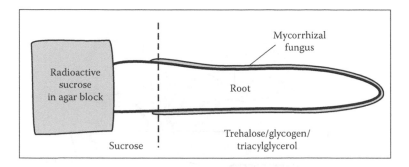

Figure 5.4 Technique to study translocation of ¹⁴C-sucrose by root and mycorrhizal tissue. (From Burnett, J.H., *Fundamentals of Mycology*, Arnold, London, 1976. With permission.)

cycle. Time-lapse photomicrography showed lipid bodies moving in both directions along hyphae (Bago *et al.*, 2003).

5.1.5 Phosphorus Transfer

Several years ago, Harley and his colleagues found that the ability of the excised beech root mycorrhiza to absorb $H^{32}PO_4^{2-}$ was greater than that of uninfected roots and that many times more radioactivity accumulated in the fungal sheath than in the core of root tissue. In short-term experiments, about 90% of the phosphate taken up is retained in the sheath. Therefore the main benefit of symbiosis to mycorrhizal plants has been thought to be in phosphorus nutrition. In an arbuscular mycorrhizal plant, phosphorus (as orthophosphate) can be absorbed both directly through the root and through the external fungal hyphae.

In a mycorrhizal plant, two routes of phosphate uptake from the soil solution are available: Directly by root epidermis and root hairs, as well as via the fungal hyphae extending into soil. To determine their contribution, Smith *et al.* (2003) used a compartmented pot system to quantitatively estimate the contribution of the mycorrhizal uptake pathway to total plant P supply. Experimental plants were grown in pots. A mycorrhizal spore inoculum covered with nylon mesh was placed in the soil, which allowed the hyphae but not roots to penetrate into soil in which ³³P-labeled orthophosphate of high specific activity was mixed (Figure 5.5). The ³³P from the soil solution could only reach the plants via the hyphae; unlabeled P could be absorbed directly. After a period of growth, comparison of specific activity of ³³P in *Glomus*-inoculated and -uninoculated plants showed that mycorrhizal plants grew better in terms of dry weight production. The pathway of phosphorus transport is thought to involve the uptake of P by fungal transporters located in external hyphae, followed by its delivery into cortical cells of the root. Mycorrhizal uptake replaced the direct uptake pathway in roots colonized by fungi, presumably due to down-regulation of plant genes encoding P transporters, indicating a molecular cross-talk between plant and fungus.

5.1.6 Nitrogen Transfer

Govindarajulu *et al.* (2005) studied the uptake, assimilation, and transfer of $^{15}NO_{3-}$ or $^{15}NH_{4+}$ to mycorrhizal cultures of carrot roots colonized by *Glomus intraradices*. Inorganic nitrogen was converted into arginine by extraradical mycelium that is translocated to intraradical mycelium but transported to the plant after breakdown and releasing ammonia without carbon. Its assimilation into arginine would allow nitrogen to be moved in a nontoxic and concentrated form (four nitrogen atoms per molecule). Movies show long-distance solute translocation by means of an extended vacuole that moved from one compartment to another, crossing through the septum and suggesting that the tubular vacuoles in the hypha act as an internal distribution system (Ashford, 1998).

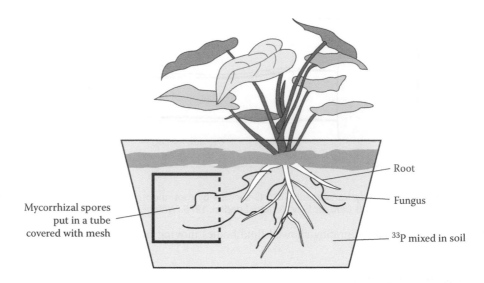

Figure 5.5 Diagram of a pot design in which only the fungal hyphae had access to phosphate and supplied it to the plant. A small tube with nylon mesh allowed only hyphae to grow in the tube containing ^{33}P. Analysis showed that mycorrhizal pathway dominated plant phosphate uptake. (Modified from Smith, S.E., Smith, F.A., and Jakobsen, I., *Plant Physiol.* 133, 16–20, 2003.)

5.1.7 Decomposition of Plant Residues

To determine if the mycelium of an arbuscular mycorrhizal plant can decompose organic substrates and translocate extracted nutrients to the plant to which it is attached, Hodge *et al.* (2001) designed a microcosm. Test plants of *Plantago lanceolata* were grown in two rows of three compartments (Figure 5.6). The plant in the central compartment in each strip was inoculated with the AM fungus *Glomus hoi*. The compartments containing plants were separated by a mesh that was permeable to hyphae but not roots. The experimental compartment contained a patch of $^{15}N/^{13}C$-labeled decomposing grass leaves. After 42 days, the ^{13}C- and ^{15}-N content of the experimental patch was less than 50% of that of controls. To examine if capture of C and N from litter was related to fungal hyphal growth, the hyphae were extracted using a membrane technique, stained and viewed using a microscope, and scored using a grid-line intercept technique. A close relationship between N capture and hyphal length density demonstrated that arbuscular mycorrhizal fungus can acquire nitrogen from organic material and transfer it to other plants. Extrapolating this to the ecosystem, mycorrhizal fungal hyphae act as an extension of the plant root, bypassing the root's P-depletion zone, and are an efficient way to explore large volumes of soil, benefiting tree growth.

5.1.8 Mycorrhiza as Conduits of Photosynthetically Fixed Carbon Compounds

An insight into the role of mycorrhiza in the ecosystem has come from a study of mycorrhized orchid plants by David J. Read. Orchids belong to the family Orchidaceae, which is one of the largest plant families, comprising 20,000–35,000 species. Its minute seeds increase in volume only in the presence of a specific ectomycorrhiza-forming fungus. Seeds put inside packets constructed of fine mesh were placed in soil (McKendrick *et al.*, 2000, 2002). At intervals of time extending to several months the contents in the recovered packets were examined microscopically. Mycorrhizal seedlings showed improved growth (dry weight) only when grown with their roots in direct contact with a photosynthetic plant. Mycorrhizal plants were labeled by exposure to $^{14}CO_2$, and the test plants examined by autoradiographic imaging. Transfer of carbon from the autotrophic plant to non-photosynthetic or partially photosynthetic plants such as orchids and the Indian pipe (*Monotropa*)

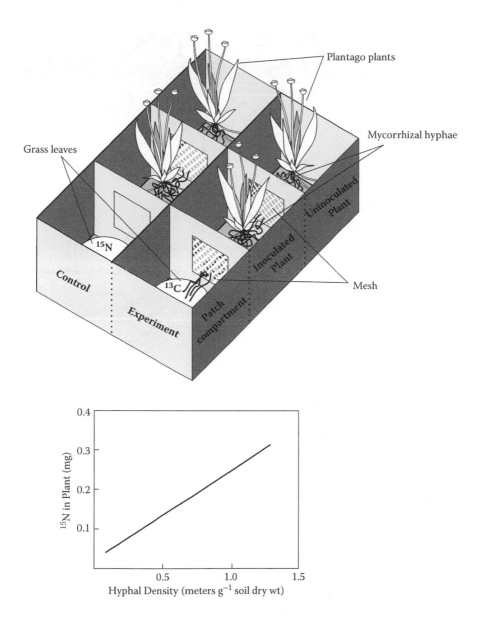

Figure 5.6 Top, design of a microcosm to demonstrate that mycorrhizal fungus decomposes organic material and transfers the extracted carbon and nitrogen compounds to the plant. For clarity plant roots are not shown. Graph shows that nitrogen captured from ^{15}N-labeled plant leaf material is related to hyphal density. (Redrawn from Hodge, A., Campbell, C.D., and Fitter, A. H., *Nature* 413, 297–298, 2001.)

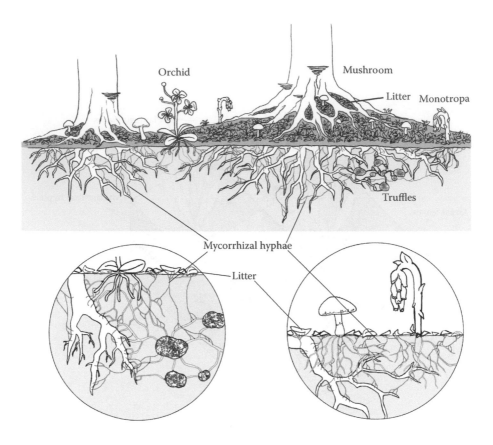

Figure 5.7   Diagram to illustrate the new concept of mycorrhiza in nutrient recycling and as conduits for transfer of carbon compounds to nonphotosynthetic plants.

growing in deep shade have provided strong evidence for a new role of mycorrhiza, i.e., mycorrhizal
hyphae connect plant roots and serve as conduits for hyphal transfer of simple carbon compounds
between plants. A mycorrhizal interconnection between trees with nonphotosynthetic plants plays
an important role in supplementing the carbon requirement of plants growing in shade under the
canopy (Figure 5.7). Mycoheterotrophic plants are unambiguous examples of plant-to-plant carbon
transfer via mycorrhizal fungi. These results suggest a mechanism for species of plants to grow in
shade in the understory of forests and reveal an important mechanism that determines biodiversity
(Bidartondo *et al.*, 2004; Bidartondo, 2005).

   Mycorrhizal association benefits both partners. The photosynthate from the green-leaved partner passes from the plant to the fungus, and in return the fungus can pass the mineral nutrients
absorbed from soil to the plant (Cameron *et al.*, 2006). The experiment that supports this view is
shown in Figure 5.8. Ramets of a forest orchid were assembled in a compartmented Petri dish. The
mycorrhizal mycelium extended from orchid shoot onto agar. A groove was cut to sever the hyphal
connection and the setup was kept in a growth chamber under light. Following feeding of C as
$^{14}CO_2$, radioactivity measurements showed plant-to-fungus transfer in the bridged agar block but no
radioactivity in the agar to which the hyphal connection was severed; fungus-to-plant transfer was
studied using double-labeled (^{13}C–^{15}N) glycine. Both ^{13}C and ^{15}N were assimilated by the fungus
and transferred to the root and shoot of the orchid, demonstrating that photosynthetically made
compounds pass from the plant to the fungus, in return for mineral nutrients (N) passing from the
fungus to the plant.

5.1.9 Cheating in Plant-Fungus Marriages

A liaison between a plant and a fungus is not without risk. A plant may emit chemical signal(s) similar to the usual host, inviting the fungus to physically link up with it (the .interloper) and to provide it carbon compounds and nutrients drawn from the host plant to which the fungus is attached (Gardes, 2002). An example is a non-chlorophyllous plant known as Indian pipe or *Monotropa*, which occurs in deeply shaded forests of pine and oak trees. It will not grow if a photosynthetic tree is not close by, or if its hyphal connections are severed. It survives only because its underground part links up through a *specific* fungus to surrounding photosynthetic plants. Apparently, the non-photosynthetic plant emits some chemical signal(s), mimicking the photosynthetic plant, resulting in mycorrhizal fungal hyphae to link up with it (Figure 5.8). Some authors call the non-photosynthetic plants cheaters because they survive at the cost of a photosynthetic plant through shared hyphal connections. The nutritional cost to the photosynthetic trees may be negligible because the trees are unlikely to be carbon-limited.

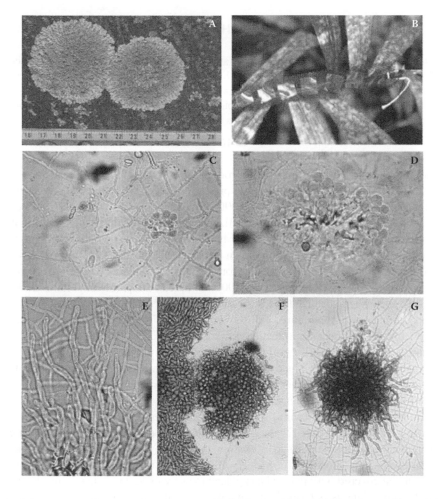

Figure 5.8   (A) Lichen thallus on the bark of a tree. (B) Plastic cover slips secured over leaves in forest; mycelium from fungal propagule. (C) Hyphae enveloping algal cells. (D) Algal cells displaced peripherally by elongating fungal hyphae. (E) Longitudinal contacts of fungal hyphae with algal filaments. (F) Two adjacent thalli of different lichens. (G) A young lichen thallus. (Photos (B)–(G) from New Phytologist Trust and author, with permission.)

5.1.10 Genome Sequence

The genome sequences of the mushroom *Laccaria bicolor* (Martin *et al.*, 2008) and the truffle *Tuber melanosporum*, which form ectomycorrhiza (Martin *et al.*, 2010), showed that symbiosis induces an increased expression of carbohydrate and amino acid transporters in both. The overall pattern of induction of genes coding for enzymes acting on polysaccharides is similar in both *L. bicolor* and *T. melanosporum* symbiotic transciptome. A striking difference is the presence of an invertase gene in *T. melanosporum*, whereas *L. bicolor* has none, raising the question of the nature of photosynthetically produced sugar that is translocated to the fungus. Another surprise is that, given the capacity of *L. bicolor* to persist in litter, the genome of this fungus revealed only a single gene encoding an endoglucanase and no genes for cellobiohydrolases—enzymes thought to be crucial for utilization of cellulose in litter. Also not detected were the conventional genes for the oxidative systems necessary for lignin degradation. A possibility is that rhizosphere microorganisms associated with mycorrhiza may have an important role in mobilizing soil nutrients.

5.2 ENDOPHYTIC FUNGI

Mutualistic interactions between fungal invaders and a host plant are deciphered as a balance, under environmental, physiological and genetic control, that results in fitness benefits for both partners. Under this view, parasitism is an unbalanced symbiosis. As biotic and abiotic stresses commonly result in the production of ROS [reactive oxygen species], rapid and strong activation and scavenging of ROS is potentially a prime mechanism in maintaining this balance.

—K.-H. Kogel, P. Franken, and R. Hückelhoven (2006)

The recognition of endophytic fungi came from observations in 1981 of toxicosis in cattle grazing on grasses. This was attributed to alkaloids produced by fungal endophyte *Neotyphodium* and its sexual stage *Epichloë* (Bacon and White, 2000). The term *endophytic fungus* has been applied to any fungus that is isolated from surface-sterilized segments of plant tissue on nutrient medium. Since recognition of endophytic fungi, there have been several reports of endophytic fungi and of the benefits they provide to the plant. For example, in Lassen Volcanic and Yellowstone National Parks, where the annual soil temperature fluctuates from 20° to 50°C, plants are colonized by an endophytic fungus *Curvularia* sp. (Redman *et al.*, 2002). Whereas the plants of *Dichanthelium lanuginosum* grown from surface-sterilized seeds in sterile soil that had been inoculated with *Curvularia* sp. survived constant soil temperature of 50°C, the nonsymbiotic plants died. Re-isolation of the fungus demonstrated that thermal protection was also provided to the fungus. A benefit to endophytic fungus could be that the plant host provides a critical environment for completing its asexual and sexual stages (Strobel, 2002). Since the discovery of the anticancer drug taxol (paclitexel) from an endophytic fungus inside yew trees, there has been great interest in the isolation of endophytic fungi and testing their metabolites for anticancer properties.

By definition, an endophytic fungus lives in mycelial form in *biological association* with the living plant, at least for some time (Maheshwari, 2006). It is essential that proof of an endophyte be given as hyphae in living plant tissue by examination of the structures on sterile tissues by scanning electron microscopy (Porras-Alfaro *et al.*, 2008). It has been suggested that before isolation work plants may be covered with plastic bags placed over twigs and new leaves used. The minimal requirement before a fungus is termed an "endophyte" is the demonstration of its *hyphae in living tissue*. For this a staining technique—for example, acridine orange—for rapid visualization of hyphae in hand-cut sections has been used. Since identification of a fungus from hyphal features alone is rarely possible, the identification techniques will require methods of immunofluorescence detection, DNA sequencing, and comparison of sequences to homologous sequences registered in

GenBank. There is a need also of comparing the biochemical activity of aseptically grown plants with those inoculated with the endophyte. The facultative endophytes would offer the most difficulty. They might be opportunistic and become biotrophic under certain environmental conditions. Both types are challenging because of their interactions with the plant, modifications of plant growth, biosynthetic capabilities, and evolutionary implications.

5.3 LICHENS

Symbiosis generates novelty.

—**Lynn Margulis**

Approximately 20,000 species of lichens are known. They (Figure 5.8) grow on tree bark, rock, and asbestos roofs in both polar regions and arid lands. Illustrations of different forms of lichens can be found in Sanders (2002) and these may superficially resemble plants. These organisms are new entities formed by the symbiotic association of a tangled mass of fungal hyphae (mycobiont) that holds a photosynthetic green or a blue-green algal partner (photobiont) resulting in a stable thallus of specific structure. Lichens are classic examples of evolutionary novelties evolved by cyanobacteria or algal genome acquisition (Margulis and Sagan, 2002). Crustose lichens are composed of a flat and crust-like thallus, whereas certain foliose lichens hang as threads from branches of trees. The credit for establishing the combination of two dissimilar organisms living together as a single entity is due to a Swiss botanist, Simon Schwendener (1872), who said

> As the result of my researches, the lichens are not simple plants, not individuals in the ordinary sense of the word; they are, rather, colonies, which consist of hundreds of thousands of individuals, of which, however one alone plays the master, while the rest in perpetual captivity prepare the nutriment for themselves and the master. This master is a fungus of the class Ascomycetes, a parasite which is accustomed to live upon other's work. Its slaves are green algae, which it has sought out, or indeed caught hold of, and compelled into its service. It surrounds them, as a spider its prey, with a fibrous net of narrow meshes, which is gradually converted into an impenetrable covering; but while the spider sucks its prey and leaves it dead, the fungus incites the algae found in its net to more rapid activity, even to more vigorous increase.

For nomenclatural purposes, names given to lichens are regarded as applying to their fungal component (mycobiont). *Peltula polyspora* is lichen (mycobiont) and its algal component (photobiont) is *Anacystis montana*, a blue-green alga. The photobiont in lichens is generally a green or yellow-green eukaryotic alga or prokaryotic blue-green nitrogen-fixing alga. If certain lichens are placed in the dark, the photobiont—usually a cyan bacterium such as *Nostoc* or green alga such as *Trebouxia*—cannot live. They do not survive in persistent all-light or all-dark conditions, nor do they live when the environment is entirely wet or dry; an alternation between environments is essential to their growth and survival.

Lichens can survive in the most extreme and severe environments where neither plants nor fungi can exist alone. In the Himalayan Mountains, they grow at altitudes up to 18,000 feet. On the Antarctic ice cap, they have been found on rocks within 300 miles of the South Pole, where it is so cold that growth is only possible for a few days in the year. At the other end of the Earth, in the Arctic tundra, lichens grow with particular luxuriance. A bushy kind, called reindeer "moss," forms ankle-deep carpets and provides the main food of reindeer in the winter. Lichens obtain their moisture from mists and find all the minerals they need dissolved in the rain. Lichens can also tolerate heat that would desiccate and kill most plants. They shrivel but remain alive and, when the opportunity comes, they take up moisture at extraordinary speed and in great quantities, absorbing as much as half their dried body weight in a mere ten minutes. Most species of lichens are extremely sensitive to sulfur dioxide in the air and are therefore indicators of air pollution.

5.3.1 Mycobiont and Photobiont

The mycobiont is mostly an Ascomycotina. The fungus is the dominant member of the partnership and determines the morphology of the lichen. Based on the chlorophyll content of alga, it was estimated that the alga *Nostoc* comprises about 5% of the lichen *Peltigera*. Beneath the upper, peripheral algal layer in which the photobiont partner is entrapped by loosely interwoven fungal hyphae is a gas-filled interior (medulla) 400–1000 μm thick. Freeze-fracture electron microscopy shows hydrophobin rodlets over the surface of fungus (Scherrer *et al.*, 2000). When the material was extracted with hot sodium dodecyl sulphate (SDS), the protein self-assembled into rodlets characteristic of hydrophobin. Since hydrophobins are typically found in the fungal cell wall, it is thought that the mycobiont shields the phycobiont from desiccation.

5.3.2 *In Situ* Study

A simple technique allows observation of microscopic phases of lichen development in nature (Sanders, 2005). Plastic microscope cover slips are attached to the surface of leaves in nature and removed at intervals for microscopic examination. Germ hyphae originating from fungus spores with food reserves in the form of large oil droplets were observed growing toward algae (predominantly *Trebouxia*) with branches surrounding the algal cells, indicating that some substance(s) that attracts hypha and induces branching was being produced by the phycobiont.

5.3.3 Synthesis

The fruiting bodies of lichens are of fungal origin. Therefore the spores that are discharged give rise only to fungi. In a technique of isolating fungal components, a small piece of lichen is fixed to the inner side of a Petri dish lid and inverted over the lower half containing an agar layer, from which the culture of the mycobiont is established (Ahmadjian, 1967). Lichen fungi grow on a variety of media, although very slowly. There is no marked preference for organic or inorganic sources of nitrogen, disaccharides, and polysaccharides. Axenically cultured fungal colonies are hard, cartilaginous, and slow-growing. Growth of *Xanthoria parietina* was accelerated by homogenized colonies every two months and incubation in fresh medium on a reciprocal shaker (Scherrer *et al.*, 2000). Sufficient material for a biochemical experiment could be obtained in about one year. SDS-insoluble, trifluoroacetic acid soluble protein with hydrophobin motif -C-X_7-CCN- was isolated. The deduced amino acid sequence was used to design primers and to clone the hydrophobin gene and study its expression by Northern blotting. The gene was expressed only when the mycobiont was growing symbiotically, not when growing alone in axenic culture. AM fungi receive carbon from the host and convert this into trehalose and glycogen.

Many lichen fungi are partially or wholly deficient in the vitamins thiamine and biotin. Some strains on solid media reach a colony size of only 1–2 mm in diameter after 9–12 months. The temperature range for maximum growth of lichen fungi is 14–28°C, pH optimum is 4.5–7.4, and light has no influence. Most isolated mycobionts do not produce spores and are therefore difficult to relate to free-living fungi. The synthesis of lichen established the controversy regarding the dual nature of the lichen thallus (Ahmadjian *et al.*, 1980). Suspension of *Trebouxia* (phycobiont) and *Cladonia* (mycobiont) was mixed on the surface of autoclaved soil in clay pots and incubated under light–dark cycle. The cultures were periodically wetted and dried. After 12 weeks ultrastructural examination of synthesized squamules showed peg-like haustorium penetrating algal cells.

5.3.4 Movement of Carbohydrate

By incubating samples of intact lichen in solutions of $NaH^{14}CO_3$ in the light and dissecting out the mycobiont at intervals, fixed ^{14}C was found in fungal medulla, showing that photosynthetic carbon compounds move from the alga to the fungus. An "inhibition technique" identified the mobile carbohydrate between the alga and the fungus. Portions of lichens were permitted to photosynthesize in the presence of $NaH^{14}CO_3$ and a high concentration (1–2%) of nonradioactive glucose was added on the basis that it would compete for entry of the photosynthetic product moving from an alga into the fungus, and this would diffuse out into the medium (Hill and Smith, 1972). Chromatographic analyses of inhibited and noninhibited lichens showed that [^{14}C] glucose was detectable after 1 min but not detected after 2 min; it was being rapidly converted into [^{14}C] mannitol. It was concluded that mannitol, ribitol, or sorbitol were the mobile forms of carbohydrate; the conversion of fixed carbon into polyol occurred rapidly. Over half of fixed carbon moved from alga to fungus. The main storage form of carbon in the mycobiont is triacylglycerol (Bago *et al.*, 2002).

5.4 CONCLUDING REMARKS

An emerging view is that almost any fungus lives in association with either the underground or the aboveground parts of a plant. A common theme in symbiosis is the molecular signals that are exchanged between partners, leading to recognition and interactions between phototroph and the fungus. The chemical nature of these signals needs to be studied. Among the unusual features of AM fungi is the presence of an unusually large number of nuclei—as many as a few hundreds to tens of thousands of nuclei. A second unusual feature is that individual cells can have very large amounts of genetic variation. The processes that generate and maintain the high intracellular genetic diversity is not known.

REFERENCES

Ahmadjian, V., Russell, L. A., and Hildreth, K. C. (1980). Artificial reestablishment of lichens. I. Morphological interactions between the phycobionts of different lichens and the mycobionts *Cladonia cristatella* and *Lecanora chrysoleuca*. *Mycologia* 72:73–89.

Akiyama, K., Matsuzaki, K., and Hayashi, H. (2005). Plant sesquiterpenes induce hyphal branching in arbuscular mycorrhizal fungi. *Nature* 435:824–827.

Ashford, A. E. (1998). Dynamic pleomorphic vacuole systems: Are they endosomes and transport compartments in fungal hyphae? *Adv. Bot. Res.* 28:119–159.

Bacon, C. W. and White, J. F. (2000). *Microbial Endophytes*, Marcel Dekker, NY.

Bago, B., Pfeffer, P. E., Abubaker, J., Jun, J., Allen, J. W., Brouilette, J., Douds, D. D., Lammers, P. J., and Schachar-Hill, Y. (2003). Carbon export from arbuscular mycorrhizal root involves the translocation of carbohydrate as well as lipid. *Plant Physiol.* 131:1496–1507.

Bago, B., Zipfel, W., Williams, R. M., Jun, J., Arreola, R., Lammers, P. J., Pfeffer, P. E., and Schachar-Hill, Y. (2002). Translocation and utilization of fungal storage lipid in the arbuscular mycorrhizal symbiosis. *Plant Physiol.* 128:108–124.

Bidartondo, M. I. (2005). The evolutionary ecology of myco-heterotrophy. *New Phytol.* 167:335–352.

Bidartondo, M. I. and Bruns, T. D. (2005). On the origins of extreme mycorrhizal specificity in the Monotropoideae (Ericaceae): Performance trade-offs during seed germination and seedling development. *Mol. Ecol.* 14:1549–1560.

Burggraff, A. J. P. and Beringer, J. E. (1989). Absence of nuclear DNA synthesis in vesicular-arbuscular mycorrhizal fungi during in vitro development. *New Phytol.* 111:25–33.

Burnett, J. H. (1976). *Fundamentals of Mycology*. Arnold, London.

Cameron, D. D., Leake, J. R., and Read, D. J. (2006). Mutualistic mycorrhiza in orchids: Evidence from plant–fungus carbon and nitrogen transfers in the green-leaved terrestrial orchid *Goodyera repens*. *New Phytol.* 171:405–416.

Finlay, R. D. and Rosling, A. (2006). Integrated nutrient cycles in boreal and forest ecosystems—The role of mycorrhizal fungi. In G. M. Gadd, ed., *Fungi in Biogeochemical Cycles.* Cambridge University Press, Cambridge, pp. 28–50.

Gardes, M. (2002). An orchid-fungus marriage: Physical promiscuity, conflict and cheating. *New Phytol.* 154:1–14.

Genre, A., Chabaud, M., Timmers, T., Bonfante, P., and Barker, D. G. (2005). Arbuscular mycorrhizal fungi elicit a novel intracellular apparatus in *Medicago truncatula* root epidermal cells before infection. *Plant Cell* 17:3489–3499.

Gianinazzi-Pearson, V., Arnould, C., Oufattole, M., Arango, M., and Gianinazzi, S. (2000). Differential activation of H^+-ATPase genes by an arbuscular mycorrhizal fungus in root cells of transgenic tobacco. *Planta* 211:609–613.

Govindarajulu, M., Pfeffer, P. E., Jin, H., Abubaker, J., Douds, D. D., Allen, J. W., Bücking, H. J., Lammers, P. J., and Schachar-Hill, Y. (2005). Nitrogen transfer in the arbuscular mycorrhizal symbiosis. *Nature* 435:819–823.

Hans, J., Hause, B., Strack, D., and Walter, M. H. (2004). Cloning, characterization, and immunolocalization of a mycorrhiza-inducible 1-deoxy-d-xylulose 5-phosphate reductoisomerase in arbuscule-containing cells of maize. *Plant Physiol.* 134:614–24.

Harley, J. L. (1968). Mycorrhiza. In *Fungi,* Vol. 3, G. C. Ainsworth and A. S. Sussman, eds., Vol. III, pp. 139–178. Academic Press, NY.

Harrison, M. J. and van Buuren, M. L. (1995). A phosphate transporter from the mycorrhizal fungus *Glomus versiforme*. *Nature* 378:626–629.

Hijri, M., Hosny, M., van Tuinen, D., and Dulieu, H. (1999). Intraspecific ITS polymorphism in *Scutellospora castanea* (Glomales, Zygomycota) is structured within multinucleate spores. *Fung. Genet. Biol.* 26:141–151.

Hill, D. J. and Smith, D. C. (1972). Lichen physiology. XII. The "inhibition technique." *New Phytol.* 71:15–30.

Hodge, A., Campbell, C. D., and Fitter, A. H. (2001). An arbuscular mycorrhizal fungus accelerates decomposition and acquires nitrogen directly from organic material. *Nature* 413:297–298.

Hodge, A. and Fitter, A. H. (2010). Substantial nitrogen acquisition by arbuscular mycorrhizal fungi from organic material has implications for N cycling. *Proc. Natl. Acad. Sci. USA* 107:13754–13759.

Ingold, C.T. and Hudson, H.J. (1993). *The Biology of Fungi*, Chapman and Hall, London.

Kogel, K.-H., Franken, P., and Hückelhoven, R. (2006). Endophyte or parasite—What decides? *Curr. Opin. Plant Biol.* 9:358–363.

Kuhn, G., Hijri, M., and Sanders, I. R. (2001). Evidence for the evolution of multiple genomes in arbuscular mycorrhizal fungi. *Nature* 414:745–748.

Leigh, J., Hodge, A., and Fitter, A. H. (2009). Arbuscular mycorrhizal fungi can transfer substantial amounts of nitrogen to their host plant from organic material. *New Phytol.* 181:199–207.

Lewis, D. H. and Harley, J. L. (1965). Carbohydrate physiology of mycorrhizal roots of beech. I. Identity of endogenous sugars and utilization of exogenous sugars. *New Phytol.* 65:224–237.

Maheshwari, R. (2006a). Plant–fungus marriages. *Resonance* 11:33–44

Maheshwari, R. (2006b). What is an endophytic fungus? *Curr. Sci.* 90:1309.

Margulis, L. and Sagan, D. (2002). *Acquiring Genomes*. Basic Books, New York. p. 240.

Martin, F., Aerts, A., Ahrén, D., *et al.* (2008). The genome of *Laccaria bicolor* provides insights into mycorrhizal symbiosis. *Nature* 452:88–92.

Martin, F., Duplessis, S., Ditengou, F., Lagrange, H., Voiblet, C., and Lepyrie, F. (2001). Developmental cross talking in the ectomycorrhizal symbiosis: Signals and communication genes. *New Phytol.* 151:145–154.

Martin, F., Kohler, A., Murat, C., *et al.* (2010). Périgord black truffle genome uncovers evolutionary origins and mechanisms of symbiosis. *Nature* 464:1033–1038.

McKendrick, S. L., Leake, J. R., and Read, D. J. (2000). Symbiotic germination and development of myco-heterotrophic plants in nature: Transfer of carbon from ectomycorrhizal *Salix repens* and *Betula pendula* to the orchid *Corallorhiza trifida* through shared hyphal connections. *New Phytol.* 145:539–548.

McKendrick, S. L., Leake, J. R., Taylor, D. L., and Read, D. J. (2002). Symbiotic germination and development of the myco-heterotrophic orchid *Neottia nidus-avis* in nature and its requirement for locally distributed *Sebacina* spp. *New Phytol.* 154:233–247.

Palmqvist, K. (2000). Carbon economy in lichens. *New Phytol.* 148:11–36.

Pfeffer, P. E., Douds, D. D., Jr., Becard, G., and Schachar-Hill, Y. (1999). Carbon uptake and the metabolism and transport of lipids in an arbuscular mycorrhiza. *Plant Physiol.* 120:587–598.

Pirozynski, K. A. and Malloch, D. W. (1975). The origin of land plants: A matter of mycotrophism. *BioSystems* 6:153–164.

Porras-Alfaro, A., Herrera, J., Sinsabaugh, R. L., *et al.* (2008). Novel root fungal: Consortium associated with a dominant desert grass. *Appl. Environ. Microbiol.* 74:2805–2813.

Quispel, A. (1951). Some theoretical aspects of symbiosis. *Antonie Van Leeuwenhoek.* 17:69–80.

Read, D. J. and Perez-Moreno, J. (2003). Mycorrhizas and nutrient cycling in ecosystems: A journey towards relevance? *New Phytol.* 157:475–492.

Redecker, D., Hijri, M., Dulieu, H., and Sanders, I. R. (1999). Phylogenetic analysis of a dataset of fungal 5.8S rDNA sequences shows that highly divergent copies of internal transcribed spacers reported from *Scutellospora castanea* are of ascomycete origin. *Fung. Genet. Biol.* 28:238–244.

Redecker, K., Kodmer, R., and Graham, L. E. (2000). Glomalean fungi from the Ordovician. *Science* 289:1920–1921.

Redman, R. S., Sheehan, K. B., Stout, R. G., Rodriguez, R. J., and Henson, J. M. (2002). Thermotolerance generated by plant/fungal symbiosis. *Science* 298:1581–1582.

Ruiz-Dueñas, F. J. and Martínez, A. T. (2009). Microbial degradation of lignin: How a bulky recalcitrant polymer is efficiently recycled in nature and how we can take advantage of this. *Microbial Biotechnol.* 2:164–177.

Sanders, I. R., Alt, M., Groppe, K., Boller, T., and Wiemken, A. (1995). Identification of ribosomal DNA polymorphisms among and within spores of Glomales: Application to studies on the genetic diversity of arbuscular mycorrhizal communities. *New Phytol.* 130:419–427.

Sanders, W. B. (2005). Observing microscopic phases of lichen life cycles on transparent substrata placed *in situ*. *Lichenologist* 37:373–382.

Sanders, W. B., and Lücking, R. (2002). Reproductive strategies, relichenization and thallus development observed *in situ* in leaf-dwelling lichen communities. *New Phytol.* 155:425–435.

Scherrer, S., De Vries, O. M. H., Dudler, R., Wessels, J. G. H., and Hoenneger, R. (2000). Interfacial self-assembly of fungal hydrophobins of the lichen-forming Ascomycetes *Xanthoria parietina* and *X. ectaneoides*. *Fung. Genet. Biol.* 30:81–93.

Schwendener, S. (1872). Erörterung zur Gonidienfrage. *Flora,* N. R. 30:161–166.

Simard, S. W., Perr, D. A., Jones, M. D., Myrold, D. D., D'urali, D. M., and Moine, R. (1997). Net transfer of carbon between mycorrhizal tree species in the field. *Nature* 388:679–682.

Smith, D. C. (1978). What can lichens tell us about real fungi? *Mycologia* 70:915–934.

Smith, S. E., Barker, S. J., and Zhu, Y.-G. (2006). Fast moves in arbuscular mycorrhizal signaling. *Trends Plant Sci.* 11:369–371.

Smith, S. E., Smith, F. A., and Jakobsen, I. (2003). Mycorrhizal fungi can dominate phosphate supply to plants irrespective of growth responses. *Plant Physiol.* 133:16–20.

Strobel, G. A. (2002). Microbial gifts from nature. *Can J. Plant Pathol.* 24:14–20.

Strobel, G. and Daisy, B. (2003). Bioprospecting for microbial endophytes and their natural products. *Microbiol. Mol. Biol. Rev.*, 67:491–502.

Viera, A. and Glenn, M. G. (1990). DNA content of vesicular-arbuscular mycorrhizal fungal spores. *Mycologia* 82:263–267.

CHAPTER **6**

Fungi as Plant Pathogens

Peace will not—and cannot—be built on empty stomachs.

—**Norman E. Borlaug**

Plants are a vastly available source of food for the fungi, provided the fungi can breach the multilayered cell wall and counter the myriad intracellular defensive chemicals produced by the plant. The necrotrophic fungi secrete plant cell-wall degrading enzymes, causing much tissue destruction known as soft rots; however, these are a relatively unimportant pathogen of plants. The most destructive fungal pathogens infect the aerial parts, producing thousands of spores that are disseminated by wind or rain. Because of mutation, newer and more virulent races of fungal pathogens keep arising, requiring constant breeding of new varieties. The breeding of rust-resistant, high-yielding dwarf varieties of wheat by the American plant breeder Norman Borlaug led Mexico to become a net exporter of wheat by 1963. Norman Borlaug was awarded the 1970 Nobel Peace Prize for his contribution to peace through preventing hunger. Presently, a rust disease on soybean caused by *Phakospora pachyrhizi* (http://www.stopsoybeanrust.com/mcbrazil.asp) is of concern. This pathogen is spreading westward from its center of origin in China.

6.1 THE RUST FUNGI

Even though there are several plant pathogenic fungi, the focus here is on the rust fungi (Basidiomycotina, Uredinales), which have been a scourge of mankind since the ancient times (Maheshwari, 2007). Rust fungi are foliar pathogens that infect the aerial parts, mostly the leaves, with infection being readily visible as spore-producing areas on the leaf surface (Figure 6.1). In experiments in which infected plants were kept in an atmosphere containing radioactive carbon dioxide, the photosynthetic rate ($^{14}CO_2$ fixed per gram fresh weight per hour) was measured by analyzing the distribution of radiolabeled and carbon compounds by autoradiography of the whole plant (Livne, 1964). Accumulation of radioactivity at the rust pusules suggests that the parasite subverts the normal pattern of translocation of carbon compounds from photosynthetic leaves to the developing grains or seeds. It has been suggested that the parasite produces some cytokinin-like substances and the infected cells draw the plant-made carbon compounds toward them and convert these into chitin, spore lipids, trehalose, pentitols, and hexitols for packaging as structural or reserve materials into countless disseminative spores. Since the rust fungi cannot yet be cultured (grown separately from the plant host), their biosynthetic potentialities cannot be experimentally assessed. As a consequence the grains or the seeds are either shrunken or not formed at all, with high infection intensity. Norman Borlaug remarked very aptly, "Without food in the stomachs, there is instability and chaos."

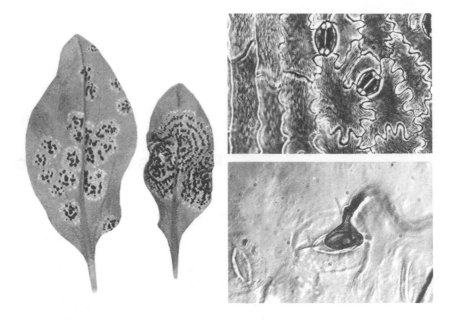

Figure 6.1   Left, snapdragon leaf infected with *Puccinia antirrhini*. Above right, snapdragon leaf showing cuticle ridges. Below right, an appressorium cell formed at the tip of urediospore germ tube over stomata of snapdragon leaf.

Rust fungi are obligate biotrophs. The mycelium grows in living plant tissue with hypha growing intercellularly and forming peg-like projections called haustoria (sing., haustorium). The haustoria penetrate through the cell wall into the cytoplasm of host cells. Electron microscopic studies revealed that the haustorium remains separated from the host cytoplasm by invagination of host cell membrane. Although the mechanisms remain to be discovered, the parasite is able to absorb nutrients from the host cell. Evidence is emerging of an additional role of the haustorium: It determines the specificity of the host–parasite interaction. Only certain races of the parasite suppress host defense responses and are able to absorb carbon compounds from the host, causing disease. The strategy of the biotrophic rust fungus is to keep feeding on living host cells, producing innumerable wind-disseminative spores for its spread and survival. The rust fungi starve the host plant but do not kill it. They are the elite among the fungal parasites.

6.1.1 Thigmotropism and Thigmomorphogenesis

Upon contacting a host, urediospores imbibe water, swell, and produce a germ tube that adheres strongly to the plant surface (cuticle). Contrary to the popular belief that the germ tube finds a portal of entry by chance, rather a thigmotropic mechanism constantly reorients the germ tube toward stomata for entry inside the host (Figure 6.2). The cuticle covering the aerial surfaces of plants is comprised of a layer of cutin and nano-sized wax crystals arranged in specific orientations in the form of ridges with a regular repeating pattern (Figure 6.3). Importantly the sequence of events from germination of urediospores to formation of appressorium is the same on peeled epidermis, on a dead leaf, and on a cellophane replica of the leaf surface. This suggested that the fungus perceives the minute topographical signals and constantly reorients the direction of germ tube elongation at a right angle to the parallel ridges on epidermal cells (Maheshwari and Hildebrandt, 1967). To test this concept, Hoch and Staples microfabricated parallel ridges on silicon wafers (Hoch *et al.*, 1987). Germ tube elongation on 0.2-μm parallel ridges was perpendicular, as on the natural leaf surface (Figure. 6.4). However, a 0.6-μm change in the elevation in the contact surface on a guard cell

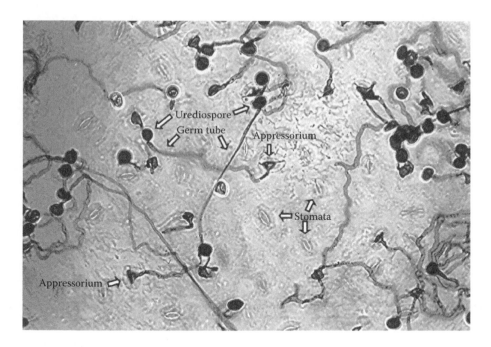

Figure 6.2 Road map of germinating urediospores of *P. antirrhini* on snapdragon leaf. (From Maheshwari, R. *Cur. Sci.* 93, 1249–1256, 2007.)

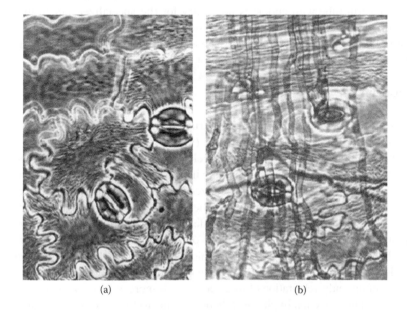

(a) (b)

Figure 6.3 (a) Light microscopy of snapdragon leaf surface and *Puccinia antirrhini* urediospore germ tubes. (b) Perpendicularly orientated germ tube on cuticle around stomata. (From Maheshwari, R., Hildebrandt, A. C. and Allen, P. J., *Nature* 214, 1145–1146.)

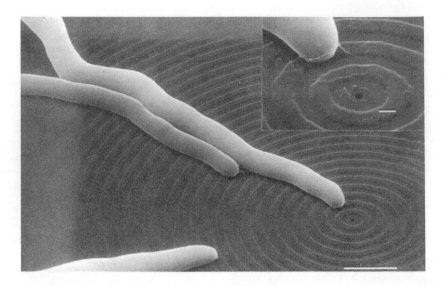

Figure 6.4   Scanning electron microscope image of perpendicularly oriented germ tubes of *Uromyces phaseoli* on microfabricated ridges. (From Richard C. Staples and Harvey Hoch, with permission.)

around the stomata pore induced the germ tube tip to form a bulbous swelling called an appressorium (Figure 6.5), accompanied by the division of nuclei in this structure and perhaps programming differentiation of other pathogenesis-related cells. The appressorium fastens the germ tube onto the surface and enables the parasite to build up the turgor pressure required to force entry of the germ tube into the host through the stomata. When urediospores were put directly on a mesophyll cell of leaf from which the epidermis had been peeled, or on host callus culture (Maheshwari *et al.*, 1967), urediospores did not germinate and no infection occurred. When experiments were repeated using pre-germinated spores, infection did not occur. This suggested that touch-induced morphogenetic differentiation of germ tube with cuticle is necessary for the specialized penetration structure to form for entry of germ tube into the host and infection to occur.

Appressorium development is a prelude to the nuclear division accompanied by the development of cellular structures, leading to the development of the haustorium (Figure 6.6).

6.1.2 Haustorium

The haustorium of rust fungi is a knob-shaped structure that invaginates the host cell plasma membrane and is surrounded by an extracellular matrix (Figure 6.7). The first haustorium is the primary site of contact between the rust pathogen and host cells, and supposedly—if reaction is compatible—the infection hyphae ramify intercellularly, forming haustoria inside other cells. Since using heterologous probes, nutrient transporters have been localized in haustorium, haustorium is considered to be the nutrient-absorbing structure for the parasite. The available data suggest that infection structures including haustoria are induced by contact of the infection hyphae with an artificial membrane incorporating lipid/waxes and carbohydrate constituents of the plant cell wall (Maheshwari *et al.*, 1967d; Heath, 1990). Haustoria are prominent in obligate biotrophs. Apparently through secretion of certain proteins or molecules, the fungus subverts the normal acropetal flow of carbon to developing grains toward itself. Concomitantly, respiration of infected tissue is increased and a reduction in the levels of chlorophyll occurs. The reduction in photosynthetic rate due to the lowering of ribulose 1,5-bisphosphate carboxylase activity affects grain production. The grains are shrunken or not formed at all.

Although rare strains of rust fungi have yielded to culturing on artificial media, it is not known whether the saprophytic mycelium ever formed haustoria. It is believed that in nature the rust

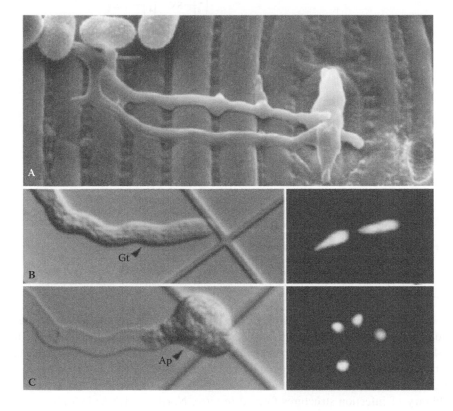

Figure 6.5 Signaling for growth orientation and cell differentiation in urediospore germ tubes by surface topography. (A) Urediospores of *Uromyces phaseoli* germinating on artificial surface on silicon wafer microfabricated with 0.2-μm ridges. This elevation does not provide an inductive stimulus for nuclear division in germ tube (Gt), fluorescing objects in the right panel. (B) Silicon wafer microfabricated with 0.5-μm ridges. The tip of the germ tube at 0 time contains two nuclei (right panel). (C) A germ tube that has crossed 15 ridges (60 min) has differentiated an appressorium (Ap) containing four nuclei (right panel). Images were viewed under scanning electron microscope or fluorescence microscope. (From Harvey C. Hoch and Richard C. Staples, with permission.)

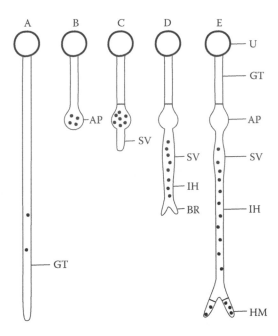

Figure 6.6   Diagram of sequential development of infection structures in urediospores of bean rust germinated on nitrocellulose membrane. (A) On nonhydrophobic membrane (control), the nuclei of dikaryotic urediospore migrate into germ tube but do not divide. (B–E) On hydrophobic membrane incorporating paraffin to simulate cuticle texture, nuclear and cell divisions occur during infection structure formation. Urediospore germinating on hydrophobic membrane undergo mitosis to generate infection structures. AP, appressorium; SV, substomatal vesicle; IH, infection hypha; BR, branch; HM, haustorium mother cell. (Based on Maheshwari, R., Hildebrandt, A. C. and Allen, P. J. (1967a). *Can J. Bot.* 45: 447–450 (1967); and Maheshwari, R., Allen, P. J. and Hildebrandt, A. C. (1967d). *Phytopathology* 57: 855–862.)

mycelium can absorb nutrition only through the haustorium. Hence knowledge of the structure and cytology of haustorium will be crucial to understanding the biotrophic mode of life of the parasite. Mendgen and co-workers succeeded in isolating haustorium from homogenates of infected broad bean leaves by affinity chromatography with lectin, Concanavalin A (Hahn and Mendgen, 1997; Voegele *et al.*, 2001). These were used for the preparation of mRNA and for construction of a cDNA library of infection structures formed *in vitro*. Northern (RNA) hybridization identified genes involved in nutrient uptake and vitamin biosynthesis. Using antibodies against a yeast hexose transporter as a heterologous probe and against a putative plant-induced amino acid transporter, nutrient transporters have been immunochemically localized exclusively in the haustorium plasma membrane of the bean rust fungus *Uromyces fabae* (Figure 6.8), suggesting that the haustorium is the cellular structure through which the fungus draws sugar and amino acids for its nutrition. Differential screening of expressed sequence tag sequences, analysis, and database searches identified about 30 genes that are up-regulated during parasitic growth (Jakupovic *et al.*, 2006). Among these genes are those encoding nutrient (glucose and amino acid) transporters, vitamin biosynthesis, ATP generation, and metallothionein (implicated in binding and detoxification of metals and scavenging of oxygen radicals).

6.2 HOST RESISTANCE AND PATHOGEN AVIRULENCE

An especially interesting aspect of rust fungi is the formation of five different types of spores or cells (pleomorphism) for a complete life cycle involving asexual and sexual stages. After the

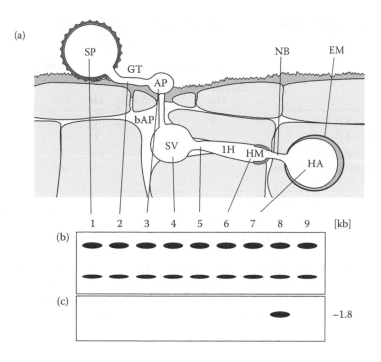

Figure 6.7 Specific expression of hexose transporter gene in haustoria. (a) Diagram of infection structure in vertical section of leaf tissue. SP, urediospore; GT, germ tube; AP, appressorium; SV, substomatal vesicle; IH, infection hypha; HM, haustorial mother cell; HA, haustorium. (b) Ethidium bromide stained DNA in denaturing agarose gel. (c) Northern blot of gel in (b). (Adapted from Voegele, R.T., Struck, C., Hahn, M., and Mendgen, K., *Proc. Natl. Acad. Sci. USA* 98, 8133–8138, 2001.)

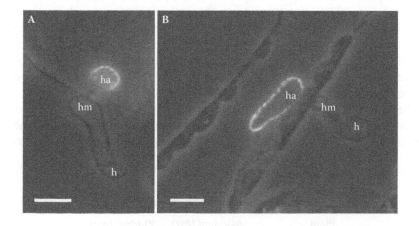

Figure 6.8 Immunofluorescence localization of amino-acid transporter in haustoria of bean rust fungus *Uromyces fabae*. (A) Early and (B) fully developed. Bar is 5 μm. h, haustorium; HM, haustorial mother cell; h, intercellular hypha. (From Kurt Mendgen., with permission.)

Table 6.1   Segregation of Rust Reaction in Crosses on Parent Wheat Varieties

| | Reaction | | | | | |
| | Parent Varieties | | F_2 Plants | | | |
Race	Ottawa 770 B *LLnn*	Bombay *llNN*	*LN*	*Lnn*	*llN*	*llnn*
Race 22 $a_L a_L A_N A_N$	S	I	I	S	I	S
Race 24 $A_L A_L a_N a_N$	I	S	I	I	S	S
Number of plants observed			110	32	43	9
Expected number of plants based on 9:3:3:1			109	36	36	12

Note: I=Immune; S= Susceptible
Source: Data from Flor, H.H., *Adv. Genet.* 8, 29–54, 1956. Table adapted from Johnson, T., Genetics of pathogenicity, in J.G. Horsfall and A.E. Dimond, eds., *Plant Pathology*, Vol. II, Academic Press, New York, 1960.

discovery in 1927 of pycnia in the rust fungi and of + (plus) and − (minus) mating types, crossing (hybridization) of rust races could be carried out by intermixing the exudates of pycnia, with each pycnium formed by infection from a single basidiospore. Studies in the 1930s by Stakman and his associates provided the groundwork that led to the formulation of what is called the gene-for-gene theory. The theory developed after the life cycle of rust fungi involving formation of different spore types was understood. Pure cultures of physiologic races, notably the autoecious flax rust *Melampsora lini*, could be crossed and the pathogenic properties studied by inoculation of dikaryotic (aeciospores) formed on a given differential host. Flor (1956) made his observations after infecting flax varieties carrying different resistant genes with the progeny of crosses between different races of rust pathogen. He crossed two different races of flax rust to obtain the F_2 cultures of the obligate parasite on living flax plants (Table 6.1). The urediospores formed on plants were then tested for the ability to infect more than 30 different varieties of flax that had previously been selected as carrying single genes for rust reaction. Flor's hybridization experiments showed that rust fungus contained genetic factors that conditioned its pathogenic behavior—a landmark conclusion considering the time when the finding was made. The results of hybridization of rust races and of their hosts showed that virulence is a dominant character and avirulence is recessive. If both parent races were virulent on a given host, the progeny was also virulent. A pathogenic trait such as infection type 1, 2, or 4 suppressed in F_1 (see Figure 15.1) reappeared in F_2. By parallel crosses between flax varieties that differed in genes governing host reaction and of crosses between rust races that differed in genes governing virulence, Flor found that flax varieties possessed resistance genes whereas their rust races possessed genes for virulence, and that the resistance and avirulence genes are numerically equal. That is, for every gene capable of mutating to give resistance in the host, there is a gene in the pathogen capable of mutating to a virulent condition that will overcome the host resistance. This is known as the gene-for-gene theory according to which for every Resistance (R) gene in the host there is a complimentary *Avirulence* (*Avr*) gene in the parasite. Confusing definitions of terms exist in the literature. Here, when the plant is a susceptible host, the pathogen is termed *virulent* (Table 6.2). When the plant is resistant, the pathogen is said to be avirulent and the

Table 6.2   Specificity of the Gene-for-Gene Resistance

| | Fungus (Parasite) | |
Plant	Avirulent (*Avr*)	Virulent (*avr*)
Resistant (*Rl*−)	Resistance	Disease
Susceptible (*rlr*)	Disease	Disease

interaction is said to be incompatible. Resistance is triggered only if the *R* gene product (R protein) in the plant recognizes a specific *Avr* gene product (Avr protein) excreted by the pathogen in the cytoplasm of the host cell (Table 6.2). In other words, disease results when *Avr*, *R*, or both are absent. A mutation either in the avirulence or resistant gene that results in loss of function will result in a change from an incompatible to a compatible reaction.

The R-Avr protein complexes are thought to move into the nucleus and bind to specific promoters, acting as transcription factors and activating expression of host defense genes resulting in hypersensitive response. The recognition mechanism in which the Avr protein binds and activates the promoter of the cognate *R* gene is based on work done on bacterial plant pathogens (Romer *et al.*, 2007). In this study mutation in *Avr* abolished pathogen recognition by the matching pepper *R* gene, which suggested that recognition involves the transcriptional activation of host genes. In this study the pepper host plant was infiltrated with *Avr* and its binding to a defined promoter region of the resistance gene in the host DNA was studied. Gene-for-gene interactions have been demonstrated between plants, biotrophic and other fungi or fungus-like *Phytophthora infestans*, and bacteria.

6.3 DUAL ROLE FOR HAUSTORIUM

Based on the (rust) gene for (host) gene relationship, the haustorium–host cell interface (Figure 6.9) is thought to be the site for uptake of nutrients from the host cell into which it has penetrated. Further it is thought that uptake of nutrients (sugars, amino acids) occurs by means of symport with protons derived from the metabolism of sucrose mechanism for the acquisition of nutrients but also for determining the specificity of the host–parasite interaction. Based on the localization of nutrient transporters at the haustorium surface, the interaction between Avr and R protein molecules is thought to take place inside the cytoplasm of the host cell. A model postulates the Avr-R protein complex in an infected plant cell signals the plant cell to respond to attack by suicide or a localized host cell death called hypersensitive response (HR), limiting the spread of the biotrophic pathogen to the infected cell. A model predicts that products of *R* genes act as receptors for pathogen-encoded Avr proteins to initiate a signal that activates plant defense responses to arrest pathogen propagation. A haustorium-specific cDNA library and antibodies against putative proteins, the transfer of protein from the rust haustorium into host was detected by immunofluorescence microscopy (Kemen *et al.*, 2005). In the plant the expression of corresponding *R* and *Avr* gene products (Figure 6.10) induce a HR-like cell death. Based on the rust gene for host gene relationship, and the theoretical reasoning that *Avr* genes are highly polymorphic, genomic DNA from germinated flax rust urediospores was analyzed for variants of AvrL 567, cloned and expressed in *E. coli* as fusion proteins, and sequenced (Dodds *et al.*, 2006). Expression of *Avr* 567 in plants induced an HR-like necrotic response in flax plants, whereas other variant alleles did not, indicating that this gene could be a member of the *Avr* locus. It was hypothesized that these are recognized by the corresponding cytoplasmic resistance proteins thought to correspond to transmembrane receptor kinase, receptors without kinase domain, cytosolic kinase, and cytosolic protein with nucleotide binding site (Catanzariti *et al.*, 2006; Dodds *et al.*, 2004).

6.4 EFFECTOR MOLECULES

While the targets of Avr/R proteins can be either membrane-spanning or cytoplasmic proteins or structural molecules such as cell wall components, several types of effector molecules are thought to control different steps of the infection process—for example, for tight adhesion of hyphae to plant cell wall, for localized dissolution of the host cell wall for haustoria to penetrate, and of cytokinin-like compounds that cause host metabolites to accumulate at the infection site and affect

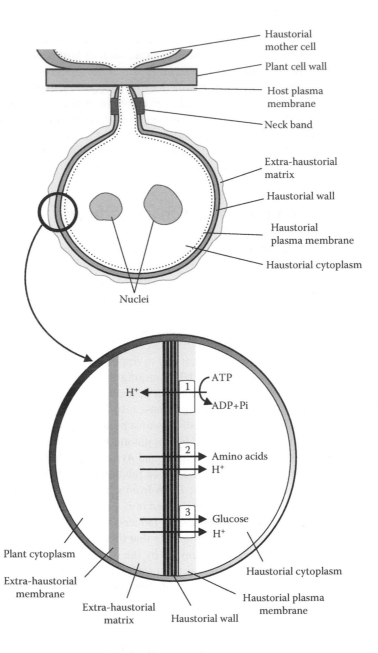

Figure 6.9   Diagram of proton-coupled nutrient uptake at haustorium surface. (Based on Voegele, R.T., Struck, C., Hahn, M., and Mendgen, K., *Proc. Natl. Acad. Sci. USA* 98, 8133–8138, 2001.)

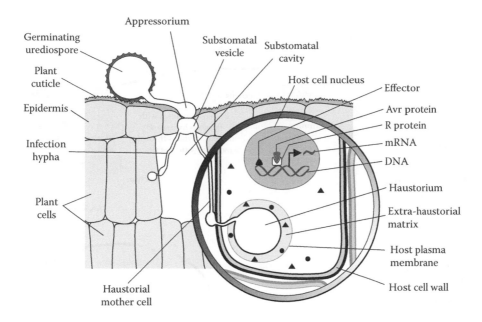

Figure 6.10 Diagram showing interaction between rust fungus and susceptible host is initiated on external surface. Interaction of haustorium-secreted Avr with host R molecules inside infected host cell and activation of host defense pathway (enlarged).

cell membrane permeability. Because the rust fungi have not been persistently cultured, the role of these compounds in pathogenesis is a matter of speculation. This is an area where knowledge of the genome sequence of rust fungus is expected to throw light. Culturing of both, the rust and the mycorrhizal fungi, is of considerable interest.

6.5 CONCLUDING REMARKS

The lifestyle of rust fungi remains an enigma. On one hand the wheat stem rust pathogen *Puccinia graminis tritici* shows extreme specificity such that a race of pathogen discriminates between two varieties of wheat, yet on the other hand this pathogen can infect a widely different plant—the barberry—to complete its life cycle (Figure 6.11), involving the asexual and sexual stages. Aeciospores germinate on wheat, produce a vast amount of urediospores, and spread infection. The pathogen produces bicelled teliospores for overwintering, which germinate to produce basidiospores that must infect barberry leaves as a secondary host to complete the life cycle. Eradicating the barberry plant is reported to reduce new races from arising due to hybridization of mixtures of races. In India barberry occurs in the cool Himalayan ranges but not in the plains where the wheat crop is grown. Alternatively, aeciospores produced on barberry may disseminate over long distances in peninsular India, causing infection of wheat on a small scale or as self-sown tillers. Second, successful entry of the parasite into the host is dependent on the ability of the parasite to sense the configurations of wax deposits on the external surface (cuticle) of the plant host. The touch-induced formation of infection structures suggests that physical surface could be very important in realizing consistent axenic cultures of rust fungi. Currently, the focus is to understand recognition events that activate a defense response in the resistant host. The knowledge may be applied to induce a resistance response in crop plants to reduce crop losses due to rust infection—for example, by engineering resistance to disease in crop plants by transfer of disease-resistance genes from one plant species to another. To

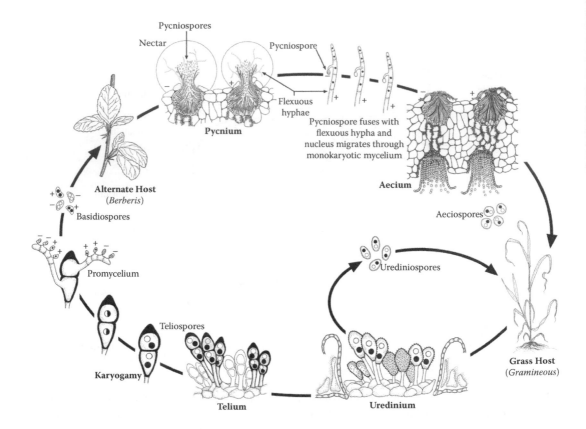

Figure 6.11 Life cycle of *Puccinia graminis tritici*. (From Jacolyn A. Morrison, USDA, Cereal Disease Laboratory, St. Paul, Minn.)

discriminate between changes in biochemical parameters in infected tissue, the genomes of both the parasite and the host are being analyzed. Using DNA extracted from 24-h germinated urediospores of *P. graminis tritici*, the genome sequence of the parasite is being analyzed. This should lead to identification of touch-inducible genes and their role in defense-related signal transduction pathway.

REFERENCES

Alexopoulos, C. J., Mims, C. W., and Blackwell, M. (1996). *Introductory Mycology*. Wiley, New York.

Braun, E. J. and Howard, R. J. (1994). Adhesion of fungal spores and germlings to host surfaces. *Protoplasma* 181:202–212.

Catanzariti, A.-M., Dodds, P. N., Lawrence, G. J., Ayliffe, M. A., and Ellis, J. G. (2006). Haustorially expressed secreted proteins from flax rust are highly-enriched for avirulence elicitors. *Plant Cell* 18:243–256.

Daly, J. M., Inman, R. E., and Livne, A. 1962. Carbohydrate metabolism in higher plant tissues infected with obligate parasites. *Plant Physiol*. 37:531–538.

Dangl, J. N. and McDowell, J. M. (2006). Two modes of pathogen recognition by plants. *Proc. Natl. Acad. Sci. USA* 103:8575–8576.

Dodds, P., Lawrence, G. J., Catanzariti, A. M., Ayliffe, M. A., and Ellis, J. G. (2004). The *Melampsora lini* AvrL 567 avirulence genes are expressed in haustoria and their products are recognized inside plant cells. *Plant Cell* 16:755–768.

Dodds, P. N., Lawrence, G. J., Catanzariti, A.-M., The, T., Wang, C.-I.A., Ayliffe, M. A., Kobe, B., and Ellis, J. G. (2006). Direct protein interaction underlies gene-for-gene specificity and coevolution of the flax resistance genes and flax rust avirulence genes. *Proc. Natl. Acad. Sci. USA* 103:8888–8893.

Ellis, J. G., Dodds, P. N., and Lawrence, G. J. (2007). The role of secreted proteins in diseases of plants caused by rust, powdery mildew and smut fungi. *Curr. Opin. Microbiol.* 10:326–332.

Flor, H. H. (1956). The complementary genic systems in flax and flax rust. *Adv. Genet.* 8:29–54.

Hahn, M. and Mendgen, K. (1997). Characterization of *in planta*-induced rust genes isolated from a haustorium-specific cDNA library. *Mol. Plant Microbe Interactions* 10:427–437.

Hahn, M., Neef, U., Struck, C., Göttfert, M., and Mendgen, K. (1997). A putative amino acid transporter is specifically expressed in haustoria of the rust fungus *Uromyces fabae*. *Mol. Plant Microbe Interactions* 10:438–445.

Heath, M. C. (1990). Influence of carbohydrates on the induction of haustoria of the cowpea rust fungus *in vitro*. *Exp. Mycol.* 14:84–88.

Hoch, H. C., Staples, R. C., Whitehead, B., Comeau, J., and Wolf, E. D. (1987). Signaling for growth orientation and cell differentiation by surface topography in *Uromyces*. *Science* 239:1659–1663.

Hu, G. and Rikenberg, F. H. J. (1998). Ultrastructural localization of cytokinins in *Puccinia recondite* f. sp. *tritici*-infected wheat leaves. *Physiol. Mol. Pl. Pathol.* 52:79–84.

Jakupovic, M., Heintz, M., Reichman, P., Mendgen, K., and Hahn, M. (2006). Microarray analysis of expressed sequence tags from the haustoria of the rust fungus *Uromyces fabae*. *Fung. Genet. Biol.* 43:8–19.

Johnson, T. (1960). Genetics of pathogenicity. In J. G. Horsfall and A. E. Dimond, eds., *Plant Pathology*, vol. II. Academic Press, New York.

Jones, J. D. G. and Dangl, J. L. (2006). The plant immune system. *Nature* 444:323–329.

Kemen, E., Kemena, A. C., Rafiqi, M., Hempel, U. Mendgen, K., Hahn, M., and Voegele, R. T. (2005). Identification of a protein from rust fungi transferred from haustorium into infected plant cells. *Mol. Plant Microbe Interact.* 18:1130–1139.

Livne, A. (1964). Photosynthesis in healthy and rust-affected plants. *Plant Phys.* 39:614–621.

Maheshwari, R., Hildebrandt, A. C. and Allen, P. J. (1967a). The cytology of infection structure development in urediospore germ tubes of *Uromyces phaseoli* var. *typica* (Pers.) Wint. *Can J. Bot.* 45:447–450 (1967).

Maheshwari, R. and Hildebrandt, A. C. (1967b). Directional growth of urediospores germ tubes and stomatal penetration. *Nature* 214:1145–1146.

Maheshwari, R., Hildebrandt, A. C. and Allen, P. J. (1967c). Factors affecting the growth of rust fungi on host tissue cultures. *Bot. Gaz.* 128:133–159.

Maheshwari, R., Allen, P. J., and Hildebrandt, A. C. (1967d). Physical and chemical factors controlling the development of infection structures from urediospore germ tubes of rust fungi. *Phytopathology* 57:855–862.

Maheshwari, R. (2007). A scourge of mankind: From ancient times into the genomics era. *Curr. Sci.* 93:1249–1256.

Mendgen, K., Struck, C., Voegele, R. T., and Hahn, M. (2000). Biotrophy and rust haustoria. *Physiol. Mol. Plant Pathology* 56:141–145.

Romer, P., Hahn, S., Jordan, T., Stauß, Bonas, U., and Lahaye, T. (2007). Plant pathogen recognition mediated by promoter activation of the pepper *Bs3* resistance gene. *Science* 318:645–648.

van den Buren, E. A. and Jones, J. G. G. (1998). Plant disease resistance proteins and the gene-for-gene concept. *Trends Biochem. Sci.* 23:454–456.

Voegele, R. T., Struck, C., Hahn, M., and Mendgen, K. (2001). The role of haustoria in sugar supply during infection of broad bean by the rust fungus *Uromyces fabae*. *Proc. Natl. Acad. Sci. USA* 98:8133–8138.

CHAPTER 7

Fungi as Chemical Factories

Fungi have evolved as chemical factories that produce various classes of enzymes for breaking down plant litter, using the solubilized compounds for their nutrition. Fungi also produce a variety a great diversity of secondary metabolites—some having antagonistic activity for survival amidst competing microflora and microfauna occurring in their habitat.

Fungi secrete several classes of enzymes into their microenvironment for breaking down organic material (plant litter) into oligomeric or monomeric compounds for their absorption inside the cells for extracting carbon, nitrogen, and phosphorus compounds from which the fungi synthesize proteins, nucleic acids, and phospholipids for growth to continue. Their absorptive mode of nutrition is facilitated by the secretion of a diversity of organic molecules called secondary metabolites, some of which possess toxic or antagonistic activity toward other organisms. For example, the yeasts that live in the skin of sugary fruits excrete ethanol—not for human consumption, as some may think, but to arrest the competing microflora and microfauna in their habitat. The same may be said for penicillin, an antibiotic first identified as a secretion from the fungus *Penicillium notatum* that acts powerfully against bacteria. Some fungi live inside the plant as endophytes in a symbiotic relationship, producing secondary metabolites that confer thermotolerance to plants and thereby perhaps mutually extending their own and that of their host plants. In the niche in which the fungi live—approximately 10 μm around the hypha—minute amounts of such compounds may suffice for the fungus. Man's quest is to obtain the potentially useful compounds in large quantities for use through applications of microbiology, biochemistry, and chemical engineering, or in one single word, by biotechnology. This chapter gives a few better-studied examples to illustrate the realized as well as the perceived uses of fungi.

7.1 FUNGAL FACTORIES

7.1.1 Penicillin

The turning point in the industrial exploitation of fungi was the chance discovery of *Penicillium notatum* by Alexander Fleming in the 1920s, followed by the isolation in 1943 of the strain NRRL 1951 of *Penicillium chrysogenum* (Raper *et al.*, 1944; Raper, 1946). To get enough penicillin, the NRRL strain was subjected to continuous processes of mutation and selection for penicillin to be produced as the "wonder drug" active against the causes of meningitis, community-acquired pneumonia, and sepsis. On the occasion of being awarded the Nobel Prize in Medicine for the discovery of penicillin, Alexander Fleming said:

> The origin of penicillin was the contamination of a culture plate of staphylococci by a mould. It was noticed that for some distance around the mould colony the staphylococcal colonies had become translucent and

evidently lysis was going on. This was an extraordinary appearance [Figure 7. 1] and seemed to demand investigation, so the mould was isolated in pure culture and some of its properties were determined.

The mould was found to belong to the genus *Penicillium* and it was eventually identified as *Penicillium notatum*, a member of the *P. chrysogenum* group, which had originally been isolated by Westling from decaying hyssop.

Having got the mould in pure culture I planted it on another culture plate and after it had grown at room temperature for 4 or 5 days I streaked different microbes radially across the plate. Some of them grew right up to the mould—others were inhibited for a distance of several centimetres. This showed that the mould produced an antibacterial substance which affected some microbes and not others [figure 7.1].

In the same way I tested certain other types of mould but they did not produce this antibacterial substance, which showed that the mould I had isolated was a very exceptional one.

Then the mould was grown on fluid medium to see whether the antiseptic substance occurred in the fluid. After some days the fluid on which the mould had grown was tested in the same way that I have already figured for lysozyme—by placing it in a gutter in a culture plate and then streaking different microbes across the plate. The result shown in [the figure] is very similar to that observed with lysozyme with one very important difference, namely that the microbes which were most powerfully inhibited were some of those responsible for our most common infections.

This was a most important difference.

By this method and by the method of serial dilution I tested the sensitivity of many of the common microbes which infect us and found…that many of the common human pathogens were strongly inhibited while many others were unaffected. This led us to our first practical use of penicillin, namely in the preparation of different culture medium. There was such a sharp distinction between the sensitive and insensitive microbes that by adding penicillin to the culture medium all the sensitive microbes were inhibited while all the insensitive microbes grew out without hindrance. This made it very easy to isolate microbes like the whooping-cough bacillus and Pfeffer's influenza bacillus which are normally found in the respiratory tract in association with large numbers of cocci which are sensitive to penicillin.

Fleming published his findings in 1929 in a paper titled "On the Antibacterial Action of Cultures of a *Penicillium*," with special reference to their use in the isolation of *B. influenza*. This attracted the attention of Ernst Chain, a biochemist refugee from Hitler's persecution who was working in Oxford. With Howard Florey, Ernst Chain isolated a few milligrams of penicillin, which was highly active against staphylococci. By this time World War II had started and an opportunity for clinical trial came Fleming's way. A policeman in the Radcliffe Hospital had acute staphylococcal blood poisoning. A course of treatment was started and a spectacular recovery got under way—but tragically the supply ran out; the policeman relapsed and died.

To Alexander Fleming, a secretion from a fungus lysing the bacteria was a scientific curiosity. He had not thought that it might be of medical value. The penicillin paradigm illustrates that fungal biotechnology generally begins with the discovery of a strain by chance, which as Pasteur had stated, favors the prepared mind. Efforts of groups of workers were required to improve the yield by cultural and genetic manipulations to obtain sufficient amounts of penicillin to enable a fungus's metabolic product to be purified and tested before its use by humans. The chemical structure of penicillin was worked out by the late 1940s. All penicillins are beta-lactam antibiotics (Figure 7.2).

Although Fleming's original culture was *P. notatum* NRRL 832 (ATCC9179) obtained as a laboratory contaminant, the industrially useful culture *Penicillium* NRRL 1951 was derived in 1943 from an infected cantaloupe in a market in Peoria. It is this strain that was used by J. F. Stauffer and Myron P. Backus in the Botany Department at the University of Wisconsin to get enough penicillin. They mutagenized this strain by treatment of conidia by x-ray, UV, and nitrogen mustard [methylbis (β-chloroethyl) amine]. The survivor colonies were selected on a honey-peptone-agar medium. Approximately 250 isolates were grown in flask cultures containing a corn-steep-lactose medium from which a further selection was made of 12 cultures. Based on time-consuming tests, three to

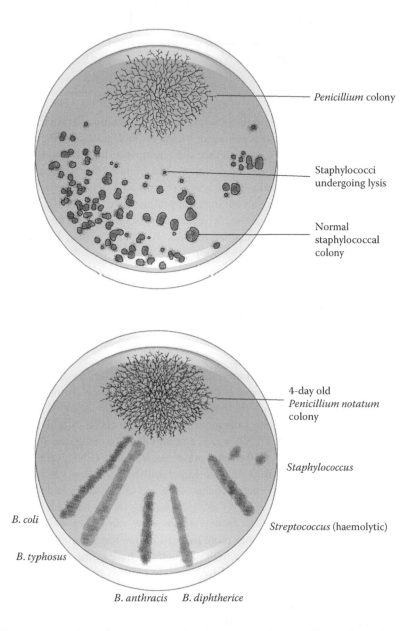

Figure 7.1 Above: Dissolution of staphylococcal colonies around *Penicillium* colony. Below: Technique used to demonstrate antagonistic activity of a four-day-old colony of *Penicillium notatum* on agar against species of bacteria streaked radially. The bacteria are (counterclockwise) *Staphylococcus; Streptococcus; B. diphtherice; B. anthracis; B. typhosus; B. coli.* (Drawings based on photographs in Alexander Fleming's Nobel lecture.)

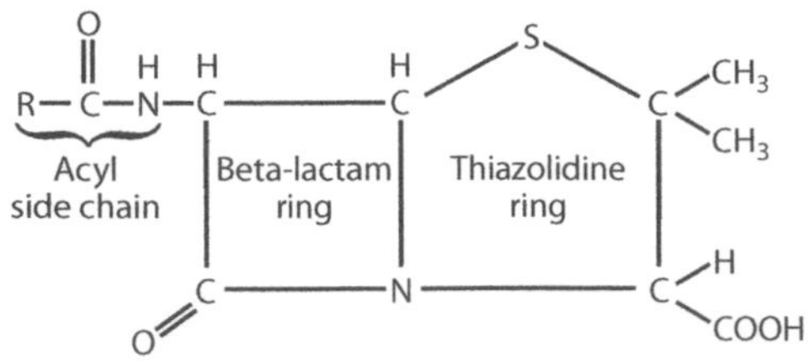

Figure 7.2 General structures of penicillins.

four finalists were selected and UV treatments were repeated to obtain a strain that did not produce an undesirable yellow pigment and with penicillin yields of 3000 units/ml. Penicillin yields were influenced by composition of medium, type of inoculum, temperature, pH, and tightness of the cotton used to plug the culture flasks (implying aeration). The yield of the Wisconsin strain was 100 times that of the original Fleming strain. There was no correlation between growth rate on agar or in liquid medium and yield. Reflecting on commercial production of penicillin, Kenneth B. Raper, a professor of bacteriology and botany at the University of Wisconsin and a key player in the development of penicillin-producing strains, recalled in 1978 (Katz, E. R. 2007; Kenneth Raper, Elisha Mitchell and *Dictyostelium. J. Biosci,* 31, 195-200):

> When subjected to untold mutations and selections during the past 35 years, it can be said, insofar as I have been able to determine, that the cultures now being used for penicillin production throughout the world stem from the cantaloupe strain isolated in Peoria in July 1943.... To achieve this end, let it be said that the penicillin saga was made possible by teamwork: teamwork within the several institutional groups and among different groups, whether in government, universities, or industry.

There are three important steps in the biosynthesis of penicillin G (benzylpenicillin) (Figure 7.3). The first reaction in a biosynthetic pathway is the condensation of three amino acids—L-a-amino-adipate, L-cysteine, and L-valine—into the tripeptide d-(L-a-amino-adipyl)-L-cysteinyl-D-valine (LLD-ACV). Subsequently, the enzyme isopenicillin N synthase (IPNS) catalyzes ring closure of LLD-ACV, yielding isopenicillin N (IPN), which contains the characteristic β-lactam ring. Both d-(L-a-amino-adipyl)-L-cysteinyl-D-valine synthetase (ACVS) and IPNS have been localized to the cytosol in *P. chrysogenum*. The final enzyme in penicillin biosynthesis is isopenicillin N:acyl CoA acyltransferase (IAT). This enzyme exchanges the L-a-amino-adipate moiety of IPN for a CoA-activated phenylacetyl or phenoxyacetyl group, resulting in the formation of the antibiotics penicillin G or penicillin V, respectively. Newer strains of *P. chrysogenum* obtained by further mutagenesis and selection produce 50 times more penicillin than the Wisconsin strain (Rodriguez-Saitz *et al.*, 2005). Two possible reasons for high productivity have come to light. One, a mutation in the *pahA* gene that encodes phenylacetate 2-hydroxylase reduces precursor availability. Second, high penicillin-producing strains contain 6 to 14 copies of three biosynthetic clustered genes.

Among the several lessons of the penicillin story, one is to carefully examine a contaminated culture before discarding it; if it is a contaminant not seen before, study the physiology of the contaminant in some detail. The penicillin story tells how a program of repeated mutagenesis and selection led to the selection of a useful strain; how cultural parameters were designed to obtain the optimum yield before the performance of the strain could be evaluated on an industrial scale; and how the efficacy and side effects of the product were tested on humans before it was marketed.

Whereas the original penicillin-producing strain was obtained by an alert mind, though by chance, we have an example of a biotechnologically important fungus that was obtained as a result of a deliberate search. Endo and his associates in Japan tested 8000 strains of microorganisms for their ability to produce an inhibitor of cholesterol biosynthesis, one of the major causes of atherosclerosis.

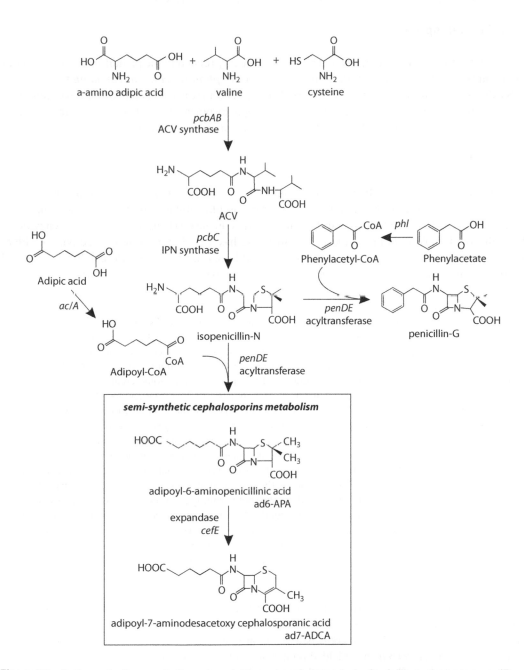

Figure 7.3 Pathway for the production of penicillin and cephalosporin in *Penicillium chrysogenum*. (From Koetsier, M.J., Gombert, A.K., Fekken, S., Bovenberg, R.A.L., van den Berg, M.A., Kiel, J.A.K.W., Jekel, P.A., Janssen, D.B., Pronk, J.T., van der Klei, I.J., and Daran, J.-M., *Fung. Genet. Biol.* 47, 33–42, 2010, with permission of Elsevier.)

Lovastatin produced by mutant strains of *Aspergillus terreus* by batch fermentation can produce nearly 2200 mg per liter within 12 days in fed-batch fermentations (Casas Lopez *et al.*, 2003).

7.1.2 Cephalosporin

Cephalosporins are structurally and pharmacologically related to the penicillins. Cephalosporin compounds were first isolated in 1948 from cultures of *Cephalosporium acremonium* from a sewer in Sardinia by an Italian scientist, Giuseppe Bronze. The first agent, cephalothin (coalition), was launched by Eli Lilly in 1964. Like the penicillins, cephalosporins have a beta-lactam ring structure that interferes with synthesis of the bacterial cell wall and are therefore bactericidal.

7.1.3 Statins

A statin is a drug prescribed for therapeutic intervention in patients who have a heritable risk factor for cholesterol accumulation leading to cardiovascular disease. Statin acts as a competitive inhibitor of 3-hydroxy-3-methylglutaryl coenzyme A reductase (HMG-CoA), the first enzyme in the pathway of cholesterol biosynthesis, and therefore has a cholesterol-lowering activity. In 1998, Akira Endo was awarded the Lasker-DeBakey Award for Clinical Medical Research. At the award ceremony, Endo was praised as (http://www.laskerfoundation.org/awards/2008_c_description.htm):

> Akira Endo, For the discovery of the statins—drugs with remarkable LDL-cholesterol-lowering properties that have revolutionized the prevention and treatment of coronary heart disease. By 1971, when Endo began his work (at Sankyo Company in Tokyo), thwarting the reductase seemed like a possible way to keep the body's cholesterol production in check. Endo had a novel idea about how to find substances that block the reductase. Aware that organisms can secrete compounds with powerful biological activities—presumably to kill their competitors—Endo proposed that some creatures might spit out chemicals that foil the reductase and thus impede a vital activity of their neighbors. Over the next two years, Endo and colleagues grew more than 6000 fungi, harvested the broth in which each had grown, and tested whether the material could interfere with an early step of cholesterol synthesis in a test tube. They then separated its components from one another, keeping track of the active material. By this and additional methods, he purified a substance from the fungus *Penicillium citrinum*, called mevastatin or compactin, that blocks the reductase. Additional analysis revealed that compactin resembles HMG-CoA and competes strongly with it for the site on the enzyme that catalyzes HMG-CoA's conversion to mevalonate. This capacity to bind specifically to the spot normally reserved for the enzyme's substrate suggested that the potential drug would not randomly clasp molecules in cells and disrupt other important activities. Endo wondered whether compactin would reduce enzyme activity in animals, as it does in test tubes. In 1979, he showed that the compound dramatically lowers blood cholesterol in dogs and monkeys with no obvious toxic effects. Statin use is increasing—30 million people worldwide are taking them—and has begun to make a dent in those numbers.

Lovastatin ($C_{24}H_{36}O_5$) (Figure 7.4) is produced as a secondary metabolite by the fungi *Penicillium citrinum* sp., *Monascus ruber*, and *Aspergillus terreus*. It catalyzes the rate-limiting step of cholesterol biosynthesis. Currently *A. terreus* is a commonly used producer of this drug. In some cases, pelleted growth of *A. terreus* has yielded higher titers of lovastatin than obtained with filamentous growth. Uncontrolled filamentous growth occurs when using rapidly metabolized substrates. The rapid increase in viscosity accompanied by filamentous growth greatly impedes oxygen transfer, and this is said to explain the low titers of lovastatin. Lovastatin productivity of wild strains of *A. terreus* is relatively low, but selected mutants can produce lovastatin titers of ≥ 2200 mg^{-1} within 12 days in fed-batch fermentations. Selection and composition optimization of a suitable medium was important for establishing a process for producing lovastatin. Experimental design strategies are widely used for optimizing fermentation media. Of the major culture nutrients, carbon and nitrogen

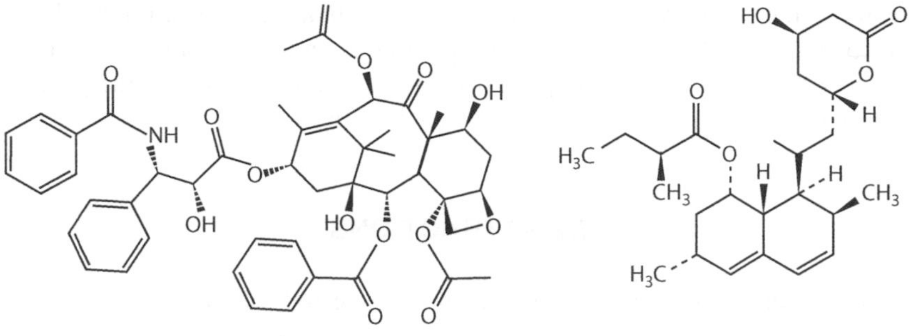

Figure 7.4 Chemical structures of (a) Paclitaxel and (b) Lovastatin.

sources generally play a dominant role in fermentation productivity because these nutrients are directly linked with the formation of the biomass and the metabolite. Also, the nature and concentration of the carbon source can regulate secondary metabolism through phenomena such as catabolic repression. Biosynthesis of lovastatin appears to depend on the carbon and nitrogen sources.

7.1.4 Anticancer Drugs

In a screening of trees for new natural products, the needles of Pacific yew trees were found to contain minute quantities of a highly derivatized diterpenoid named taxol that had shown anticancer properties in assays using mammalian cell cultures. Arguing that there may have been a transfer of genes from host to the fungus living inside (endophytic), Stierle *et al.* (1993) isolated a hyphomycetous (fungi Imperfecti) fungus, *Taxomyces adreanae*, from Pacific yew trees. From 21-day-old culture filtrates of this fungus grown in a medium that included glucose, sodium acetate, and sodium benzoate, 24–50 ng taxol per liter could be isolated. The discovery of the fungus was a result of a search of uncommon fungi of potential industrial importance. Taxol has a complex structure consisting of several rings—a four-member ring, a six-member ring, and an eight-member ring—and peripheral functionalities (Figure 7.4) that bind to polymerized tubulin, stabilizing it against disassembly and consequently inhibiting mitosis and showing anticancer properties. Possibly taxol is produced as a result of a dynamic interaction between the host and the fungus. Endophytic fungi are culturable on artificial media.

7.1.5 Defensins

The basis of a fungal factory is the isolation and investigation of the characteristics of new isolates and a knowledge of the physiology and biochemistry of the strain. In 2005, the fungus *Pseudoplectania nigrella* was found in soil in northern European pine forests and identified in screens for secretory proteins that are active against bacteria in plate assays (Mygind *et al.*, 2005; Jing *et al.*, 2010). Following culturing of the fungus, messenger RNA was extracted from mycelia of laboratory-grown *P. nigrella*, converted into complementary DNA, and mutagenized with transposons encoding a β-lactamase gene that lacks a signal peptide. Plasmids were introduced into *Escherichia coli*, and bacteria secreting β-lactamase were selected on ampicillin plates from which plasmids were isolated and the fungal cDNA inserts were sequenced. Sequence similarity against public and private databases identified a nucleotide sequence containing an open reading frame encoding a peptide of 95 amino acids in length that exhibited a 50–55% sequence identity to several defensins of invertebrates. The cDNA was transferred into an *Aspergillus oryzae* high-efficiency protein expression system that yielded a primary low molecular weight secreted and antimicrobial compound plectasin. *In vitro*, the recombinant peptide was especially active against *Streptococcus pneumoniae*, including strains resistant to conventional antibiotics. Plectasin showed extremely low toxicity in mice and cured them of experimental peritonitis and pneumonia caused by *S. pneumoniae* as efficaciously as vancomycin and penicillin. These findings illustrate the potential of fungi as a novel source of antimicrobial substances called defensins.

7.2 ATTRACTIONS

The features of fungi that have established them as preferred cellular factories over animal and plant cells are:

1. The raw material (nutritional) requirements for a fungus-based factory is relatively simple: A carbohydrate as a source of carbon and energy; inorganic ions as sources of nitrogen, phosphorus, and sulfur; and trace minerals as micronutrients and sometimes vitamins (mostly thiamine, nicotinic acid, and pyridoxine). All these are easily supplied as commercial-grade compounds available in bulk such as sugar cane or beet root molasses, corn-steep liquor, starch, and soya bean flour. In penicillin fermentation, corn-steep liquor added to growth medium following the initial discovery increased the penicillin yield from 20 unit cm^{-3} to 100 unit cm^{-3} of sucrose, urea, and phosphate. The stimulatory effect of corn-steep liquor is due to phenyl ethylamine, which is preferentially incorporated into the penicillin molecule to yield benzyl penicillin (Penicillin G). Thus, it became a standard practice to add side-chain precursors to the culture medium, in particular, phenylacetic acid.
2. The diversity of fungi occurring in freshwater, marine, or terrestrial habitats as psychrophile, mesophile, or thermophile species offers scope of finding and selecting an appropriate strain more suited to prevailing conditions of temperature. For example, thermophilic or thermotolerant molds may be specially suitable for fermentation in the tropics where the water required for removal of heat generated during fermentation in the growth vessel itself has a temperature well above the optimal range (20–30°C) of common molds. Moreover, a process at 45–50°C based on thermophilic molds will have a reduced risk of contamination. This is an important criterion considering that large volumes (100–1000 liters) of culture medium need to be maintained free from contamination for a time period extending to several days. (In a biochemical sense, the term *fermentation* is an energy-generation process wherein the organic compounds act as both electron donors and electron acceptors. However, in biotechnology, fermentation simply means culture of any cell system on a large scale for the production of chemical products.) Their adaptability to diverse environmental conditions provides greater opportunity of finding and selecting an appropriate strain for use. The thermophilic fungus *Thermomyces lanuginosus*, which forms thin mycelial suspensions, has found use in the production of thermostable lipase for use in the manufacture of detergents for hot-water machine washing.
3. The rapid growth rates of fungi make the scaling up of the process possible. For example, in liquid medium the budding yeast *Saccharomyces cerevisiae* has a doubling rate of 0.5 h^{-1}, almost matching the bacterium *E. coli*. The molds *Aspergillus* and *Neurospora* double their mass in about 2.2–2.7 hours at 25°C. Thus a fungus-based process is completed in usually 1 week or less, compared to an animal- or plant-based process, which takes months.
4. Selection of exceptionally high-producing variants among hundreds or thousands of colonies is facilitated by color tests or replica plating on selective media. Special media also allow colonial growth of some filamentous fungi for plate assays.
5. Molecular transformation tools for strain improvement are well developed. In yeast, the transforming DNA is integrated at a homologous site in the chromosome, although it is commonly ectopic in filamentous molds.
6. Expression cloning (Figure 7.5) combines the advantageous features of unicellular yeast and the multicellular molds, facilitating construction of strains secreting higher levels of a desired enzyme:
7. Posttranslational processing—e.g., glycosylation by addition of oligosaccharide $GlcNAC_2Man_{2-6}$—to asparagine residues in newly formed protein act as a secretion signal, allowing the protein to enter the secretion pathway. Glycosylation, phosphorylation, or acetylation allow active heterologous proteins to be produced. Chaperonin molecules allow correct folding of polypeptide.

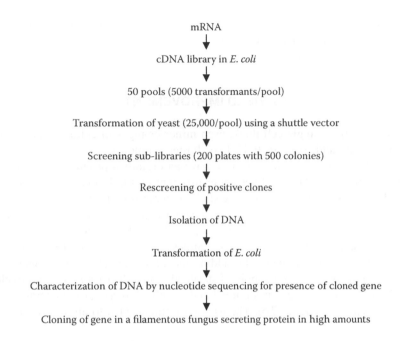

mRNA
↓
cDNA library in *E. coli*
↓
50 pools (5000 transformants/pool)
↓
Transformation of yeast (25,000/pool) using a shuttle vector
↓
Screening sub-libraries (200 plates with 500 colonies)
↓
Rescreening of positive clones
↓
Isolation of DNA
↓
Transformation of *E. coli*
↓
Characterization of DNA by nucleotide sequencing for presence of cloned gene
↓
Cloning of gene in a filamentous fungus secreting protein in high amounts

Figure 7.5 Scheme for expression cloning.

8. Inducible synthesis of some proteins allows their production with minimum background proteins.
9. The large surface area of cells provides for increased sites for protein secretion.
10. For certain situations, spontaneous hyphal fusions between genetically related strains allow construction of heterokaryotic mycelium and may offer advantages—for example, large-scale production of therapeutic antibody.
11. Depending on the ecological niche, a fungus tolerating oxygen or temperature can be selected by enrichment culture technique.
12. A useful strain can be permanently preserved as spores. The spores also serve as material for mutagenesis for further strain improvement.
13. The septate hypha is of special significance in fermentation. For culturing air has to be continuously bubbled into the medium with concomitant mechanical stirring of growing culture mass. However, when subjected to shear, the septate hyphae—like rungs in a ladder—provide strength to hyphae, limiting shearing and "bleeding" of protoplasm from severed hyphal compartments. The approximately 0.5-μm diameter central pore in the septum allows protoplasm to move forward and backward at a rate of about 4–6 cm per hour, bringing all compartments in cytoplasmic continuity. Furthermore, with short lateral branches (bridges) connecting adjacent hyphae, the entire mycelium is an interconnected and intercommunicating unit wherein metabolic activities may be synchronized. Dysfunctional organelles can be replaced by the migration of functional organelles from other hyphal compartments.
14. Woronin bodies may seal off parts of mycelium by plugging the septal pore, allowing the isolated compartments to take up specialized functions by generating and maintaining a reducing environment for the redox-sensitive enzymes and formation of special secondary metabolites by shunt pathways. An indirect evidence for this function is the accumulation of pigments and differentiation of spores in parts of the colony.
15. The mycelial growth form allows collection of extracellularly produced products by simple filtration. The brewing industry is attempting to genetically engineer yeast strains that will aggregate and settle down, yielding clear beer or wine.
16. Germinating spores or living mycelia can be immobilized in an agar matrix and used as beads for the production of biologically active molecules. Examples of species and strains that have mycelia suitable for reuse as agar beads are: *Rhizopus oryzae* (ATCC 1145), *Mucor plumbeus* (ATCC

4740), *Cunninghamella echinulata* var. *elegans* (ATCC 8688a), *Aspergillus niger* (ATCC 9142), and *Phanerochaete chrysosporium* (ATCC 2475).

17. Fungi are generally safe organisms for large-scale use.

7.3 YIELD IMPROVEMENT

Since a hypha is only a single cell thick, even minor changes in culture conditions can influence its physiology and yield. Back in 1947, J. W. Foster—a pioneer investigator of fungal metabolism—noted, "Indeed it is a common event to have an organism produce no detectable amount of a particular metabolic product, and yet under different cultural conditions, produce that very substance abundantly." In 1955 in a pioneering study of penicillin production, Backus and Stauffer noted antibiotic yield by surface-grown cultures to be markedly affected by factors that the worker may be ignorant of, such as the tightness of the cotton plug in the culture flasks/bottles, implying aeration. Fungi are therefore grown in glass or stainless-steel vessels fitted with control modules for *in situ* sterilization of culture media, for adding inoculum and antifoaming agents, for adding nutrients at desired rate and time, for supply of sterile air and exhaust for carbon dioxide, as a stirrer for good-mixing, for automatic control of pH and dissolved oxygen, of temperature, and for the aseptic removal of samples (Figure 7.6). The use of a fermenter led to an increased production of recombinant glucoamylase by *Aspergillus niger*. Continuous dilution of the medium with a mineral salt medium and slow addition of peptone and glucose was required for increased yield.

Fungus growth on a glucose-ammonium medium generally results in the acidification of the medium due to absorption of NH_4^+ and counterrelease of H^+ from the cell. In a fermenter, pH is continuously monitored by pH electrodes and an automatic addition of alkali or acid as required.

Figure 7.6 Fermenter with equipment for online control of culture parameters. (From Spinco/Bioengineering, with permission.)

Production of recombinant glucoamylase was enhanced over tenfold when pH was maintained (O'Donnell *et al.*, 2001). Several *pacC* genes have been identified. A *pacC* gene encodes a transcription factor; the *pal* genes encode components of a pH signal transduction pathway.

7.4 HETEROLOGOUS PROTEIN PRODUCTION

Recombinant DNA techniques allow regulatory properties of a strain to be modified for overproduction of a particular protein—e.g., more than 10–15% of total cellular protein—thereby allowing it to be obtained for practical needs.

7.4.1 Chymosin

According to legend, an Arab nomad with a saddlebag of milk to sustain him on a journey was crossing the desert on a horse. After several hours when he stopped to quench his thirst, the milk had separated into a pale watery liquid and a solid white lump (cheese). Taking a cue from this, the ancient Romans recognized a link between saddlebags made from the stomach of a suckling calf and the curdling of milk. Therefore a preparation made from the lining of the stomach of suckling calves began to be added. The preparation, rennet, contains a specific protease enzyme, chymosin (rennin), that causes casein in milk to clump into a solid gel. Protests by animal rights activists that it was inhumane to kill calves led to a search for substitutes. Japanese microbiologists perceived that since microorganisms secrete a variety of enzymes, the screening of soil microorganisms could provide a substitute for rennet. The fungus *Mucor pusillus*, having high milk-clotting activity, was obtained. They transferred the gene into a safe fungus, the Koji mold (mostly *A. oryzae*, *A. sojae*, *A. awamori*, and *A. kawachii*), which had long been used in Japan for the production of fermented food and beverages. A recombinant host strain produced heterologous *Mucor* (*Rhizomucor*) *miehei* protease in excess of 3 g/liter. This yield compares favorably with that of recombinant proteins (2–10 g per liter) in milk produced by transgenic animals. Chymosin identical to calf rennet is being produced commercially by yeast or filamentous fungi by transformation with a plasmid containing an artificially synthesized chymosin gene. This pure form of "vegetarian cheese" or Chy-max® was the first commercial product of recombinant DNA technology in the U.S. food supply.

Mucor rennin has been used as a model for investigations of the expression of fungal proteins in heterologous host cells (cells of other species), the processing of inactive precursor proteins into active precursor proteins into active enzymes, the effects of glycosylation on secretion, and the activity and stability of proteins. The *Mucor* rennin gene was expressed in yeast cells that secrete the foreign proteins at concentrations exceeding 150 mg/liter. Recombinant *Aspergillus oryzae* strains produce *M. miehei* acid protease in excess of 3 g/liter. A strain of *Penicillium duponti* can produce a highly thermostable acid protease, and *Malbranchea pulchella* var. *sulfurea* and *Humicola lanuginosa* produce thermostable alkaline proteases that can be concentrated without loss in activity simply by vacuum evaporation of the culture medium at 45°C.

7.4.2 Lipases

Lipases stable at pH 10–11 and temperatures from 30°C to 60°C were obtained from *H. lanuginosus* (syn. *Thermomyces lanuginosus*) and *Rhizomucor miehei* for use in detergents to remove oil stains. A lipase gene from *H. lanuginosus* was cloned and expressed in *A. oryzae*. The *R. miehei* lipase was the first lipase whose three-dimensional structure was deduced by x-ray analysis. Although the overall structures of the lipases are quite different, the lipase catalytic center always has the same three amino acids (serine-histidine-aspartic acid) as the serine proteases. The lipase catalytic site is covered by a short α-helical loop that acts as a "lid" that moves when the enzyme is adsorbed at the oil–water interface, allowing the substrate access to the active site. Mutation of an active-site serine causes an alteration in the motion of the lid, which affects the binding affinity of the enzyme.

7.4.3 Lactoferrin

Lactoferrin is an iron-binding glycoprotein present in human milk that plays a protective role against microbial and viral infections. Expression of human lactoferrin (hLF), a 78-kD glycoprotein, was achieved by placing cDNA under the control of the *A. oryzae* amylase promoter that secreted it at levels up to 25 mg/liter. Subsequently a modification of this production system combined with a classical strain-improvement program further increased production in excess of 2 g/liter on *Aspergillus awamori* as a glucoamylase fusion polypeptide and processed into mature hLF by an endogenous peptidase. The recombinant lactoferrin was indistinguishable from human milk lactoferrin in size, immunoreactivity, and iron-binding capacity. The recombinant protein had potent antimicrobial activity. Lactoferrin is the largest heterologous protein and the first mammalian glycoprotein expressed in a fungus.

7.4.4 Human Vaccine

Mammalian cell cultures require expensive media of undefined composition for production of therapeutic antibodies. The average antibody yield from hybridoma cells is about 100 mg/liter. If transgenic fungi could be used, the production may become possible as these fungi possess Golgi for processing secretory proteins by the addition of N-glycan structures (a short oligosaccharide chain linked to asparagine residue in a protein molecule). Single-chain antibody fragments have been produced in *Aspergillus* and *Trichoderma* as fusion proteins with a *Trichoderma* cellulase *cbh1* promoter (Ward *et al.*, 2004). The idea being explored by Neugenesis is to construct two separate vectors of light and heavy antibody chains as fusion proteins with a secretory protein such as glucoamylase or cellobiohydrolase. A specific amino acid sequence is engineered between the secretory enzyme clips off the antibody by a host protease during the secretory process in the Golgi. The two transformed fungal strains—i.e., one producing the light chain and the other producing the heavy chain (Figure 7.7)—are fused into a heterokaryon. Expectedly, the heterokaryon would produce both the antibody subunits and process them into intact monoclonal antibody molecules in a synthetic medium of defined composition (Allgaire *et al.*, 2009).

7.5 METHYLOTROPHIC YEASTS

During the 1970s Phillips Petroleum developed *Pichia pastoris* (yeast) that can be grown on a large scale in completely defined medium containing methanol as the sole carbon source for producing single-cell proteins. *Pichia* has alcohol oxidase (AOX1) linked to a strong promoter that is inducible by methanol and can be grown to high cell densities. Since it ferments glucose and related sugars even in the presence of air (absence of the Crabtree effect), it is called a nonconventional yeast. The alcohol oxidase—the enzyme that catalyzes the first step in the metabolism of methanol—constitutes as much as 35% of the soluble protein in the cell. In a gene-fusion approach, *Pichia* expresses foreign genes to gram quantity of protein (Cereghino *et al.*, 2002). The posttranslational folding and disulfide bond formation and glycosylation are similar to that in mammalian proteins. The production of nearly 15 g/liter of animal protein (gelatin) has led many companies to adopt the *Pichia* expression system for production of heterologous proteins.

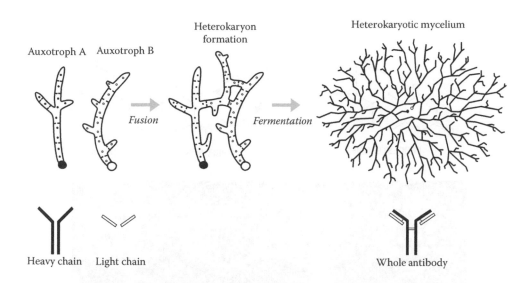

Figure 7.7 Scheme for production of humanized antibody by heterokaryotic mycelium of *Neurospora crassa*. Germinating spores of two auxotrophic strains (A and B) fuse to produce a heterokaryotic mycelium that can grow on a minimal medium lacking supplements because of complementation of non-allelic mutant genes. The mycelium secretes the heterologous whole antibody molecule. (Based on a figure kindly provided by W. Dorsey Stuart, Novozyme-Neugenesis, Davis.)

7.6 NEW METHODS OF YIELD IMPROVEMENT

The examples described in earlier sections show that gene products from unrelated fungi have been successfully produced using fungi as host organisms. However, yields of heterologous proteins are several orders of magnitude lower than of homologous proteins. Currently attempts are being made to identify the factors concerned with the production of proteins.

7.6.1 Increase in Gene Copy Number

The results of inserting extra copies of the desired gene have been disappointing. For example, in *Coprinus cinereus* inserting as many as eight copies of isocitrate lyase gene gave only 25% activity compared to untransformed fungus (Mellon and Casselton, 1988). *Aspergillus niger* transformed with the glucoamylase *glaA* gene also showed no correlation between the copy number of the transforming gene and its level of glucoamylase. Since journals publish negative results reluctantly, it appears that generally increasing yields through increases in copy numbers has been a failure. In view of processes of gene-silencing due to RIP and quelling (Chapter 8), it is now understood why increasing the copy number has failed to overexpress a gene in the fungi, or in the plants.

7.6.2 Manipulation of Morphology

The colony morphology of fungi in different species in liquid shake cultures is broadly of three types—freely dispersed, clumped, or compact hyphae—and depending on the conditions of culture (viz., composition of medium, inoculum size, pH, temperature, stirring rate), the addition of inert material to the culture medium in the same species is quite variable and affects yields.

A use of some species of fungi is in the 11α-hydroxylation of progesterone for obtaining an active steroid for medicinal use, for example, in alleviating pain and inflammation (Figure 7.8). Whereas progesterone can be synthesized chemically from plant sterols such as diosgenin, the regio- and stereospecific insertion of OH is carried out by incubation with harmless, living fungi having

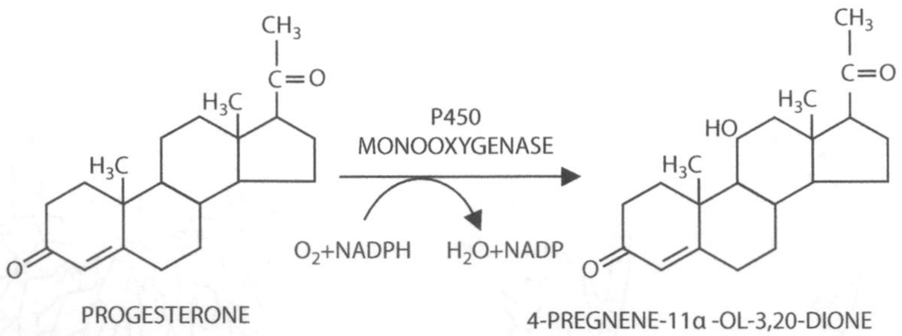

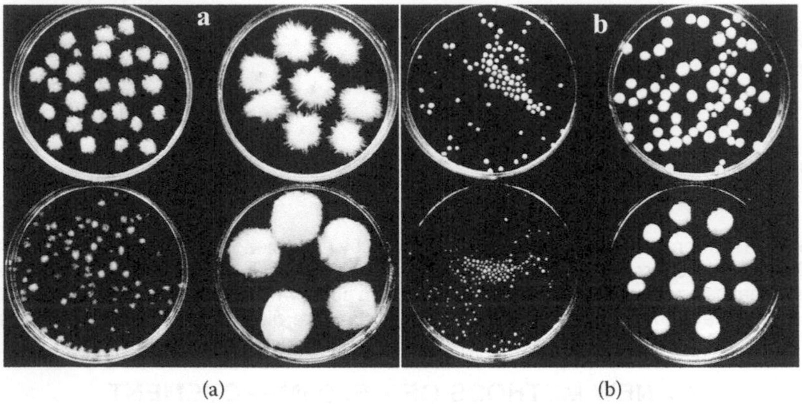

Figure 7.8 Influence of fungal morphology on biocatalytic activity. *R. nigricans* pellets, obtained in shake flasks at various inoculum concentrations and shaking frequency of (a) 100 rpm and (b) 225 rpm. The same position of Petri dishes in both figures represents pellets from cultures with identical inoculum, which was the highest in the cultures at the left bottom and decreased clockwise. (Original photograph provided by Polona Znidarsic-Plazl.)

membrane-bound steroid hydroxylase. The highest induction activity is dependent on morphology. In *Rhizopus nigricans* the highest hydroxylation rate correlated with lower diameters (pellets with diameters from 1.6 ± 0.3 to 13.9 ± 1.6 mm) and thus larger specific surface area of agglomerates harvested from the intensive growth phase of batch cultivation (Žnidaršič-Plazl and Plazl, 2010).

The filamentous form of *A. niger* was better for pectic enzyme synthesis, whereas the pellet form was optimum for citric acid production (Papagianni, 2004). Pellet form was also required for penicillin production by *Penicillium chrysogenum*. The wood-degrading fungus *Trichophyton rubrum* has different morphologies depending on its cultivation in baffled or unbaffled flasks. The pellet size in baffled flasks was small and gave higher yield of MnP ligninase. In *Phanerochaete chrysosporium* the ligninolytic activity (manganese peroxidase) was produced during differentiation of spores. These cells disappeared coincident with the time of enzyme secretion, suggesting that the differentiated cells acted as an enzyme reservoir, releasing their contents by the autolysis process. *Penicillium urticae* produced the antibiotics patulin and griseofulvin following conidiation. External parameters of culture affect the morphology of mold, its physiology, and its yield.

7.6.3 Hyperbranching Mutants

With their large surface area, filamentous fungi have high protein-secreting abilities, but the region of the hypha that actually secretes protein is contentious. In Chapter 1 we saw that protein synthesis occurs throughout the growing hypha but glucoamylase is secreted only from the growing apical region where the nascent wall is laid down, presumably because this region is more porous.

An approach of obtaining higher protein yield may therefore be through mutants with increased branching intensity. Hyperbranching *A. niger* mutants produced higher amylase and protease on solid substrates (Biesebeke *et al.*, 2005).

7.6.4 Modification of Cell Wall

Structure of the cell wall may be one of the most important determinants controlling the secretion of protein. *N. crassa* produces light and large forms of invertase enzyme. Trevithick and Metzenberg (1966) found that if the invertase is of a size that can pass through 40–70 Å pores in the cell wall it was secreted externally as a light form. On the other hand, the large form of enzyme remained in the periplasm or was bound to the cell wall. The observation suggests that structural features of protein and/or the cell wall determine the amount of protein secreted. Structural changes in the cell wall with altered chemical and physical properties can occur in response to carbon sources affecting secretion (Mishra and Tatum, 1972; Pessoni *et al.*, 2005). Culture morphology of *Sporotrichum thermophile* was strikingly different when grown with cellulose or in its depolymerized form (cellobiose) (Gaikwad and Maheshwari, 1994). In a cellulose medium, the mycelium autolysed and the cell-wall-bound β-glucosidase was released, whereas in a cellobiose medium the hyphal cells remained intact and β-glucosidase was not released into the medium.

7.6.5 Other Molecular Manipulations

After a protein is translated in the ribosomes it is transported into the endoplasmic reticulum where subunits are assembled by correct disulfide bonds and the protein molecule correctly folded by folding enzymes and chaperones, and other modifications such as disulfide bond formation and glycosylation (Sagt *et al.*, 2000) take place catalyzed by an array of proteins. Following quality control checks, the carrier vesicles transport protein to Golgi and finally to the plasma membrane for release into the periplasm between the plasma membrane and cell wall, which acts as a sieve allowing proteins of a particular size to pass through depending on the porosity. The low protein secretion can be due to a defect in any of the steps. At least 23 genes with roles in secretion have been identified in yeast. From the yeast genome sequence, about 20% of genes control functions related to cell wall biogenesis (de Groot *et al.*, 2001), indicating that remodeling cell wall structure for enhancing protein secretion may be tricky; it might make the organism vulnerable to osmotic bursting due to a lack of or altered cross-linking of wall polymers.

To understand reasons for the low yield of foreign proteins (Table 7.1), Gouka *et al.* (1996) integrated a single copy of a number of heterologous genes at a defined locus, controlled by the expression signal of *Aspergillus awamori* endoxylanase gene. With stability of mRNA, translation efficiency by construction of a synthetic gene with codon usage was optimized. Effects of other manipulations on protein secretion studied included overproduction of protein disulfide isomerase (Ngiam *et al.*, 2000; Valkonen *et al.*, 2003), of chaperone (Conesa *et al.*, 2002), and the use of a protease-deficient strain. However, success was limited. For example, lignin peroxidase of *Phanerochaete chrysosporium* was not overproduced when the encoding gene was fused to a gene of well-secreted cellobiohydrolase or glucoamylase in a protease-deficient strain.

7.6.6 New Expression Hosts

The species of fungi currently used do not fulfill all of the requirements, such as the production of enzymes stable at broad pH and temperature range, or a culture morphology giving nonviscous growth and thereby reducing energy cost in operation in large-scale fermentation, free of undesirable pigments and proteases. From a screening of more than 1000 fungi, Novozymes Inc. (www. novozymes.com) selected a filamentous fungus, *Fusarium venenatum*, as a new expression host that

Table 7.1   Yields of Some Proteins and Metabolites from Fungi

Product/Source	Method	Host Fungus	Yield (per liter)
Calf chymosin	Heterologous	*Aspergillus niger*	1.3 g
Cellulase	Homologous	*Trichoderma reesei*	30–40 g
	Homologous	*Chrysosporium lucknowense*	80 g
Glucoamylase	Homologous	*Aspergillus niger*	20 g
Lipase			
Thermomyces lanuginosus	Heterologous	*Aspergillus oryzae*	
Rhizomucor miehei	Heterologous	*Aspergillus oryzae*	
Plectasin			
Pseudoplectania nigrella	Heterologous	*Escherichia coli*	3.5 mg
Protease (alkaline)	Heterologous	*Acremonium chrysogenum*	4 g
Fusarium			
Protease (acid)	Heterologous	*Aspergillus oryzae*	20 mg
Mucor pusillus	Heterologous	*Aspergillus oryzae*	5 mg
Lactoferrin (human)	Heterologous	*Aspergillus niger*	25 mg
Interleukin-6	Heterologous	*Aspergillus niger*	150 mg
Human Chorionic Gonadotropin	Heterologous	*Pichia pastoris*	3 mg
Human insulin	Heterologous	*Pichia pastoris*	1.5 mg
Citric acid		*Aspergillus niger*	130–150 g

secretes low protease, low total spectrum of proteins, high heterologous expression, has favorable fermentation and morphology, and is safe (generally regarded as safe, or GRAS). Using this fungus, the first microbe-produced recombinant trypsin, which has better stability than animal-derived trypsin, was commercialized in 2002. Dyadic International has patented a novel-expression system based on *Chrysosporium lucknowense* that grows in a pH range from 4.5 to 9 and forms dispersed mycelia that produce 200-fold more neutral cellulase for an application in softening denim used in the manufacture of jeans.

7.7 BIOFUEL ETHANOL

In the 1970s, when the supply of petroleum for manufacturing petrol as transportation fuel became acute, bioethanol derived from fermentation of cellulosic biomass became an attractive alternative. A worldwide program started for screening and selecting fungi that secreted mixtures of exo- and endocellulase and β-glucosidase in a culture medium. Strains of *Trichoderma* selected from hundreds of species were developed through mutagenesis and selection that, under optimized conditions, secreted up to around 30 grams cellulase enzyme per liter of the culture medium, generating the possibility of an industrial process for converting cellulosic material into glucose and ethanol by a two-step process (Figure 7.9), generating much euphoria (Figure 7.10). Iogen, a private biotechnology firm based in Ottawa, announced the first shipment of 100,000 liters (26,417 gallons) of cellulosic ethanol processed from wheat straw to the Shell Company (http://www.iogen. ca/news_events/press_releases/2008_10_ 25.html). The protocol of the process has not been made public. The genome of *T. reesei* has been sequenced but the reasons for its high cellulase productivity are not yet known (Martinez *et al.*, 2008). The *T. reesei* genome (33.9 Mb) contains 9129 genes, compared to 10,620 in the model filamentous fungus *Neurospora crassa* (38.7 Mb). Surprisingly, *T. reesei* has only seven genes encoding endo- and exoglucanases (cellobiohydrolase)—crucial components of the cellulase system. It is assumed that the filamentous hyphae, though only 2–4 μm in

Trichoderma   Yeast

Cellulase   Anaerobic fermentation

Cellulose + H$_2$O ⎯⎯⎯⎯⎯⎯→ Glucose ⎯⎯⎯⎯⎯⎯⎯⎯⎯⎯⎯→ Ethanol + CO$_2$

Step 1   Step 2

Figure 7.9 Scheme for manufacturing biofuel ethanol from cellulosic raw material.

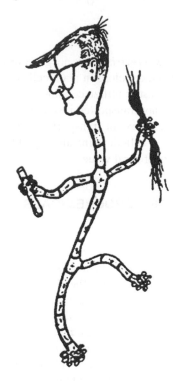

Figure 7.10 Cartoon depicting Elwyn T. Reese in the garb of *Trichoderma reesei*, a cellulolytic fungus (doi:10.1186/1754-6834-22; http://www.biotechnologyforbiofuels.com/content/2/1/22).

diameter, are packed with endoplasmic reticulum, ribosome, and Golgi (see Figure 1.8), which bud off ultramicroscopic vesicles filled with cellulase at astonishing speed; the vesicles fuse with the plasma membrane and the cell wall is rendered leaky to allow the exit of cellulase molecules into the surrounding aqueous medium. *T. reesei* has the smallest set of enzymes (approximately half of the number in other species) discovered thus far among the plant cell-wall-degrading fungi. The 10 gene-encoding enzymes involved in cellulose degradation and the 16 gene-encoding enzymes involved in hemicellulose degradation occur as different sets of genes in separate regions of the genome, suggesting their induction by different inducers (sophorose, cellulose, lactose), identified by scientists in Natick, Massachusetts. As complete hydrolysis of cellulose and hemicellulose requires multiple enzymes acting synergistically, *T. reesei* alone may not provide the mixture of different classes of enzymes that would be required for conversion of the plant biomass to fermentable sugars. For example, molecular data confirm that *T. reesei* lacks the ability to degrade lignin, the refractory compound enwrapping cellulose in biomass, thereby limiting the access of cellulase to β-(1,4)-glycoside bonds in cellulose for the release of glucose. The genomic data, however, allow a prediction of the lifestyle of *T. reesei* in nature. Since *T. reesei* is a specialist decomposer of cellulose, it cannot be a primary colonizer of biomass that contains a mix of polymers—hemicellulose, lignin, and cellulose. Rather, *T. reesei* can exist in nature primarily as a secondary colonizer of dead organic matter comprised mostly of the plant cell walls.

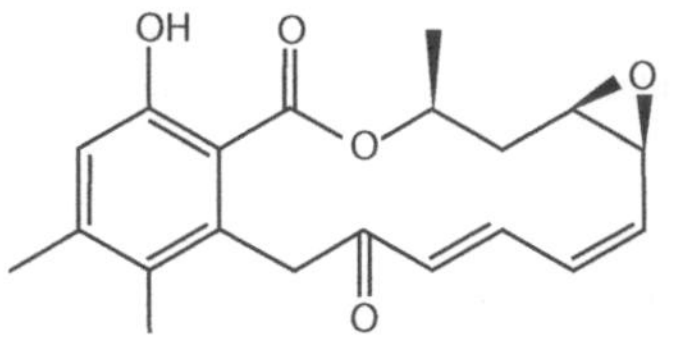

Figure 7.11   Structure of monocillin.

Although the yeast *Saccharomyces cerevisiae* does not grow on cellulose, a prospect is to develop strains that would release glucose from cellulose in plant cell walls and simultaneously ferment it to produce alcohol. The inability of yeast to utilize cellulose is because yeast does not express cellulase. Additionally cellodextrins are not transported inside the cell for catabolism (Galazka *et al.*, 2010). Genetic engineering of the cellodextrin transporter and cellodextrin hydrolyzing enzyme resulted in efficient growth of yeast on cellodextrins and a yield of 0.4 gram ethanol per gram of glucose, which is about 86% of the theoretical value.

7.8 AGROCHEMICALS

Early in evolution the immobile plants may have evolved a mechanism for adaptation and survival in environments that subject them to wide variations in temperature, both seasonally and diurnally. For plants growing in deserts, the problem of surviving high temperatures is more acute. This manifests in their morphology—for example, they have reduced leaves to conserve water, the stems are flattened and succulent to absorb and store water from the little rain they receive, or they have thorns to discourage a thirsty desert animal in search of liquid from nibbling. Taking a cue from endophytic fungi, a rhizosphere fungus *Paraphaeosphaeria quadriseptata* was isolated from about 1 mm^3 of soil surrounding the roots of a Christmas cactus, *Opuntia leptocaulis*, growing in the Sonora Desert, where temperature can be 55°C at 1400 h (McLellan *et al.*, 2007). Cultivation of a test plant, *Arabidopsis*, with the fungus enhanced its thermotolerance and enabled it to survive otherwise lethal temperatures. Chemical analysis of ethyl acetate extracts of this fungus led to identification of a C18 polyketide monocillin (MON) as the major metabolite (Figure 7.11). Application as low as 0.1 µM MON to seedlings enhanced their heat tolerance. MON application may trigger a heat-shock response mediated by a family of proteins called the heat-shock proteins (HSPs). Some promote the degradation of misfolded proteins; others promote the activation of proteins already aggregated (as when denatured by heat), thereby preserving the three-dimensional shape of the cell's proteins, i.e., they are chaperone proteins that prevent other proteins from taking conformations that would be inactive. MON can interact with HSP90—a conserved molecular chaperone that links to many proteins, including HSP101, allowing the plant to immediately recover from heat shock. *Arabidopsis* mutant seedlings that lacked HSP101 did not survive heat stress.

7.9 CONCLUDING REMARKS

The message is clear: The isolation and identification of fungi and their maintenance in culture, and knowledge of their physiology, provides information on how fungal species exist in nature among competing organisms and is vital for their exploitation. This is underscored by the search for the fungus *Pseudoplectania nigrella* and the discovery of a peptide antibiotic from it (Mygind *et al.*, 2005). Hitherto, improvement via mutagenesis and selection was the method to improve kinetics of product formation. Discoveries in functional genomics and DNA microchip technology hold promise for strain improvement.

REFERENCES

Allgaier, S., Taylor, R. D., Brudnaya, Y., Jacobson, D. J., Cambareri, E., and Stuart, W. D. (2009). Vaccine production in *Neurospora crassa*. *Biologicals* 37:128–132.

Backus, M. P. and Stauffer, J. F. (1955). The production and selection of a family of strains of *Penicillium chrysogenum*. *Mycologia* 47:429–463.

Barredo, J. L., Diez, B., Alvarez, E., and Martin, J. F. (1989). Large amplification of a 35-kb DNA fragment carrying two penicillin biosynthetic genes in high penicillin producing strains of *Penicillium chrysogenum*. *Curr. Genet.* 16:453–459.

Bermek, H., Gülseren, I., Li, K., Jung, H., and Tamerler, C. (2004). The effect of fungal morphology on ligninolytic enzyme production by a recently isolated wood degrading fungus *Trichophyton rubrum* LSK-27. *World J. Microbiol. Biotechnol.* 20:345–349.

Burlingame R. P. and Chandra, R. (2005). Gene discovery and protein production technology. *Industr. Microbiol.* 1:35–37.

Casas Lopez, J. L., Sanchez Perez, J. A., Fernández Sevilla, J. M. F. G., Acién Fernández, F. G. E., Molina Grima, E. M., and Chisti, Y. (2003). Production of lovastatin by *Aspergillus terreus*: Effects of the C:N ratio and the principal nutrients on growth and metabolite production. *Enzyme Microbiol. Technol.* 33:270–277.

Casas López, J. L., Sánchez Perera, J. A., Fernández Sevilla, J. M., Rodríguez Porcel, E. M., and Christi, Y. (2005). Pellet morphology, culture rheology and lovastatin production in cultures of *Aspergillus terreus*. *J. Biotechnol.* 116:61–77.

Cereghino, G. P. L., Cereghino, J. L., Ilgen, C., and Cregg, J. M. (2002). Production of recombinant proteins in filamentous cultures of the yeast *Pichia pastoris*. *Curr. Opin. Biotechnol.* 13:329–332.

Christgau, S., Kofod, L. V., Halkier, T., Andersen, L. N., Hockauf, M., Dörreich, K., Dalbøge, H., and Kauppinen, S. (1996). Pectin methyl esterase from *Aspergillus aculeatus*: Expression cloning in yeast and characterization of recombinant enzyme. *Biochem. J.* 318:705–712.

Conesa, A., van den Hondel, C. A. M. J. J., and Punt, P. J. (2000). Studies on the production of fungal peroxidase in *Aspergillus niger*. *Appl. Env. Microbiol.* 66:3016–3023.

Contreras, R., Carrez, D., Kinghorn, J. R., van den Hondel, C. A. M. J. J., and Fiers, W. (1991). Efficient KEX-2 like processing of a glucoamylase-interleukin-6 fusion protein by *Aspergillus nidulans* and secretion of mature interleukin-6. *Nature Biotechnol.* 9:378–381.

de Groot, P. W. J., Ruiz, C., Vazquez de Aldana, C. R., *et al.* (2001). A genomic approach for the identification and classification of genes involved in cell wall formation and its regulation in *Saccharomyces cerevisiae*. *Comp. Funct. Genom.* 2:124–142.

De Terra, N. D. and Tatum, E. L. (1963). A relationship between cell wall structure and colonial growth in *Neurospora crassa*. *Am. J. Bot.* 50:669–677.

Díez, B., Barredo, J. L., Alvarez, E., Cantoral, J. M., van Solingen, P., Groenen, M. A., Veenstra, A. E., and Martín, J. F. (1989). Two genes involved in penicillin biosynthesis are linked in a 5.1 kb SalI fragment in the genome of *Penicillium chrysogenum*. *Mol. Gen. Genet.* 218:572–576.

Elander, R. P. (2003). Industrial production of β-lactam antibiotics. *Appl. Microbiol. Biotechnol.* 61:385–392.

Endo, A. (1985). Compactin (ML-236B) and related compounds as potential cholesterol-lowering agents that inhibit HMG-CoA reductase. *J. Med. Chem.* 28:405–407.

Fleming, A. Penicillin. http://www.nobelprize.org/nobel_prizes/medicine/.../1945/fleming-lecture. Available on the Internet May 7, 2001.

Gaikwad, J. S. and Maheshwari, R. (1994). Localization and release of β-glucosidase in the thermophilic and cellulolytic fungus, *Sporotrichum thermophile*. *Exp. Mycol.* 18:300–314.

Galazka, J. M., Tian, C., Beeson, W. T., Martinez, B., Glass, N. L., and Cate, J. H. D. (2010). Cellodextrin transport in yeast for improved biofuel production. *Science* 330:84–86.

Gouka, R. J., Punt, P. J., Hessing, J. G., and van den Hondel, C. A. (1996). Analysis of heterologous protein production in defined recombinant *Aspergillus awamori* strains. *Appl. Env. Microbiol.* 62:1951–1957.

Gouka, R. J, Punt, P. J., and van den Hondel, C. A. M. J. J. (1997). Efficient production of secreted proteins by *Aspergillus*: Progress, limitations and prospects. *Appl. Microbiol. Biotechnol.* 47:1–11.

Grimm, L. H., Kelly, S, Krull, R. and Hempel, D. C. (2005). Morphology and productivity of filamentous fungi. *Appl. Microbiol. Biotechnol.* 69:375–384.

Huge-Jensen, B., Andreasen, F., Christensen, T., Christensen, M., Thim, L., and Boel, E. (1989). *Rhizomucor miehei* triglyceride lipase is processed and secreted from transformed *Aspergillus oryzae*. *Lipids* 24:781–785.

Jiménez-Tobon, G., Kurzatkowski, W., Rozbicka, B., Solecka, J., Pocsi, I., and Penninckx, M. J. (2003). *In situ* localization of manganese peroxidase production in mycelial pellets of *Phanerochaete chrysosporium*. *Microbiology* 149:3121–3127.

Jing, X.-L., Luo, X.-G., Tian, W.-J., Lv, L.-H., Jiang, Y., Wang, N., and Zhang, T.-C. (2010). High-level expression of the antimicrobial peptide plectasin in *Escherichia coli*. *Curr. Microbiol.* 61:197–202.

Keller, N. P., Bennett, G., and Turner, G. (2011). Secondary metabolism: Then, now and tomorrow. *Fung. Genet. Biol.* 48:1–3.

Koetsier, M. J., Gombert, A. K., Fekken, S., Bovenberg, R. A. L., van den Berg, M. A., Kiel, J. A. K. W., Jekel, P. A., Janssen, D. B., Pronk, J. T., van der Klei, I. J., and Daran, J.-M. (2010). The *Penicillium chrysogenum aclA* gene encodes a broad-substrate-specificity acyl-coenzyme A ligase involved in activation of adipic acid, a side-chain precursor for cephem antibiotics. *Fung. Genet. Biol.* 47:33–42.

Maheshwari, R. (2006). Fungi as cell factories: Hype, reality and hope. *Indian J. Microbiol.* 46:307–324.

Macauley-Patrick, S., Fazenda, M. L., McNeil, B., and Harvey, L. M. (2005). Heterologous protein production using the *Pichia pastoris* expression system. *Yeast* 22:249–270.

Martinez, D., Berka, R. M., Henrissat, B., *et al.* (2008). Genome sequencing and analysis of the biomass-degrading fungus *Trichoderma reesei* (syn. *Hypocrea jecorina*). *Nature Biotechnol.* 26:553–560.

McLellan, C. A., Turbyville, T. J., Kithsiri Wijeratne, E. M., *et al.* (2007). A rhizosphere fungus enhances Arabidopsis thermotolerance through production of an HSP90 inhibitor. *Plant Physiol.* 145:174–182.

Mellon, F. M. and Casselton, L. A. (1988). Transformation as a method of increasing gene copy number and gene expression in the basidiomycete fungus *Coprinus cinereus*. *Curr. Genet.* 14:451–456.

Mishra, N. C. and Tatum, E. L. (1972). Effect of L-sorbose on polysaccharide synthetases of *Neurospora crassa*. *Proc. Natl. Acad. Sci. USA* 69:313–317.

Müller, C., McIntyre, M., Hansen, K., and Nielsen, J. (2001). Metabolic engineering of the morphology of *Aspergillus*. *Adv. Biochem. Eng. Biotechnol.* 73:103–128.

Mygind, P. H., Rikke, L., Fischer, R. L., Schnorr, K. M., Hansen, M. T., *et al.* (2005). Plectasin is a peptide antibiotic with therapeutic potential from a saprophytic fungus. *Nature* 437:975–980.

Ngiam, C., Jeenes, D. J., Punt, P. J., Van Den Hondel, C. A. M. J. J., and Archer, D. B. (2000). Characterization of a Foldase, protein disulfide isomerase A, in the protein secretory pathway of *Aspergillus niger*. *Appl. Environ. Microbiol.* 66:775–782.

Nyssönen, E., Penttilä, M., Harkki, A., Saloheimo, A., Knowles, J. K. C., and Keränen, S. (1993). Efficient production of antibody fragments by the filamentous fungus *Trichoderma reesei*. *Nature Biotechnol.* 11:591–595.

O'Donnell, D., Wang, L., Xu, J., Ridgway, D., Gu, T., and Moo-Young, M. (2001). Enhanced heterologous protein production in *Aspergillus niger* through pH control of extracellular protease activity. *Biochem. Eng. Jour.* 8:197–193.

Papagianni, M. (2004). Fungal morphology and metabolite production in submerged mycelial processes. *Biotechnol. Adv.* 22:159–255.

Pessoni, R. A. B., Freshour, G., and de Cássia, F. R. R. (2005). Cell wall architecture and composition of *Penicillium janczewskii* as affected by inulin. *Mycologia* 97:304–311.

Punt, P. J., van Biezen, N., Conesa, A., and Albers, A. (2002). Filamentous fungi as cell factories for heterologous protein production. *Trends Biotechnol.* 20:200–206.

Raper, K. (1946). The development of improved penicillin-producing molds. *Ann. N.Y. Acad. Sci.* 48:41–56.

Raper, K. B., Alexander, D. F., and Coghill, R. D. (1944). Penicillin. II. Natural variation and penicillin production in *Penicillium notatum* and allied species. *J. Bacteriol.* 48:639–659.

Rodríguez-Sáiz, M., Díez, B., and Barredo, J. L. (2005). Why did the Fleming strain fail in penicillin industry? *Fung. Genet. Biol.* 42:464–470.

Romano, N. and Macino, G. (1992) Quelling: Transient inactivation of gene expression in *Neurospora crassa* by transformation with homologous sequences. *Mol. Microbiol.* 6:3343–3353.

Sagt, C. M. J., Kleizen, B., and Verwaal, R. (2000). Introduction of an N-glycosylation site increases secretion of heterologous proteins in yeast. *Appl. Environ. Microbiol.* 66:4940–4944.

Selker, E. U. and Garrett, P. W. (1988). DNA sequence duplications trigger inactivation in *Neurospora crassa*. *Proc. Natl. Acad. Sci. USA* 85:6870–6874.

Stierle, A., Strobel, G., and Stierle, D. (1993). Taxol and taxane production by *Taxomyces andreanae*, an endophytic fungus of Pacific yew. *Science* 260:214–216.

Strobel, G. and Daisy, B. (2003) Bioprospecting for microbial endophytes and their natural products. *Microbiol. Mol. Biol. Rev.* 67:491–502.

Sen Gupta, C. and Dighe, R. R. (1999). Hyper expression of biologically active human chorionic gonadotropin using the methyl tropic yeast *Pichia pastoris. J. Mol. Endocrinol.* 22:273–283.

te Biesebeke, R ., Record, E., van Biezen, N., Heerikhuisen, M., Franken, A., Punt, P. J., and van den Hondel, C. A. M. J. J. (2005). Branching mutants of *Aspergillus oryzae* with improved amylase and protease production on solid substrates. *Appl. Microbiol. Biotechnol.* 69:44–50.

Trevithick, J. R. and Metzenberg, R. L. (1966). Molecular sieving by *Neurospora* cell walls during secretion of invertase isozymes. *J. Bacteriol.* 92:1010–1015.

Valkonen, M., Ward, M., Wang, H., Penttilä, M., and Saloheimo, M. (2003). Improvement of foreign-protein production in *Aspergillus niger* var. *awamori* by constitutive induction of the unfolded-protein response. *Appl. Environ. Microbiol.* 69:6979–6986.

Van den Hondel, C. A. M. J. J., Punt, P. J., and van Goucom, R. F. M. (1991). Heterologous gene expression in filamentous fungi. In J. W. Bennett and L. L. Lasure, eds., *More Gene Manipulations in Fungi*, Academic Press, San Diego, pp. 396–428.

Wang, Y., Liang, Z.-H., Zhang, T.-S., Yao, S.-Y., Xu, Y.-G., Tang, Y.-H., Zhu, S.-W., Cui, D.-F., and Feng, Y. (2001). Human insulin from a precursor over expressed in the methylotrophic yeast *Pichia pastoris* and a simple procedure for purifying the expression product. *Biotechnol. Bioeng.* 73:74–79.

Ward, M., Lin, C., Victoria, D. C., Fox, B. P., Fox, J. A., Wong, D. L., Meerman, H. J., Pucci, J. P., Fong, R. B., Heng, M. H., Tsurushita, N , Gieswein, C., Park, M., and Wang, H. (2004). Characterization of humanized antibodies secreted by *Aspergillus niger. Appl. Env. Microbiol.* 70:2567–2576.

Ward, P., Piddington, C. S., Cunningham, G. A., Zhou, X., Wyatt, R. D., and Conneely, O. M. (1995). A system for production of commercial quantities of human lactoferrin: A broad spectrum natural antibiotic. *Nature Biotech.* 13:498–503.

Withers, J. M., Swift, R. J., and Wiebe, M. (1998). Optimization and stability of glucoamylase production by recombinant strains of *Aspergillus niger* in chemostat culture. *Biotechnol. Bioeng.* 59:407–418.

Žnidaršič-Plazl, P. and Plazl, I. (2010). Development of a continuous steroid biotransformation process and product extraction within microchannel system. doi:10.1016/j.cattod.2010.01.042.

Gene Silencing

CHAPTER 8

Transformation and Discovery of Gene-Silencing Phenomena

As is the way of scientific progress, once one issue is resolved, other, more intriguing, unexpected problems are revealed.

—John. H. Burnett

In 1928, F. Griffiths discovered an ingredient in heat-killed cells of the pathogenic bacterium *Streptococcus pneumoniae* that could transform a live, nonpathogenic mutant strain of the bacterium into a pathogenic strain. In 1944, O. T. Avery and his colleagues identified the transforming principle in the cell-free extracts as DNA. The introduction of a piece of homologous or foreign DNA molecule into an organism such that it is stably integrated into the host genome and stably alters its character is one of the most powerful tools in biology. It allows genes to be cloned through their ability to complement mutant phenotypes and to study their functions by selective knockout, the replacement of an endogenous gene with an engineered derivative.

8.1 TRANSFORMATION PROCEDURE

Methods were developed for transferring DNA into fungal cells and to select the transformed cells that have taken up this DNA. In the prototype experiment done by Mishra (1979) in Tatum's laboratory, wild-type DNA was introduced into an inositol-requiring mutant (auxotroph) of *Neurospora crassa*. As a result of transformation with a functional allele, the rare prototroph (non-inositol-requiring cell) could be directly selected on minimal media lacking the nutritional supplement. The reason for choosing the *inositol* (*inl*) strain for the experiment was that it was not known to revert spontaneously. Moreover, the *inl* mutant has an altered composition of phospholipids and cell wall and it was thought that this would allow entry of DNA inside the cells. The prototrophic selection method requires that the nutritional mutation be introduced into the recipient (host) strain by a prior sexual cross. This time-consuming difficulty is overcome by the use of a dominant selectable marker. For example, the mutant β-tubulin gene (Bml^R), which confers resistance to the fungicide benomyl, and the hygromycin resistance gene (*hph*) allow transformants to be selected even in the wild-type background. The DNA-treated conidia are plated on a medium that contains benomyl or hygromycin on which the untransformed cells cannot grow, allowing the transformed colonies to be selected.

The entry of DNA into cells is facilitated by removal of the cell wall. The cell wall can be removed by treating cells with a commercial mixture of β-(1,3)-glucanase and chitinase enzymes (Novozyme) obtained from a soil fungus *Trichoderma viride* (see Davis, 2000). The protoplasts (spheroplasts), freed from the rigid cell wall, become round, are prone to bursting in a hypertonic environment, and therefore need to be osmotically stabilized to regenerate a mycelial culture. Because the commercial preparations of the lytic enzymes vary from batch to batch, the proto-plast method yields variable results; it was therefore replaced by the electroporation or the particle bombardment methods for introducing DNA. In the now commonly used electroporation method, conidia are placed in a solution of DNA and subjected to strong electric pulses to facilitate the entry of DNA into the cell, presumably by transiently opening holes in the membrane.

In practice the DNA to be used for transformation is cut into fragments by DNA-cutting enzymes called restriction enzymes. A fragment is joined into a small circular DNA molecule called a plas-mid (vector) containing a point (restriction site) for joining the foreign DNA, one of which will have the gene of interest. The vector DNA is one that contains a selectable marker and regulatory sequences necessary for the expression of the transgene. The ligated, recircularized DNA fragment containing the gene of interest is introduced into the fungal cell (generally conidia) by transforma-tion. Typically, 1–20 μg of DNA per 10^7 cells in 50 μl of pH 7.5 buffer containing 1 M sorbitol or polyethylene glycol 4000 and 50 mM $CaCl_2$ are incubated before plating the conidia on selective agar medium to select the rare, specific colonies of transformed cells. The transformation frequency is about 1 to 20 per μg of DNA. The specific fragment of DNA from a single chosen transformed colony can be reisolated (cloned) by taking advantage of the fact that it is now tagged with DNA sequences of the transformation vector.

8.2 HOMOLOGOUS VS. ECTOPIC INTEGRATION OF TRANSGENE

In contrast to yeast, where the transforming DNA molecule commonly integrates at the related sequence by crossover (homologous recombination), integration in the filamentous fungi is com-monly at unrelated sequences (heterologous or ectopic recombination). As an example, Figure 8.1(a) shows the partial restriction map of a circular plasmid carrying the selectable marker gene, hygro-mycin phosphotransferase gene (*hph*) from *Escherichia coli*, and the control regions (promoter and terminator). This vector was used to transform *Ascobolus immersus* (Rhounim *et al.*, 1994). On average, the transformation frequency is 25 per μg of plasmid DNA. The recipient may be trans-formed by one or more ectopic integrations of the plasmid DNA at random sites in the genome (ecto-pic integration). The mode of integration is determined by Southern blot analysis of transformants [Figure 8.1(b)]. Transformants, which have integrated only one copy of the transgene, will exhibit only one *XbaI* band (lane1 and lane 4). Multiple integrations are seen in transformants in lane 2 and in lane 3, which have three and four integrations, respectively. Note that the *XbaI* site is located in the 3′ flanking sequence of the *hph* gene. The number of *XbaI* fragments will correspond to the number of transgenic copies if the multiple integrations are dispersed in the genome. This will also be seen if the multiple copies tandemly integrate at a single site.

The structure of transformant DNA is consistent with a model that suggests end-to-end joining of the input DNA, followed by integration into the genome (Watson *et al.*, 2008). The end-joins typi-cally have from one to five nucleotides in common and are near or within the original cleavage site of the plasmid. Ectopic integrations occur by attaching linear DNA to two ends of genomic DNA via the same joining mechanism.

8.3 PURIFICATION OF TRANSFORMANT

Purification of the transformed nucleus is facilitated in those fungi that form uninucleate conidia—for example, *Aspergillus* (Chapter 11)—or that can be manipulated to selectively produce uninucleate cells, as in *N. crassa*. In *N. crassa*, multinuclear macroconidia are easily obtained and are therefore used for transformation. However, as these cells are usually multinucleate, the

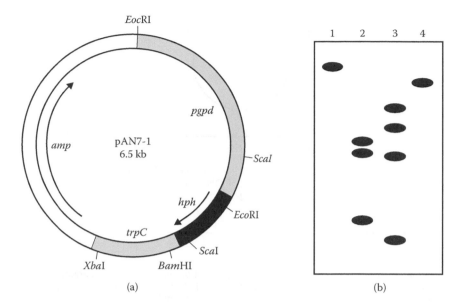

(a) (b)

Figure 8.1 (a) Map of plasmid used to obtain Hyg^R transformants of *Ascobolus immersus*. Black thick lines correspond to the *hph* gene, dotted lines are regions containing regulatory elements *gpd* promoter and *trpC* terminator from *Aspergillus nidulans* and restriction sites for *XbaI* (Xb), *BamHI* (B), *EcoRI* (E), *ScaI* (Sc), and ampicilllin resistance (*amp*). (b) Southern hybridization analysis of transformants using an *hph* probe corresponding to *ScaI-ScaI* fragment. Transformants (lanes 1 and 4) have integrated one copy of the transgene. Transformants (lanes 2 and 3) have integrated three and four copies of the transgene, respectively. (Based on Rhounim, L., Grégoire, A., Salama, S., and Faugeron G., *Curr. Genet.* 26, 344–351, 1994.)

transforming DNA may integrate at random locations in the genome and in different numbers of copies in different nuclei. Since not all nuclei in the same cell are transformed similarly, the transformants are commonly heterokaryotic. Thus a single transformed nuclear type needs to be purified. During microconidia formation, nuclei from the mycelium enter into microconidia in varying numbers, generally between one and four. The primary transformants are heterokaryotic, which requires the purification of a transformant having a single nuclear type. This is done by the rather time-consuming method of repeatedly plating a dilute suspension of macroconidia in series and picking up single colonies showing the transformed phenotype. Alternatively, a transformed nucleus from a heterokaryotic transformant may be purified by crossing it to an untransformed strain and selecting the transgene-bearing segregants among the meiotic progeny. However, purification of the transformed nucleus by the crossing method has generally been unsuccessful because often the transgene is not transmitted to the progeny.

8.4 GENE-SILENCING PHENOMENA

8.4.1 Silencing by Mutation

Selker *et al.* (1987) investigated the underlying reason for the noninheritance of transgenes in *Neurospora*. Specifically, they determined the fate of transforming DNA carrying single and duplicated DNA sequences and sequences that are normally methylated—the rationale being that methylated sequences are generally inactive (silent). A vector (Figure 8.2) was constructed that had a single copy of the *am*$^+$ gene and a duplicated *am*$^+$ flank region. This plasmid vector containing the wild type (*am*$^+$) was used to transform an *am* (glutamate dehydrogenase) deletion mutant of *Neurospora*

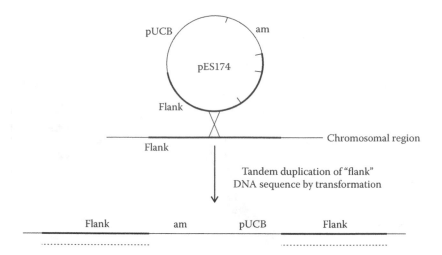

Figure 8.2 Selker *et al.*'s (1987) method of generating linked duplication of DNA sequence in *N. crassa* by homologous recombination of a plasmid. The only homologous sequence between pES174 and host was a 6-kb "flank" region (indicated by heavy line). The fate of this duplicated sequence between fertilization and nuclear fusion was studied.

crassa to study the fate of a single copy of the *am*+ gene and a duplicated *am*+ flank region. The transformant was crossed to a strain lacking the *am* gene and the ascospore progeny were analyzed by Southern hybridization using the *am* gene probe to determine if they had inherited the transgene. The progeny showed a 10-kb band that was not present in the primary transformant (Figure 8.3). This novel band could result only if the *Bam*HI restriction sites *b*, *c*, and *d* in the duplicated region had become modified. (*Bam*HI is a restriction enzyme obtained from *Bacillus amyloliquifaciens*; it differs, for example, from *Eco*RI, a restriction enzyme from *Escherichia coli*.) By using a pair

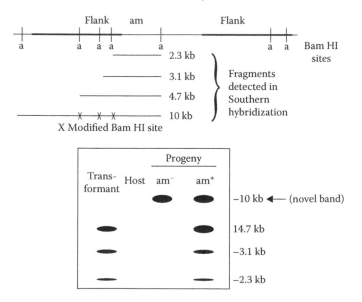

Figure 8.3 Discovery of RIP phenomenon. Detection and modification of transforming sequences in the *am*+ progeny from the cross of transformant × host DNA was digested with *Bam*HI, fractionated and probed with *am* region. The appearance of a novel 10-kb band in the *am*− progeny indicated modification or deletion of *Bam*HI sites b, c, and d in the duplicated regions of the progeny. The *am*+ progeny retained all four regions of the plasmid.

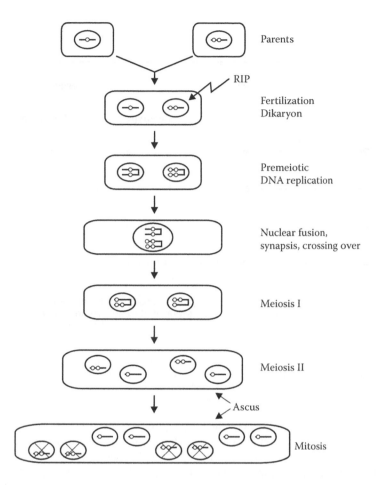

Figure 8.4 A diagram showing timing and consequence of RIP.

of restriction enzymes that distinguish methylated and unmethylated DNA (isoschizomers), the cut DNA upon electrophoresis shows a new band due to a change in the restriction sites by the modification of the cytosine residues in the DNA sequence by methylation of only the duplicated *am* flank region. The process that altered the extra DNA sequences by mutations and epigenetic modification of the cytosine residues in DNA by methylation was named RIP, an acronym for repeat-induced point mutations. By analyzing ascospore meiotic tetrads (eight-spored progeny asci), the timing of RIP was determined. It was inferred that RIP detects and mutates duplicated sequences during pairing of homologous chromosomes prior to karyogamy in the initial ascus (Figure 8.4). RIP depends upon the capacity of premeiotic cells to recognize the presence of duplicated sequences in an otherwise haploid genome. The only copies of genes immune to RIP are the tandemly repeat units in rDNA.

8.4.2 Meiotic Silencing by Unpaired DNA (MSUD)

In *Neurospora*, although strains carrying spontaneous deletions or extra DNA segments have a normal phenotype, abnormalities of translocation or duplication or deficient regions in chromosome can be manifested in the sexual phase, i.e., when these strains are crossed to normal strains and the ascospore progeny examined (Figure 8.5). Because of the extra DNA coming from the transformed strain, the region of homologous chromosome lacking the insertion will balloon out. A consequence of the lack of close pairing is that ascospores may fail to mature. The mutant designated *ascus*

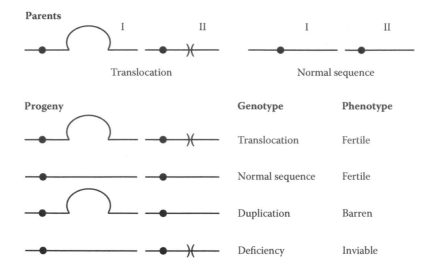

Figure 8.5 Diagram of genetic crosses using *Neurospora* strains that led to discovery of MSUD. Only two linkage groups are shown. The translocation is indicated by an inverse parenthesis that is indicated by looping out. Crosses between normal sequence strains and translocation strains yield three kinds of progeny: Normal sequence (like one parent), translocation sequence (like the other parent), and a novel class that is duplicated for the translocated segment. The duplication progeny arise when one component of the translocation segregates with a normal sequence chromosome; deletion progeny resulting from segregation of the complementary component are inviable. Duplication strains produce perithecia that lack ascospores (barren phenotype). (From Kasbekar, D.P., *J. Biosci.* 27, 633–635, 2002, with permission.)

maturation-1 (*asm-1*) produces ascospores lacking pigmentation, and these germinate very poorly. The mutant is also female sterile—it does not make female sexual structures (protoperithecia)—but is fertile when crossed as a male, although almost all eight ascospores produced in these crosses with the wild type are white, immature, and inviable. The *Asm* gene was cloned and used to construct a deletion mutant (denoted by superscript delta symbol) by targeted gene replacement. With availability of *Asm-1*$^\Delta$ in which the *Asm-1* gene had been deleted, crosses could be made using strains of opposite mating types. (Note that in the three-letter abbreviation for the gene symbol the first letter in *Asm* is capitalized; this is because of semidominant nature of ascus maturation mutation). The key observations were as follows:

1. Homozygous crosses between mutants in whom the wild-type gene is deleted—i.e., *Asm-1*$^\Delta$ *A* × *Asm-1*$^\Delta$ *a*—are barren because neither parent has the *asm-1* gene.
2. A cross of the *Asm-1*$^\Delta$ to wild-type *asm-1*$^+$ produces the normal numbers of eight ascospores, but all unpigmented and inviable, showing the dominant nature of this gene. Hence the genotype of *ascospore maturation* strain is written with the first letter capitalized, as *Asm-1*$^\Delta$.
3. Transformation of *Asm-1*$^\Delta$ strain by cloned copy of *Asm-1*$^+$ gene corrects this defect.
4. The cross of two deletion mutants yields normal numbers of eight *black* ascospores (fertile cross).

The results were explained on the assumption that *Asm-1* acts after fertilization of protoperithecia trichogyne with a male element, between stages in which the zygote nucleus immediately undergoes meiosis as a result of which four haploid nuclei are produced; each of these divides again by mitosis, giving eight nuclei that become sequestered in eight mature dark, ellipsoid ascospores. The presence of an unpaired gene (unpaired loop of DNA) activates some process in the ascogenous hyphae after karyogamy silences any unpaired DNA. Silencing extends to all homologous DNA sequences. This process is called meiotic silencing of unpaired DNA, or MSUD in short. We will note that MSUD differs from RIP, wherein the duplicated sequences are inactivated before karyogamy by mutation of both copies of duplicated sequences. A clue to the mechanistic basis of MSUD came from the isolation of a mutation *Sad-1* (*suppressor of ascus dominance-1*) that suppresses MSUD. The analysis made use of another dominant mutation called *round-spore, R*. This mutation causes all ascospores in the ascus to be round instead of ellipsoid, making it easy to detect in crosses of putative suppressor × *R* (Shiu and Metzenberg, 2002). Cloning and sequencing of the *sad-1* gene showed that it encodes a putative RNA-dependent RNA polymerase that synthesizes a double-stranded RNA from a single-stranded DNA. This revealed a connection between DNA pairing and RNA interference (RNAi) process—a finding that had been earlier implicated in quelling and also in posttranscriptional silencing of specific genes (e.g., flower color) in transgenic plants (Cogoni and Macino, 1999; Nakayashiki *et al.*, 2005).

Another example of the use of mutants in analysis of silencing phenomenon is the mutant named *banana*, which forms a single and very large ascospore shaped as a banana fruit. Using the green fluorescent protein (GFP) fused to the *histone H1* gene in a strain, a cross was made between wild type (no hH1-GFP locus) and the hH1-GFP and the heterokaryotic giant ascospores visualized to determine if the nuclei glow under blue-light excitation (Figure 8.6). Initially, only 8 of the 16 nuclei at one end of the young ascospore showed fluorescence, but approximately 12–24 h later all 16 nuclei showed hH1-GFP. Apparently, the mRNA from *hH1-GFPect* was translated in the cytoplasm surrounding the *hH1-GFPect* nuclei, and the protein is first incorporated into the nearby *hH1-GFPect* nuclei. (In this nomenclature, *ect* or EC denotes that a copy of the gene inserted by transformation is in some region other than its normal location, i.e., at ectopic location.) Since in the progeny ascospore from a cross *hH1-GFPect* × *Ban*, one half of the nuclei in the ascospore are *hH1-GFPect* and the other one half are wild type, initially only the 50% of nuclei containing *hH1-GFPect* at one end of the ascospore fluoresced. Over time, the green fluorescence spreads toward the opposite end until all nuclei of the giant ascospore fluoresced. There is no evidence so far that MSUD involves methylation.

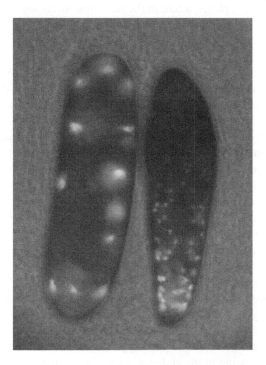

Figure 8.6 Visualization of meiotic silencing using GFP-tagged *histone H1* and β-*tubulin* in giant ascospore of banana mutant. The ascospore at left shows GFP in all 16 nuclei (8 *hH1-GFPect* + 8 non-*hH1-GFP*ect). The more mature ascospore at the right is highly multinucleate following several mitoses, and the hH1-GFP gradient is clearly visible from one end of the ascospore to the other. (Photograph courtesy of N. B. Raju, Stanford University.)

The *banana* or the *round-spore* mutants show the utility of mutants in fungal biology research. If yeast, *Neurospora*, and *Aspergillus* are the best model fungi, it is because a large repertoire of well-characterized mutants are available.

8.4.3 Silencing by DNA Methylation (MIP)

In *Ascobolus immersus* the transformation of a methionine auxotroph (*met-2*) by a plasmid carrying the *met-2+* allele resulted in its integration (Goyon and Faugeron, 1989). When the transformant carrying an extra copy of gene was crossed to wild type, both the normal and the ectopic copies were inactivated; the inactivation must occur before the premeiotic chromosome division. The frequency of inactivation of the transforming DNA was doubled if the *met-2-* genes were tandemly repeated than if the duplicated genes were at the ectopic site. No point mutations are associated with methylation. Gene inactivation is spontaneously reversible; where the progeny have a *met-2-* phenotype (Faugeron *et al.*, 1990), the reversion rate was increased by growing the fungus in the presence of 5-azacytidine, an analogue of cytidine that prevents cytosine methylation, suggesting that methylation plays a major role in this inactivation and accompanying gene silencing. This phenomenon was called MIP for *m*ethylation *i*nduced *p*remeiotically.

In another study *N. crassa* was transformed with the hygromycin phosphotransferase (*hph*) gene (Pandit and Russo, 1992). When the primary (heterokaryotic) transformant with more than one copy of the transgene was grown in the absence of hygromycin, the expression of the hygromycin (*hph*) transgene was silenced as determined by the very low percentage of colonies formed from conidial plating on hygromycin-supplemented media. Silencing is reversed if the transformant is grown in the presence of hygromycin (Figure 8.7) or 5-azacytidine, an inhibitor of methylation. This observation suggested that methylation played a role in the reversible silencing of the *hph* gene.

8.4.4 Quelling

The characteristic orange color of *N. crassa* conidia (see Figure 8.8) is due to carotenoid pigments produced from mevalonic acid by a series of reactions that involve dehydrogenation, cyclization, and *cis-trans* isomerization (Figure. 8.9). Three mutations are known to block different steps in carotenoid biosynthesis. Any of these three mutations blocks the production of pigment, resulting in an albino (white) phenotype. The three genes were named *al-1*, *al-2*, and *al-3* in the order of their discovery and all three genes have been cloned and sequenced.

In an attempt to overexpress carotenoid, Romano and Macino (1992) transformed a wild-type pink-orange strain with the wild *al-1+* gene. Unexpectedly, the typical pink-orange color of *Neurospora* was extinguished or suppressed in the transformants, which were white or pale yellow. Similarly, the duplication of the *chalcone synthase* gene (*chs*), involved in anthocyanin in flower petals, results in the loss of flower color in plants where this phenomenon is known variously as co-suppression, repeat-induced gene suppression, or homology-dependent gene silencing. These terms all mean the same thing: The mutual inactivation of gene expression by homologous sequences. The phenomenon has received more attention and publicity in plants because the expression of transgenes is necessary for crop improvement. However, *Neurospora* has provided the best clues because of the opportunity of investigating the involvement of a trans-acting diffusible molecule using heterokaryon. The albino genes of *Neurospora* provide a visual reporter system for identifying the silenced strains that have an albino phenotype (Romano and Macino, 1992).

The phenomenon of silencing of the endogenous (resident) and the transgene copies of gene in the vegetative phase was termed *quelling*. Transformation of the wild-type *Neurospora* strain with different portions of the *al-1* gene showed that a minimum of about 130 bp of coding region can induce quelling. As this size would be insufficient to code for a functional protein, the requirement of transgene protein for quelling was ruled out. If the transgene was *al-3*, Northern blot analysis shows

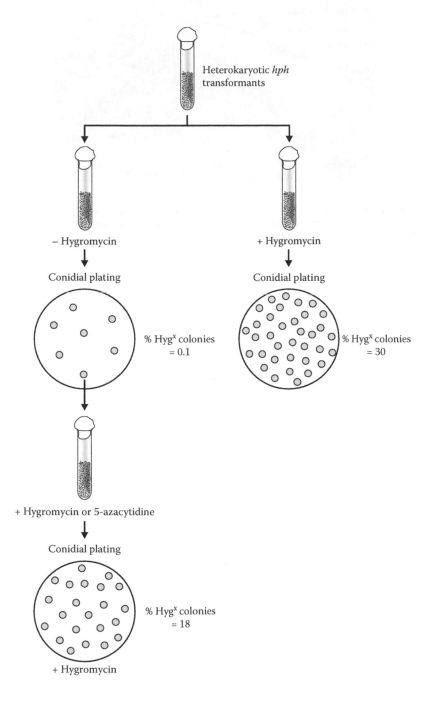

Figure 8.7 Silencing of *hph* transgene in *Neurospora crassa*. (Based on Pandit, N.N. and Russo, V.E.A., *Mol. Gen. Genet.* 234, 412–422, 1992.)

Figure 8.8 **(See color insert.)** *Neurospora crassa* growing on a sugarcane factory waste dump. The genes encoding pink-orange color have been used as a visual reporter system in gene-silencing experiments. (From P. Maruthi Mohan, with permission.)

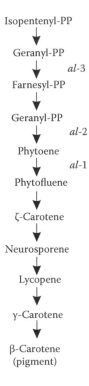

Figure 8.9 Pathway of carotene biosynthesis in *Neurospora crassa* showing the steps blocked in the *albino* mutants.

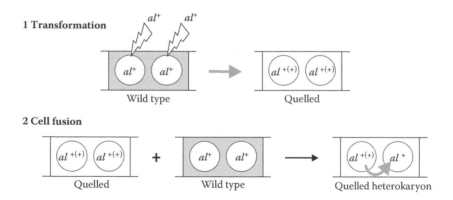

Figure 8.10 Diagram of quelling in heterokaryon. Gene introduced by transformation is shown in parenthesis. Only one nucleus of each type is shown.

that *al-3* mRNA is absent, not *al-1* or *al-2* mRNA. The transcriptional silencing of homologous endogenous genes is the hallmark of gene-silencing events in fungus or plants. But whereas in plants where almost all cases of gene silencing are associated with the repression of transcription due to the methylation of cytosine residues in the resident as well as the transgenes, quelling in *Neurospora* does not depend on methylation. This conclusion is based on results of Southern blot experiments using pairs of methylation-sensitive restriction enzymes, using an inhibitor of methylation, or using a mutant defective in methylation. To distinguish whether quelling is due to transcriptional inactivation or a posttranscriptional process such as RNA turnover, an RNAse protection assay was done. A labeled RNA probe complementary to the *al-1* gene was prepared and incubated with cellular RNA. After the unhybridized portion was removed by nuclease digestion, the size of the protected fragments was analyzed by gel electrophoresis. The amount of primary transcript (precursor mRNA) in quelled transformants was unchanged but the level of specific mRNA for the duplicated gene was reduced, suggesting that quelling is due to posttranscriptional gene silencing (PTGS).

To understand the components of the machinery by which posttranscriptional silencing is brought about, a quelled strain (albino) was mutagenized and mutants (orange phenotype) were isolated that were impaired in quelling (Cogoni and Macino, 1997b). By transformation of an albino-quelled strain with a plasmid (insertional mutagenesis), quelling deficient (*qde*) mutants were isolated that were orange in color. The rescued plasmid contained the putative *qde* gene whose sequence showed homology to RNA-dependent RNA polymerase. The experiments identified *qde* genes as a component of the silencing machinery. It was postulated that the *qde* product is RNA dependent RNA polymerase (RdRP) and makes an antisense mRNA, causing the loss of transformed phenotype.

To determine whether the presence of a transgene and endogenous gene in the same nucleus is required for silencing, a heterokaryon was constructed between quelled (albino phenotype) and the wild-type strain (orange color) containing both *al-1* silenced and nonsilenced nuclei. The white color of the heterokaryon demonstrated that quelling is dominant and the presence of a transgene and endogenous gene in the same nucleus is not a prerequisite for silencing. Rather, it suggested that silencing could occur through a mobile trans-acting molecule (Figure 8.10).

8.5 RNA SILENCING

A heterokaryon constructed between a *qde* mutant that produced no transgenic sense RNA and a wild-type strain had orange color, demonstrating that transgenic sense RNA is essential for silencing in heterokaryon. The cloning of QDE-1 after the landmark study by Fire and Mello (awarded

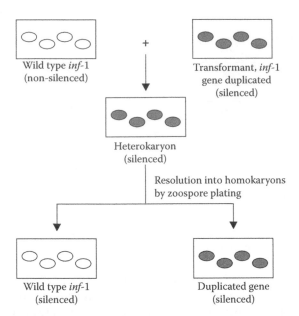

Figure 8.11   Diagram of infectious gene silencing in *Phytophthora infestans*. Hyphae are represented as rectangles. Wild-type and transformed nuclei are denoted as open and filled ovals, respectively.

a Nobel Prize in Physiology or Medicine in 2006) provided experimental evidence that an RdRP
is involved in PTGS and suggested that RNAs produced from transgenes are used as templates by
an RdRP to produce dsRNA. The wide presence of QDE-1 homologues in plants, fungi, and *C.
elegans* indicate that a conserved PTGS mechanism involving RdRP exists in all these organisms.
The *qde* genes were subsequently cloned and were found to encode key components in the quelling
pathway and demonstrated that quelling is an RNAi-related phenomenon. The discovery of quelling
in *Neurospora* has set the stage for research on RNA silencing machinery.

A case of gene silencing was discovered in the potato pathogen *Phytophthora infestans*
(Straminipila) that is maintained in the progeny nuclei even in the absence of the transgene (Van
West *et al.*, 1999) (Figure 8.11). The *inf-1* gene encodes a secretory protein elicitin—a hydrophobin
that induces necrosis in the plant, thereby restricting the spread of the pathogen. By transforming
P. infestans with plasmid containing *inf-1* gene, mutants were produced that were silenced in the
production of elicitin. The protoplasts of silenced and nonsilenced strains were fused to obtain a
heterokaryotic (silenced) strain that was resolved into homokaryotic component strains by nuclear
separation using the uninucleate zoospores of this "fungus" (see the Appendix). The individual
nuclear types multiplied mitotically in regenerating mycelium. The homokaryotic strains produced
from uninuclear zoospores were silenced; that is, once gene silencing was induced, it was maintained in the homokaryotic strain. Though both internuclear gene silencing (IGS) and quelling are
dominant, IGS differs from quelling in being infectious—i.e., being transmitted from nuclei to
nuclei—whereas quelling is not transmitted. Moreover, whereas the presence of transgene is essential for quelling, it is not for IGS (Figure 8.12).

8.6 CONCLUDING REMARKS

DNA is an inherently stable molecule, having been extracted from 18-million-year-old fossil-
ized leaves dating to the Miocene period. This suggests that DNA molecules from dead organisms or viruses in the environment can enter live cells and recombine with nuclear DNA, causing

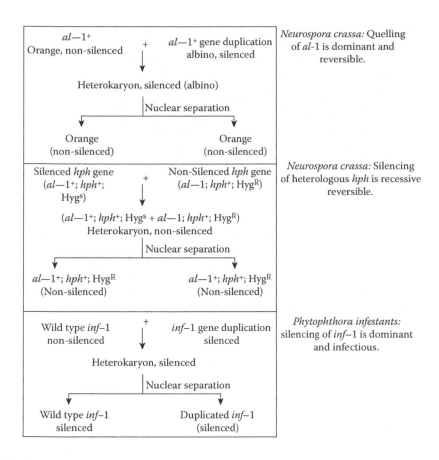

Figure 8.12 Summary diagram of transgene silencing in heterokaryons.

detrimental changes in genome structure and function. Early in evolution, cells therefore evolved defense mechanisms to protect nuclear DNA from foreign DNA sequences. Transformation experiments have led to the recognition that cells have multiple mechanisms that detect foreign DNA and silence them in the vegetative or the sexual phases, thereby maintaining the fidelity of the genome against onslaughts by extraneous DNA in the environment due to viruses or other parasitic elements. The genome defense phenomena in fungi involve the repression of sequences with similarity to RNA silencing in *Caenorhabditis elegans* and other metazoans where it is known as RNA interference (RNAi).

REFERENCES

Aramayo, R. and Metzenberg, R. L. (1996). Meiotic transvection in fungi. *Cell* 86:103–113.

Cogoni, C. and Macino, G. (1997a). Conservation of transgene-induced post-transcriptional gene-silencing in plants and fungi. *Trends Plant Sci.* 2:438–443.

Cogoni, C. and Macino, G. (1997b). Isolation of quelling-defective (*qde*) mutants impaired in post-transcriptional transgene induced gene silencing in *Neurospora crassa*. *Proc. Natl. Acad. Sci. USA* 94:10233–10238.

Cogoni, C. and Macino, G. (1999). Gene silencing in *Neurospora crassa* requires a protein homologous to RNA-dependent RNA polymerase. *Nature* 399:166–169.

Davis, R. H. (2000). *Neurospora. Contributions of a model organism.* Oxford University Press.

Dev, K. and Maheshwari, R. (2003). Silencing of hygromycin phosphotransferase (*hph*) gene during sexual cycle and its reversible inactivation in heterokaryon of *Neurospora crassa*. *Curr. Microbiol.* 47:220–225.

Faugeron, G., Rhounim, L., and Rossignol, J. (1990). How does the cell count the number of ectopic copies of a gene in the premeiotic inactivation process acting in *Ascobolus immersus*? *Genetics* 124:585–591.

Fire, A., Xu, S., Montgomery, M. K., Kostas, S. A., Driver, S. E., and Mello, C. C. (1998). Potent and specific genetic interference by double-stranded RNA in *Caenorhabditis elegans*. *Nature* 391:806–811.

Goyon, C. and Faugeron, G. (1989). Targeted transformation of *Ascobolus immersus* and *de novo* methylation of the resulting duplicated DNA sequences. *Mol. Cell. Biol.* 9:2818–2827.

Grotelueschen, J. and Metzenberg, R. (1995). Some property of the nucleus determines the competence of *Neurospora crassa* for transformation. *Genetics* 139:1545–1551.

Kasbekar, D. P. (2002). Sex and the single gene: Meiotic silencing by unpaired DNA. *J. Biosci.* 27:633–635.

Lee, H.-C., Li, L., Gu, W., Xue, Z., Crosthwaite, S. K., Pertsemlidis, A., Lewis, Z. A., Freitag, M., Selker, E. U., Mello, C. C., and Liu, Y. (2010). Diverse pathways generate microRNA-like RNAs and Dicer-independent small interfering RNAs in fungi. *Mol. Cell* 38:1–12.

Li, L., Chang, S.-S., and Liu, Y. (2010). RNA interference pathways in filamentous fungi. *Cell. Mol. Life Sci.* doi 10.1007/s00018–010–0471-y.

Mishra, N. C. (1979). DNA-mediated genetic changes in *Neurospora crassa. J. Gen. Microbiol.* 113:255–259.

Nakayashiki, H., Hanada, S., Bao Quoc, N., Kadotani, N., Tosa, Y., and Mayama, S. (2005). RNA silencing as a tool for exploring gene function in ascomycete fungi. *Fung. Genet. Biol.* 42:275–283.

Pandit, N. N. and Russo, V. E. A. (1992). Reversible inactivation of a foreign gene, *hph*, during the asexual cycle in *Neurospora crassa* transformant. *Mol. Gen. Genet.* 234:412–422.

Raju, N. B. (2009). *Neurospora* as a model fungus for studies in cytogenetics and sexual biology at Stanford. *J. Biosci.* 34:139–159.

Rhounim, L., Grégoire, A., Salama, S., and Faugeron G. (1994). Clustering of multiple transgene integrations in highly-unstable *Ascobolus immersus* transformants. *Curr. Genet.* 26:344–351.

Romano, N. and Macino, G. (1992). Quelling: Transient inactivation of gene expression in *Neurospora crassa* by transformation with homologous sequences. *Mol. Microbiol.* 6:3343–3353.

Selker, E. U., Cambareri, E. B., Jensen, B. C., and Haack, K. R. (1987). Rearrangement of duplicated DNA in specialized cells of *Neurospora. Cell* 51:741–752.

Shiu, P. K. T. and Metzenberg, R. L. (2002). Meiotic silencing by unpaired DNA: Properties, regulation and suppression. *Genetics* 161:1483–1495.

Van West, P., Kamoun, S., van't Klooster, J. W., and Govers, F. (1999). Internuclear gene silencing in *Phytophthora infestans. Mol. Cell* 3:339–348.

Watson, R. J., Burchat, S., and Bosley, J. (2008). A model for integration of DNA into the genome during transformation of *Fusarium graminearum. Fung. Genet. Biol.* 45:1348–1363.

PART IV

Model Organisms

CHAPTER 9

Yeast
A Unicellular Paradigm for Complex Biological Processes

Amitabha Chaudhuri
Genentech Inc., South San Francisco, California

In short order, yeast became a supermodel, challenging even *E. coli* in the visibility of its contributions.

—**Rowland H. Davis (2003)**

Since antiquity, yeast has been domesticated unwittingly or purposefully for the conversion of grape juice into wine by a process called fermentation. In the 18th century, the French chemist Antoine Lavoisier (1734–1794) showed that sugar in grape juice was transformed into ethanol during fermentation. Theodor Schwann (1810–1882) and Charles Cagniard-Latour (1777–1859) microscopically examined fermentation mixtures and advanced the view that the "force" that drove fermentation is "a mass of globules that reproduces by budding" and is a consequence of the growth of yeast—an idea that was quickly rejected by the influential German chemist Justus von Liebig, who maintained that the murkiness in fermenting liquid was not due to a living organism. Based on controlled experiments, chemical analyses of broth, and microscopic examinations of the sediment from successful and "diseased" fermentation vats, the versatile French scientist Louis Pasteur (1822–1895) concluded that yeast cells did not spontaneously arise from fermenting liquid but from preexisting cells. He identified yeast as the causative agent of alcoholic fermentation. The brewer's yeast *Saccharomyces cerevisiae* has today become a supermodel—"an organism that reveals and integrates many diverse biological findings applying to most living things" (Davis, 2003). A number of investigators, in approximately 700 laboratories around the world, have joined hands to make this fungus (though rather atypical) a model of all model organisms. Its advantages for the study of physiology and eukaryotic gene functions are:

- Its unicellular nature, making it a eukaryotic counterpart of *E. coli*
- Its amenability for mass culture in a simple minimal medium with a doubling time of about an hour
- The stability of its haploid and diploid phases (Figure 9.1)
- The ease of generating and detecting mutants, including conditional-lethal mutations for the study of indispensable gene functions
- The availability of a large diversity of mutant stocks
- Its growth under both anaerobic and aerobic conditions, making it ideal for the study of mitochondrial biogenesis
- Its small genome—smallest of any eukaryote
- The highly efficient cloning of genes by simple complementation of mutant genes

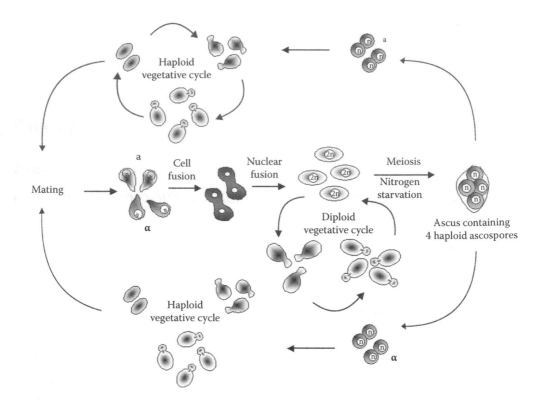

Figure 9.1   Life cycle of *Saccharomyces cerevisiae*. Both haploid and diploid cells multiply by budding. Diploid (2n) cell undergoes meiosis to form four haploid (n) cells, which are enclosed in a cell called an ascus.

- The homologous integration of transforming DNA, allowing disruption, deletion, or replacement of a gene
- Its two-hybrid method to generate a protein-interaction map for a system biology modeling of multicellular organisms (Giot *et al.*, 2003)

In his Nobel lecture, Leland H. Hartwell tells how *Saccharomyces cerevisiae* was pivotal to his life in science: "My research career has been motivated by a desire to understand cancer. Each time I have identified an intriguing aspect of the cancer problem, I have found that it could be approached more effectively in the simpler eukaryotic cell, *Saccharomyces cerevisiae*, than the human cell" (Hartwell, 2002). This chapter gives some remarkable examples of yeast in the study of biological processes in the eukaryotes and the likely further developments.

9.1 MOLECULAR MECHANISMS OF DNA REPLICATION AND CELL DIVISION

One of the most significant contributions of *Saccharomyces cerevisiae* (brewer's, baker's, or budding yeast) and the fission yeast *Schizosaccharomyces pombe* to biology is in the understanding of the eukaryotic cell cycle. Unlike unicellular bacteria and fission yeast that coordinate input from cell geometry and size in their decision to divide, the budding yeast *Saccharomyces cerevisiae* uses a different set of cell-intrinsic cues to control the timing of cell division (Moseley and Nurse, 2010). It has been proposed that the cell surveys the integrity of the bud neck and proper organization of the cytoskeleton before initiating cell division (Keaton and Lew, 2006). The positioning of the spindle at the bud neck requires that the neck is properly organized and has a correctly organized actin cytoskeleton. The organized actin structure ensures that the spindle formed in the mother cell is pulled into the bud neck by actin cables emanating from the daughter cell (Siller and Doe, 2009).

The mitotic cell division is a temporal sequence of events in which the cell first duplicates the DNA (S phase), followed by precise separation of the sister chromatids into daughter nuclei (M phase), leading finally to the separation of the two daughter cells. Each phase is separated from the preceding phase by an interval of time. The gap G1 is the time interval between mitosis and the initiation of new DNA synthesis; G2 is the gap between completion of DNA synthesis from the initiation of mitosis. In the late 1960s, Leland H. Hartwell took a genetic approach to understand the biochemical basis behind the orderly sequence of events of the mitotic cycle. He capitalized on the genetic advantages of *S. cerevisiae* to generate cell division cycle (*cdc*) mutants (Hartwell *et al.*, 1974). The *cdc* mutants were conditional temperature-sensitive alleles (*ts* alleles), which could be maintained stably by growing the mutant cells at a permissive temperature at 23°C and exhibited the mutant phenotype of cell cycle arrest only at 37°C. The growth-arrested *cdc* mutants could be observed under a light microscope to determine accurately the position within the cell cycle at which they were arrested. The mapping of the cell cycle position with growth is possible in yeast by observing the ratio of the size of the bud (daughter cell) to the mother (Figure 9.2). The technique used by Hartwell for generating yeast *cdc* mutants is shown in Figure 9.3.

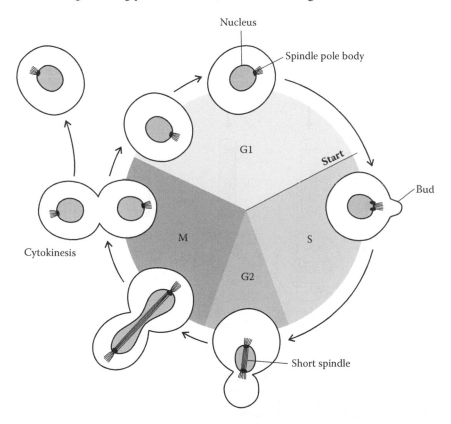

Figure 9.2 The *S. cerevisiae* cell cycle. The shape of a cell shows its position in its division cycle. The position START within G1 is the point at which the cell is committed to complete the cell cycle. The bud emerges at the beginning of the S phase and enlarges during G2 and M. The spindle pole bodies in yeast are embedded on the nuclear membrane. Yeast, like other fungi, has a closed mitosis— the nuclear envelope never breaks down. (Adapted from Watson, J.D., Gilman, M., Witkowski, J., and Zoller, M., *Recombinant DNA*, Scientific American Books, New York, 1992.)

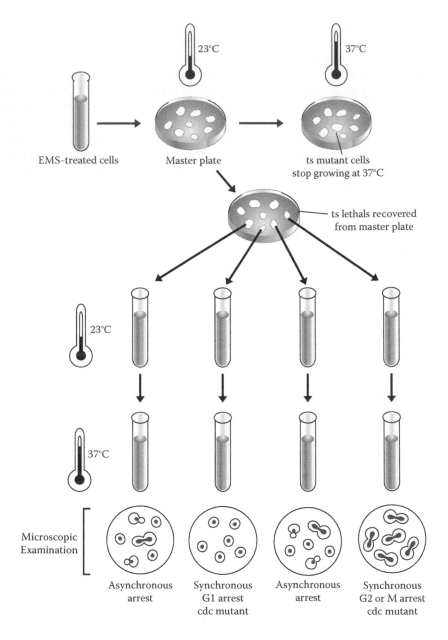

Figure 9.3   Isolation of *cdc* mutants. Cells, mutagenized with the DNA-methylating agent ethyl methanesul-
fonate (EMS), were grown at 23°C and then spread on agar at 23°C (the permissive temperature).
The colonies that were formed were replica plated and grown at 37°C (the nonpermissive temperature). Individual colonies that failed to grow at 37°C were noted as temperature sensitive (*ts*)
mutants. The *ts* mutants from the master plate were grown at 23°C, shifted to 37°C for a few hours,
and examined by microscopy. Cells in most cultures arrested at random points in the cell cycle
(asynchronous arrest), but some cells arrested at the same point of the cycle (synchronous arrest).
The latter class of cells was called *cdc* mutants. (Based on Watson, J.D., Gilman, M., Witkowski,
J., and Zoller, M., *Recombinant DNA*, Scientific American Books, New York, 1992.)

Early genetic work with yeast accelerated the understanding of the eukaryotic cell cycle, first by forging a link between the events in the cycle with specific genes, and second by establishing that the orderly sequence of events during cell cycle is a result of biochemical dependency, meaning that a prior event needed to be completed before the initiation of the next. Finally, the availability of the *cdc* mutants allowed the cloning and identification of the genes regulating cell cycle by complementation.

In all, 67 genes in the *Saccharomyces* genome database are annotated as *cdc* genes, of which 32 genes are directly involved in cell division. The rest regulate cell polarity and bud growth without taking part directly in the process of cell division. Many of these 32 genes have been cloned by complementation. For example, *cdc28* mutation in *S. cerevisiae* arrests growth of cells at a point "Start" in G1, when the cell commits to enter the cell cycle. To clone this gene, a *cdc28 ts* mutant was transformed with a genomic library from wild-type (*CDC28*) yeast strain, the cells plated and incubated at nonpermissive temperature (37°C), and the rare colonies in which *CDC28* gene was incorporated survived. (We shall follow the convention of naming a gene with three letters and a number. For *S. cerevisiae*, the wild-type alleles are written in uppercase letters in italics, and mutant recessive alleles are written in lowercase letters in italics.) The plasmid carrying *CDC28* gene was recovered, sequenced, and from the DNA sequence the encoded protein was identified as a protein kinase, an enzyme that transfers the terminal phosphate group from ATP onto a serine/threonine residue of another protein. Cloning of several *cdc* genes revealed that many of them are protein kinases, whereas others are phosphatases (enzymes that remove phosphate groups from protein substrates). Paul Nurse showed that the human Cdc2 kinase was able to complement the loss of function mutation in yeast (Lee and Nurse, 1987), underscoring the evolutionary conservation of the yeast and the human kinase. Work by Tim Hunt with sea urchin eggs led to identification of a 45-kDa protein, whose levels in dividing cells fluctuated at a regular interval, which he called *cyclin*. Biochemical analysis in different organisms revealed that binding of cyclin to the protein kinase regulated the enzyme activity. Two important conclusions from these studies were that the cyclical nature of the eukaryotic cell cycle is controlled by phosphorylation/dephosphorylation of *cyclin-dependent kinases* (CDKs) and by the targeted degradation of cyclin protein, as shown in Figure 9.4. The second important conclusion was that the mechanism of cell cycle regulation is highly conserved in all organisms from yeast to human. Leland Hartwell, Paul Nurse, and Tim Hunt were awarded the Nobel Prize in Medicine in 2001 for their contributions to understanding the fundamental process of cell division.

9.2 BUD GROWTH AND POLARITY

A basic problem in development is to understand the principles by which multicellular organisms determine the time and correct positions of new cells generating shape (patterning). This is the principle of polarity—seen at the level of an organism in the organization of the head and foot structures of humans and animals, and in the shoot and root system of plants, etc. At the cellular level, polarity is observed in the plane of cell division as in the case of the *Fucus* zygote, where the division plane of the nucleus in the rhizoid cell lies perpendicular to that in the thallus cell; in the organization of the nerve cells with dendrites at one end and axon at the other end; and in the structure of transporting epithelial cells lining our stomach and intestine with an apical and a basolateral end that separates two distinct compartments. The concept of polarity is equally prevalent in fundamental biological processes such as embryogenesis, where asymmetrical cleavage of the fertilized egg creates cells that follow distinct fates; during communication between immune cells such as B- and T-lymphocytes; or during neurogenesis, where nerve cells are actively guided so that they reach and synapse at specific regions in the brain. Studies in the last four decades have revealed that the underlying mechanism behind all manifestations of polarity in unicellular or

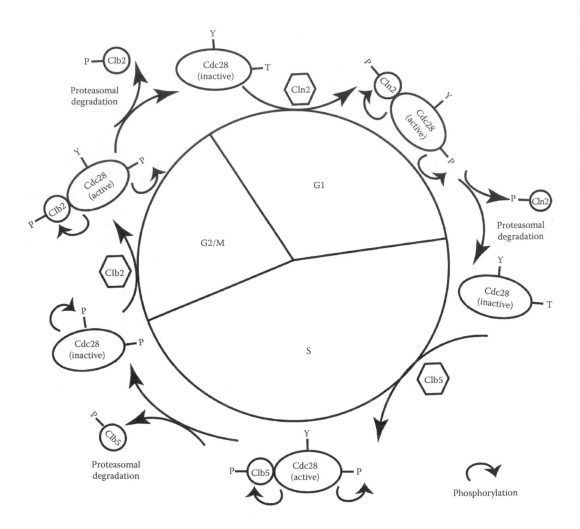

Figure 9.4 A biochemical model of cell cycle regulation in eukaryotes. Each phase of the cycle is regulated by the timely expression of a distinct cyclin protein. Entry of cells into mitosis is controlled by successive phosphorylation by kinases and dephosphorylation by phosphatases. The yeast cyclin-dependent kinase Cdc28p is activated by phosphorylation of threonine-18 (T-18) and inhibited by phosphorylation of tyrosine-19 (Y-19). Cln2p commits the cell into START by activating Cdc28. Once activated, Cdc28p phosphorylates Cln2p resulting in dissociation of the complex. Cdc28 is inactivated by dephosphorylation and phosphorylated Cln2p is degraded. Clb5 and Clb2 proteins initiate S phase and M phase, respectively. The end of the M phase is marked by rapid degradation of Clb2 and dephosphorylation of Cdc28 kinase.

multicellular organisms lies in the generation of cellular asymmetry, and a conserved mechanism involving cytoskeleton reorganization guides this process (Nelson, 2003).

The budding and the fission yeast are attractive systems to understand cellular asymmetry at the molecular level. First, *S. cerevisiae* shows polarized growth at every cell division by taking a decision where to produce the bud. Second, the complex process is genetically tractable. The budding pattern is easily observed by staining yeast cells with a fluorescent dye calcofluor, which fluoresces after binding to cell wall chitin. Bud scars are especially rich in chitin and fluoresce brightly (Figure 9.5).

At every division cycle, *S. cerevisiae* selects the site of a new bud in a spatially distinct pattern (Freifelder, 1960). Haploid **a** or α cells choose bud sites in an axial pattern in which mother and daughter cells bud adjacent to their prior mother–bud junction. On the other hand, diploid **a**/α cells

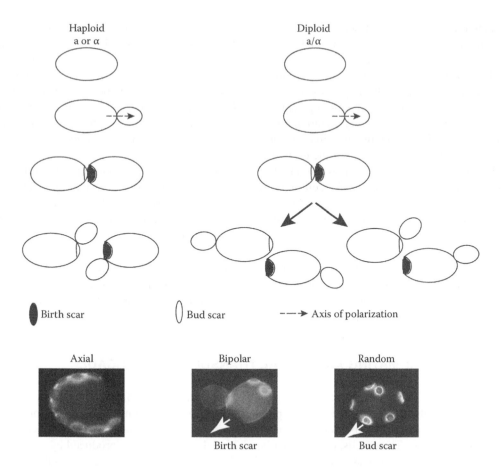

Figure 9.5 Bud-site selection in *Saccharomyces cerevisiae*. The pattern of budding is characterized by the orientation of the bud scars with respect to the birth scar. In calcofluor staining, the birth scars appear as unstained areas whereas the bud scars are brightly fluorescent. In axial pattern the bud scars always lie juxtaposed to the birth scar and to each other, forming a continuous line of scars. In bipolar budding the bud scars are seen opposite to the birth scar and also adjacent to it as shown in the figure. In random budding the bud scars are arranged randomly on the yeast cell.

choose sites in a bipolar pattern in which the mother cell buds either adjacent to the last daughter or at the pole opposite the last daughter. The daughter, however, always buds at the pole opposite its mother. The two distinct patterns of budding are shown in Figure 9.5.

In the original screen (Chant and Herskowitz, 1991), haploid yeast cells were mutagenized and plated on soft agar plates. Mutants defective in axial budding pattern were selected by observing microcolonies formed after the mother cell has undergone one to three cell divisions under a microscope. The pattern of budding is revealed by the arrangement of the daughter cells with respect to the mother cell in each microcolony. The original genetic screen identified four genes (*BUD1*, *BUD2*, *BUD3*, and *BUD4*) required for the specification of the axial budding pattern. A fifth gene, *BUD5*, was identified by molecular approaches (Chant *et al.*, 1991). Analysis of the function of the genes in haploids and diploids revealed that *BUD1*, *BUD2*, and *BUD5* are required by both types of cells to select bud sites at correct positions, whereas mutation in *BUD3* and *BUD4* affected only haploid cells. Further work from various laboratories identified other genes that affected bud site selection in haploids and diploids distinctly. The mechanism of bud site selection follows a two-step process. First, the yeast cell integrates intrinsic spatial information to define the site for the growth of the new bud. Proteins that persist at the site of the previous bud from one cell division to the next produce these spatial cues. Localization experiments using green fluorescent protein (GFP) tagged proteins support the persistence of Bud3, Bud4, and Bud10 at previous bud sites after each division cycle. Once the site is selected, the next step involves recruitment of a guanosine-5′-triphosphate (GTP) binding protein Bud1 and its regulators Bud2 and Bud5 to the site previously marked by Bud3, Bud4, and Bud10, leading to the localized activation of Bud1. It is believed that active Bud1 activates Cdc42 locally by interacting with its GTP-exchange factor Cdc24. Active Cdc42 then induces actin cytoskeleton reorganization by regulating the activity of actin-binding proteins. As a result of cell polarization, the protein transport machinery delivers membranes and components of cell wall biosynthesis at the site of the growing bud.

Bud site selection in diploids is mediated by a different group of landmark proteins Bud8 and Bud9, which mark the poles of the cell by recruiting Rax2p, which persists at the poles for many generations and is believed to activate Cdc42p via Bud1p. In the absence of Bud8 or Bud9, Rax2p fails to localize correctly, causing the diploid to bud in a random pattern. A schematic representation of the molecular machinery that guides the axis of polarization during budding is shown in Figure 9.6.

Although it may appear that generation of polarity in unicellular yeast may be different from that in epithelial cells of multicellular organisms, the basic core mechanism of organizing the cytoskeleton using Cdc42 and actin-binding proteins had evolved very early during evolution to be used universally by all eukaryotic organisms.

9.3 MATING AND SIGNAL TRANSDUCTION CASCADE

Yeast cells exist in **a-** and α-mating types (Chapter 12). Mating of cells of opposite mating types is coordinated by the release of small peptide hormones (pheromones). The **a**-cells produce a 12-amino acid peptide, the a-pheromone, and respond to the 13-amino acid peptide α-pheromone, produced by α-cells. The α-cells, on the other hand, bind the a-pheromone produced by the **a**-cells. Reception of the pheromone signal triggers a series of events that include changes in cell shape and arrest of cell growth. How these events are triggered by the pheromone signal is of general interest because signaling events must occur in unicellular organisms seeking nutrients in the surrounding environment or in a fungal pathogen searching for an entry point into a host plant, or in a higher organism responding to growth factors, hormones, neurotransmitters, and other sensory input.

Pheromone signaling is the most well-characterized eukaryotic signaling pathway. The hunt for genes in this pathway was facilitated by a selection scheme based on the growth arrest of cells in response to pheromone signaling as schematically depicted in Figure 9.7. Briefly, streaking yeast cells of opposite mating types on an agar plate in the form of a cross (cross-streaking) results in the appearance of a "zone of growth inhibition" in both strains as a result of pheromone-induced growth arrest. The zone of growth inhibition is eventually populated by growth of diploid cells, which arise as a result of mating and which are nonresponsive to pheromones [Figure 9.7(a)]. However, if mutant **a**-cells defective in producing **a**-pheromone are cross-streaked on a plate with wild-type α-cell, a zone of growth inhibition is observed around the mutant only because α-pheromone produced by wild-type α-cells arrests the growth of **a**-cells [Figure 9.7(b), top]. Conversely, if mutant **a**-cells defective in responding to α-pheromone are plated with wild-type α-cells, a zone of growth inhibition is observed only around the α-cells [Figure 9.7(b), bottom]. Finally, if both mutant **a**- and α-cells are plated together, no zone of growth inhibition is observed [Figure 9.7(c)].

Using such a simple visual selection procedure, mutants defective in sending or responding to pheromone signals were identified, and the genes were cloned by functional complementation. The pheromone pathway has been extensively analyzed by biochemical and molecular genetics techniques, providing us with a wealth of knowledge of a eukaryotic signal transduction pathway (Dohlman and Thorner, 2001).

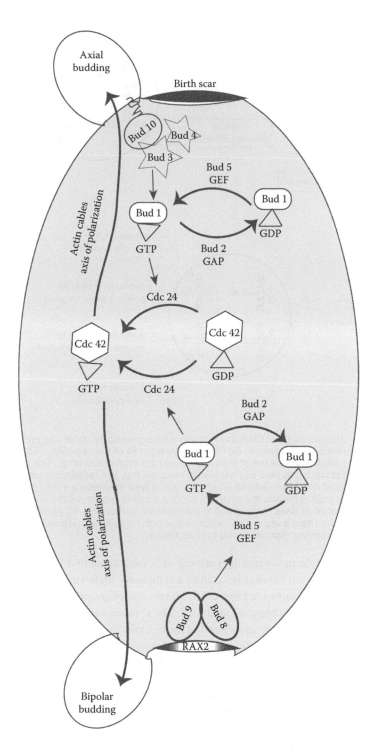

Figure 9.6 Mechanisms that generate cell polarity during bud-site selection in *S. cerevisiae*. The spatial landmark proteins on the cell surface guide the cellular machinery to build the axis of polarization through the GTP-binding proteins Bud1 and Cdc42. (Reproduced from Chant, J., *Ann. Rev. Cell Dev. Biol.* 15, 365–391, 1999, with permission from Annual Reviews Inc.)

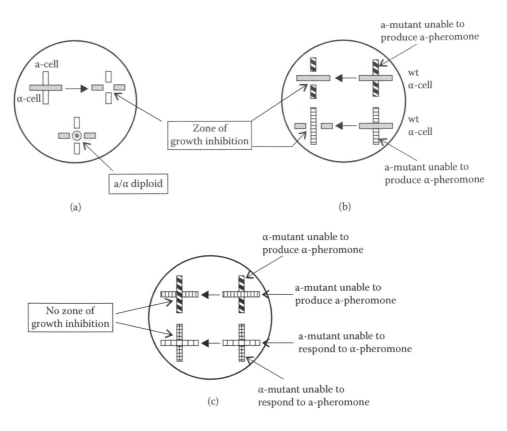

Figure 9.7 Scheme to isolate mutants defective in pheromone signaling by observing growth inhibition of cells in the presence of pheromone. (a) Wild-type yeast cells of the opposite mating type show a zone of growth inhibition where the streaks of growing cells cross each other. The diploid cells produced as a result of mating appear as new growth because they are resistant to pheromones. (b) Mutant showing growth inhibition with no effect on the wild type indicates a defect in the production of the pheromone (top), whereas the appearance of a zone of growth inhibition around wild type indicates a mutant that does not respond to the pheromone (bottom). (c) Absence of a zone of growth inhibition with mutant **a** and α cells indicates that both mutants are deficient in either producing (top panel) or responding (bottom panel) to pheromones.

   The biochemical cascade in the mating pathway of *S. cerevisiae* involves coordinated function of many proteins that work harmoniously, akin to a symphony orchestra. The multiprotein signaling complex is schematically shown in Figure 9.8. The receptors are proteins with seven transmembrane segments embedded in the plasma membrane. These receptors are named G-protein-coupled receptors (GPCRs) because they are associated at the cytoplasmic side to a heterotrimeric-protein complex, the G-proteins. The three protein subunits forming the trimeric complex are named α, β, and γ. The pheromone receptors are activated by pheromone binding at the extracellular face, which induces a conformational change in the receptor, resulting in the dissociation of the α subunit from the β–γ heterodimer. The β–γ complex functions as an adaptor protein to recruit other signaling molecules. Pheromone signaling activates two separate signaling complexes. The first complex, immediately downstream from the adaptor complex, activates Cdc42p by exchanging GDP with GTP, catalyzed by Cdc24p (GTP cycle). Active Cdc42p regulates two separate cellular processes: Reorganization of the actin cytoskeleton, leading to the appearance of mating projections, and activation of the mitogen-activated protein kinase (MAPK) module. The MAP kinase module regulating mating in *S. cerevisiae* is a multiprotein complex of three separate protein kinases, Ste11, Ste7, and Fus3, that successively phosphorylate and activate each other to turn on

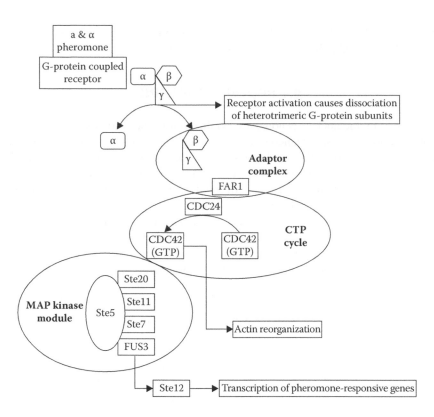

Figure 9.8   Mechanism of pheromone signaling in *S. cerevisiae*. Binding of pheromones **a** or α to the G-protein coupled pheromone receptors on α or **a** cells, respectively, initiates a kinase cascade that culminates in the expression of pheromone responsive genes driven by the transcription factor Ste 12.

the MAP kinase cascade. These kinases are tethered together by the signaling adaptor protein, Ste5. This is the first example of a signaling module characterized genetically and biochemically in a eukaryotic organism and resembles in structure and function the signaling modules found in higher organisms. Early genetic and biochemical analyses suggested that Ste5 is absolutely required for the activation of the mating pathway and participates in this process by tethering the three kinases together to facilitate efficient phosphorylation. Recent studies have revealed a more complex role of the Ste5 adaptor protein. Besides tethering the three kinases and bringing them into close proximity to facilitate phosphorylation, Ste5 has a second important function: Binding Fus3 and changing its conformation such that it becomes a good substrate for Ste7 (Good *et al.*, 2009). This complex allosteric regulation of Fus3 by Ste5 serves a very important function. As it turns out, *S. cerevisiae* uses Ste11 and Ste7 to regulate two separate genetic programs: The pheromone-induced mating pathway and the filamentous growth pathway. Ste7 phosphorylates Fus3 to induce mating or another kinase Kss1 to induce filamentous growth, as shown in Figure 9.8. What prevents signaling from crossing over from one pathway into another? Or in other words, what prevents a mating pathway from being activated in response to nitrogen starvation? The precise order of binding of Ste7 and Fus3 kinases to the Ste5 scaffolding protein insulates against signal crossover. Ste5 has a low affinity binding site for Fus3, which becomes available for Fus3 binding only when activated Ste7 is bound to the adaptor protein. This ensures that when Fus3 occupies the low affinity binding site on Ste5 it is phosphorylated by Ste7, resulting in its translocation to the nucleus, where it phosphorylates the transcription factor Ste12 to turn on the pheromone-responsive genes. The active Ste7 kinase is bound to Ste5 only in response to nutrient starvation and not in response to nitrogen starvation, thereby preventing misfiring of the pathways. The molecular mechanisms governing pheromone signaling reflect in composition and characteristics the fundamental nature of signaling pathways found in all higher organisms.

9.4 PROTEIN TARGETING

Proteins coded by the nuclear genome are synthesized in the cytoplasm and must be delivered to different membrane-bound organelles within the cell or secreted outside. How does the cell accomplish this? *S. cerevisiae* played a leading role in identifying the major players in this cellular choreography, and early conceptual breakthroughs came from mutant hunts and genetic analysis.

To identify the genes that, when mutated, would cause a defect in protein secretion, Peter Novick and Randy Schekman used two marker proteins, the enzymes phosphatase and invertase, whose secretion is easily detected by simple colorimetric assays (Novick and Schekman, 1979; Novick *et al.*, 1980). Yeast cells were mutagenized and plated. Colonies were selected that were defective in the secretion of both enzymes at the restrictive temperature 37°C, but not at permissive temperatures of 22–24°C. The mutants obtained were organized into 23 complementation groups, suggesting participation of at least 23 genes involved in the secretory pathway. The *ts* mutants are tools to examine the function of essential genes. In addition, they facilitate capturing and analyzing intermediate steps in complex biological processes by imposing the defect at the restrictive temperature. For example, at the restrictive temperature, a *SEC18* mutant accumulates 50 nm vesicles containing proteins with a pattern of glycosylation specific to the endoplasmic reticulum (ER). This suggests that the *SEC18* gene product functions in the fusion of the ER-derived vesicles with the Golgi membrane. Many of the proteins involved in vesicular transport first identified by biochemical studies of mammalian cell-free systems have been confirmed by genetic approaches using yeast.

The collection of secretion-defective mutants was further analyzed to define the order of events in the secretory pathway by the method of double mutant analysis first employed by L. H. Hartwell in 1974 to describe the sequence of events in the yeast cell cycle. Briefly, electron microscopic analysis of the mutants belonging to each of the 23 complementation groups indicated that the mutants specifically accumulated three different membrane-enclosed structures when shifted to the nonpermissive temperature. They were either endoplasmic reticulum structures, or cup-shaped structures called "Berkeley body," or 80–100-nm vesicles. Very rarely a single mutant showed overrepresentation of more than one structure, suggesting that each mutant is blocked at a discrete step in the process. In double mutant analysis, two mutants that accumulate different structures are combined and its organelle structure is analyzed. It is expected that a double mutant would accumulate a structure that corresponds to the earliest block. Using this method, the secretory mutants (*sec* mutants) were placed along a linear pathway reflecting the major steps in the secretory process (Figure 9.9).

9.5 MITOCHONDRIAL BIOGENESIS

Mitochondria are energy-generating organelles of eukaryotic cells believed to have originated from a symbiotic association between an oxidative bacterium and a glycolytic proto-eukaryotic cell. The endosymbiotic origin of mitochondria is reflected in its bilayered membrane structure, ~86-kb circular genome, organelle-specific transcription and translation, and protein assembly systems (Tzagoloff and Myers, 1986). However, during the stabilization of the symbiotic association, a majority of the mitochondrial genes were transferred to the nuclear genome. Recent analysis has revealed that ~477 proteins are required for mitochondrial function in yeast, of which only 17 are coded by the mitochondrial genome; the remaining are nuclear encoded. This partial genetic autonomy of the mitochondria was borne out from the analysis of yeast mit⁻ mutants that had point

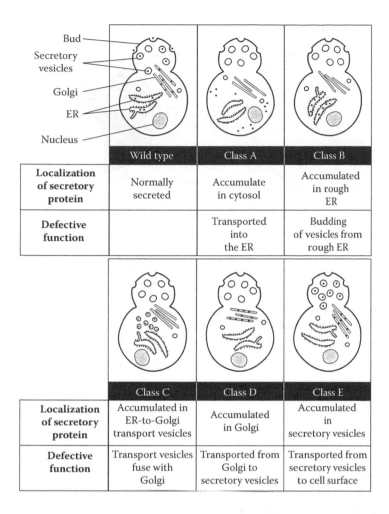

Figure 9.9 Use of yeast mutants in identification of steps in protein secretion. The mutants can be classified into five different complementation groups based on the accumulation of secretory proteins in cells. (Modified from Lodish, H., *et al., Molecular Cell Biology,* Scientific American Books, New York, 1995.)

mutations or small deletions in the mtDNA. A second class of mutants was isolated that had a petite phenotype (*pet* mutants forming small colonies) when grown on a nonfermentable source of carbon such as glycerol, ethanol, and lactate. Genetic analysis of these mutants revealed that unlike mit⁻ mutants, *pet* genes were nuclear. Hundreds of *pet* genes were identified by exhaustive genetic screens and were shown to regulate mitochondrial transcription, translation, and assembly of the electron transport chain. More recently, yeast deletion mutants were used for the identification of new *pet* genes (Steinmetz *et al.*, 2002). Briefly, 4706 homozygous deletion mutants were grown in nine different growth media, and their growth characteristics in nonfermentable carbon sources were analyzed to identify *pet* mutants. Altogether, 341 *pet* genes belonging to different classes were identified. The function of 185 mitochondria-specific genes is shown in Figure 9.10. About half of the genes participate in protein synthesis, which is reasonable considering the fact that about 95% of mitochondrial proteins are coded by the nuclear genome.

That both nuclear and mitochondrial genes function together in mitochondrial biogenesis came from the work of Schatz's group (Schatz, 2001). They demonstrated that when yeast cells are grown under anaerobic conditions, the mitochondria are devoid of cytochromes and several other proteins

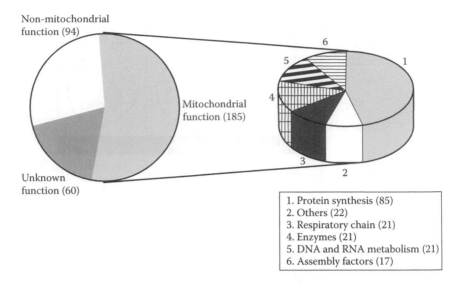

Figure 9.10   Participation of *pet* genes in biological processes. Genes were annotated using Gene Ontology
(GO) resources (www.geneontology.org).

and are difficult to detect by electron microscopy. These structures, called proto-mitochondria, are
converted into functional mitochondria when cells are shifted to aerobic conditions. This reversible
process became a useful system to demonstrate that the assembly of the electron transport chain
on the mitochondrial inner membrane required the participation of both nuclear and mitochondrial
genes. Especially noteworthy was the finding that in the absence of mitochondrial protein translation, cytochrome-c1 encoded by the nucleus became highly susceptible to proteolysis (Ross and
Schatz, 1976). The increased susceptibility was not due to increased synthesis of proteases by the
petite mutant but due to improper assembly and incorporation of cytochrome-c1 on the mitochondrial membrane. This finding led to the discovery and characterization of an elaborate system by
which proteins coded by the nuclear genes finally reach the mitochondrial matrix after traversing
the double bilayered membrane of the organelle.

   In the early 1990s a breakthrough in the area of mitochondrial protein import was made by the
research groups of Gottfried Schatz and Walter Neupert, who established methods for the isolation of mitochondria from yeast cells and set up *in vitro* mitochondrial translocation assays using
labeled proteins (Sollner *et al.*, 1989). Briefly, mitochondria from yeast cells are mixed with proteins labeled with ^{35}S-methionine by *in vitro* transcription translation in rabbit reticulocyte lysates.
Following incubation, mitochondria are isolated by density gradient centrifugation and the fate of
the labeled proteins is analyzed by biochemical and microscopic techniques. A common method
involves the treatment of the incubation mixture with proteases and examination of the fate of the
labeled proteins by gel electrophoresis. Complete translocation of the protein into the mitochondria
renders them resistant to proteolytic digestion. Three significant conclusions were reached from
these early studies: First, the mitochondrial outer membrane bears the import receptors that bind
proteins destined to the mitochondria. Second, the proteins destined to the mitochondria carry
import signals that are recognized by the import receptors. Third, mitochondrial protein import is
an energy-driven process requiring adenosine-5′-triphosphate (ATP) and a potential gradient across
the mitochondrial membranes. In the last decade the biochemical components of this complex process have been identified. Two specific receptor complexes have been characterized that reside on
each of the two membranes. The "Tom" complex (*t*ranslocase of *o*uter *m*embrane) and the "Tim"
complex (*t*ranslocase of *i*nner *m*embrane) form the translocation pore through which proteins are
translocated from cytosol into mitochondria. About 30 proteins are present in these two complexes.

Proteins destined to the mitochondria are unfolded before they can be recognized by the components of the Tom complex. The heat-shock protein hsp70 and chaperone protein MSF mediate the process of unfolding and the stabilization of the unfolded protein by using the energy of ATP hydrolysis. Once the protein traverses the outer membrane, the Tim complex threads it in. Proteins residing in the mitochondrial lumen carry a mitochondrial import signal at their N-terminus, which is processed by luminal peptidases (Wiedemann *et al.*, 2004). A simplified view of the mitochondrial protein import process is shown in Figure 9.11.

Like the eukaryotic cell cycle, mitochondrial protein import is a conserved biochemical event. This is supported by functional complementation of yeast mitochondrial import mutants by mammalian genes and also from sequence similarities between yeast and mammalian mitochondrial

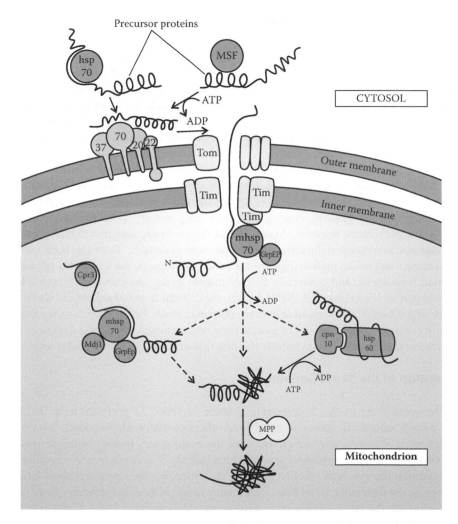

Figure 9.11 Components of the mitochondrial protein import machinery. Precursor proteins undergo unfolding in the presence of heat-shock protein-70 and MSF (mitochondrial import stimulating factor) proteins in an energy-dependent process. The mitochondrial import receptor complex binds the positively charged N-terminal signal sequence before the protein is transferred to the Tom complex in the outer membrane (OM). The protein is transferred to the components of the Tim complex in the inner membrane (IM), which pulls it into the matrix, where it is folded in the presence of chaperone proteins (mitochondrial heat-shock protein-70 and Gro-related protein-E). Once the protein is correctly folded, matrix-processing peptidase (MPP) cleaves the signal peptide. (Modified from Schatz, G., *J. Biol. Chem.* 271, 31763–31766, 1996.)

import proteins. Significantly, function of several human disease genes has been revealed by the function of their yeast homologues (Steinmetz *et al.*, 2002). As an example, when the gene associated with human deafness dystonia syndrome was cloned by positional cloning, no function could be assigned to the DDP1 peptide except that it had an N-terminal mitochondrial localization signal. Soon, however, a family of proteins from yeast bearing striking homology to DDP1 was characterized as components of the Tim complex. The human protein has been shown to function as a part of the yeast protein import machinery (Foury and Kucej, 2002). So far about 102 human diseases have been attributed to defects in mitochondrial function. The yeast system offers a great tool to analyze the function of these human genes (Koutnikova *et al.*, 1997).

9.6 FUNCTIONAL GENOMICS

In 1996, 92 collaborating laboratories in the United States, Canada, the United Kingdom, and Japan published the complete sequence of the *S. cerevisiae* genome (Goffeau *et al.*, 1996). This opened up research on how genetic instructions specify a eukaryotic cell, heralding the era of comparative genomics (comparing sequences between different species) to learn about the basic mechanisms of life that could lead to understanding how genes cause diseases and of finding rationales for treatment. Some examples of spin-offs from the yeast genome sequence are given below.

The 12 million bases in the yeast genome are packaged into 16 well-characterized chromosomes, 6466 protein-encoding genes, of which 3470 genes (56%) have homologues in other organisms. The remaining 2743 genes lack identifiable homologues in any organism. When these genes were compared with genes from 13 species of yeast, 1712 genes showed homologues, indicating their common function in this group of organisms. Among the conserved genes are those involved in cell wall biosynthesis and the pheromone response. About 78% of the genes are assigned to biological processes based on sequence homology and genetic and biochemical evidences. Yeast has been used successfully to investigate and model mammalian diseases and pathways. At the least, 31% of yeast proteins have a human homologue, and conversely 50% of human genes associated with heritable diseases have a counterpart in yeast (Hartwell, 2004). In fact, the cellular target of rapamycin was first discovered in yeast and subsequently verified in humans (Heitman *et al.*, 1991). Therefore, for the first time, the availability of a complete sequence of a eukaryotic organism accelerated the discovery and use of novel experimental approaches to investigate the function of genes at the molecular level.

9.6.1 Evolution of the Yeast Genome

Since the sequencing of the *S. cerevisiae* genome in 1996, 27 genomes from different yeast species have been sequenced, some completely and others partially. The sequence information provides a rich source of comparative data to examine the evolutionary history of the group as a whole (Dujon, 2006). The hemiascomycetous yeasts, which include *S. cerevisiae*, share similar morphology and a common lifestyle with other members of the group, yet their genomes show a high degree of divergence at the molecular level. The evolutionary range of hemiascomycetes is as large as the chordates, although the latter exhibit more diverse morphological features. Conspicuously, yeast genomes have undergone extensive chromosomal reorganization following their divergence from each other. By contrast, genomes of humans, mice, rats, and chickens retain extensive homologous segments in their chromosomes, although phylogenetically they are as distant as some of the yeast species. Additionally, yeast genomes have undergone whole-genome duplication followed by extensive gene loss. Overall genome redundancy ranges from 30–50% in five completely sequenced yeast genomes (44% in *S. cerevisiae*). Yeast genes also lack intron (< 5% in most species) in contrast to other multicellular eukaryotes. A clear understanding of the selective pressures that has led to molecular divergence within the hemiascomycetes yeasts is currently unknown.

9.6.2 Functional Analysis of the Yeast Genome

Hitherto, the total number of functional genes in an organism was estimated by saturation muta-genesis. What this means is that any gene that is associated with a function can be identified by the loss of function after mutagenesis. Therefore, if an organism is mutagenized and enough mutants are screened, the total number of mutants can give a rough estimate of the total number of functional genes. The yeast genome sequence revealed about 6400 genes, two times more than that predicted from saturation mutagenesis. A second approach to analyze gene function is by targeted disruption of genes, which is possible only after the genes have been identified and their sequences determined. R. W. Davis and his colleagues disrupted 5916 genes (91.4% of total) and analyzed the behavior of mutants under a variety of nutritional and environmental conditions (Giaever *et al.*, 2002). In their experiments, each gene was precisely deleted from the start to the stop codon and replaced with a "deletion cassette" containing a selectable marker, kanamycin. The kanamycin gene in each deletion cassette was flanked on either side by two unique 20-nucleotide sequences. These unique sequence tags can be viewed as a barcode, a permanent identifier of each deletion mutant. This clever trick of tagging every deleted gene permitted the researchers to carry out growth experiments with many deletion mutants in parallel. In a typical experiment the relative contribution of genes for growth on galactose was examined with the objective of assigning functions to as-yet uncharacterized genes (Figure 9.12). Twelve deletion mutants were grown in a medium with galactose as the sole carbon source and the relative proportion of each mutant cell in the mixed culture was determined by quan-titating the relative amount of each tag present in the culture by microarray technology (described in

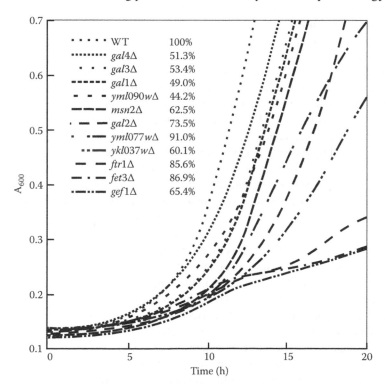

Figure 9.12 Measuring fitness of deletion strains of *S. cerevisiae* in galactose medium. Twelve different dele-tion mutants were grown together in the same tube. The growth of each strain was measured by quantifying the barcodes associated with each mutant using an oligonucleotide array (microar-ray) as described in the text. (Redrawn from Giaever, G., Chu, A.M., Connelly, C., *et al.*, *Nature* 418, 387–391, 2002.)

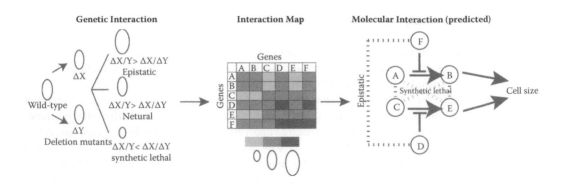

Figure 9.13 Schematic representation of the use of synthetic genetic arrays to map molecular interactions associated with a phenotype (cell size). Haploid mutants deleted for one unique gene (*X* or *Y*) are intercrossed and the cell size of the double mutant is compared with the single mutants to examine genetic interaction. An interaction map is created by examining the interaction between six separate genes. The phenotype from the interaction map can predict molecular interaction between the genes as shown in this hypothetical example. This approach can be extended to investigate genetic interaction between any number of nonessential genes (see text for details).

Section 9.6.3). Analysis of growth characteristics of each mutant in a galactose medium revealed the impact of individual genes on the utilization of galactose. By this method, the functions of two novel genes, YML090W and YML077W, in the utilization of galactose were revealed. The collection of deletion mutants is being used as a functional genomic strategy for the analysis of complex cellular processes and metabolic pathways.

The library of single-gene deletion mutants covering ~6400 open reading frames (ORFs) indicated that the vast majority of the genes in the yeast genome are individually dispensable. Only ~20% of the genes are required for cell viability (Dixon *et al.*, 2009), suggesting that the yeast genome is well insulated from genetic perturbations by the incorporation of redundancy in gene functions. The contribution of redundancy in gene functions can be mapped by systematically probing genetic interactions between genes using the library of single-deletion mutants as schematically represented in Figure 9.13. In this example, interaction between two genes *X* and *Y* that are by themselves nonessential can be examined by creating a double mutant that lacks the function of both the genes. The phenotype of the double mutant (size of the yeast cell, in this example) is then compared with the single mutants. If the cell size of the double mutant is similar to the single mutants, there is no interaction between the genes (neutral). If the double mutant is smaller than the single mutants, it indicates negative interaction between the genes (synthetic lethal). Finally the double mutant may be larger than the single mutants, suggesting positive interaction between the genes (epistatic). A pair-wise interaction between six genes is depicted as an interaction map in which the sizes of yeast cells are shown in different shades of gray. The genetic interaction map can reveal how each gene contributes to the pathway that regulates cell size, resulting in the generation of a molecular interaction map. The molecular interaction depicted in Figure 9.13 is the beginning of a hypothesis and the basis for further experimentation. This approach was used to dissect the functional organization of the early secretory pathway (Schuldiner *et al.*, 2005; Collins *et al.*, 2007). More recently, synthetic genetic arrays (SGAs) were used to query genetic interaction between 1712 yeast genes. The 5.4 million gene pairs were scored for cell size and compared with single mutants to construct genetic interaction maps (Costanzo *et al.*, 2010). The genetic interaction network highlights the functional link between genes that participate in common biological processes. In addition, it revealed connections between diverse biological processes, creating a global functional linkage map of a yeast cell.

A limitation of genetic interaction mapping is that it cannot be used for examining the relationship between essential genes since single mutants exhibit lethal phenotype. To circumvent this, molecular techniques that rely on inducible promoters to shut off gene transcription (Mnaimneh *et al.*, 2004), conditional destabilization of protein (Kanemaki *et al.*, 2003), and generation of hypomorphic alleles for essential yeast genes (Breslow *et al.*, 2008) have been developed. These studies mapped the function of 1033 essential genes, providing a clearer picture of the dispensability of the yeast genome; ~57% of the genome is indispensable, in contrast to ~16–20% determined by traditional methods.

9.6.3 Expression Pattern of Genes Using DNA Microarray

Paradigm shifts in science always go hand in hand with technological breakthroughs. The program by which a complex body develops from a single fertilized cell requires understanding of the spatial and temporal regulation of gene transcription in response to external and internal cues. In 1995 Patrick O. Brown and his colleagues at Stanford University introduced a DNA microarray technique for analyzing global expression of genes (Schena *et al.*, 1995). The yeast cell cycle was one of the first biological processes to be interrogated by this new technique. The commercial yeast microarray contains 6200 genes covalently linked on glass slides as ~100-μm dots in grids of 96 or 384 spots. Each gene is amplified by polymerase chain reaction (PCR) and the purified DNA is used for printing the microarray. Microarray technology (Figure 9.14) is fully automated, carried out by robots to increase throughput and minimize error. The microarray experiment measures the relative abundance of messages in a cell, revealing a picture of a cell's transcriptome. It has been

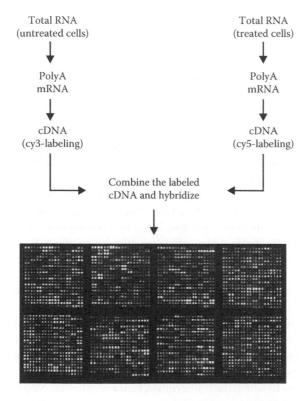

Figure 9.14 A schematic representation of a microarray experiment. Labeled cDNA prepared from untreated and treated cells was mixed and hybridized on a microarray slide spotted with 6200 yeast open reading frames (DNA sequences with initiation and termination codons). The white spots represent higher abundance of cyanin5-labeled genes and darker spots represent higher abundance of cyanin3-labeled genes. Gray spots are genes present in equal abundance (blending of white and dark color in equal proportion). In reality, the spots are detected as red (higher cyanin5 label), green (higher cyanin3 label), or yellow (equal abundance of cyanin5 and cyanin3 labels) pseudo-colors.

applied successfully to identify induction and repression of genes during specific cellular processes. The first gene expression analysis examined the genes that are regulated in a cell-cycle dependent manner in yeast (Spellman *et al.*, 1998) by comparing the relative abundance of messenger RNA (mRNA) as cells progressed through the cell cycle. Messenger RNA was harvested at defined time points to capture different phases of the cell cycle and was converted into complementary DNA (cDNA) using appropriate primers and the enzyme reverse transcriptase in the presence of red (Cy5) and green (Cy3) fluorescently labeled nucleotide precursors. Labeled cDNA (Cy3) from an asynchronous culture (control) was mixed with labeled cDNA from the synchronous culture (Cy5) and hybridized to a DNA microarray containing the yeast genes. The cDNA sequences representing individual transcripts hybridized specifically to corresponding gene sequences on the array. The fluorescence associated with each spot was quantitated using a microscope (microarray reader) that illuminated each spot with a laser beam and measured the fluorescence associated with each dye separately, to estimate the relative abundance of the transcript by the ratio of the red to green fluorescence. About 800 genes were found to be cell cycle regulated, which could be clustered into groups whose expression correlated with a specific phase of the cell cycle. For example, ~100 genes were found to be co-regulated in G1. Many of these genes function in establishing cellular polarity and initiation of bud growth, providing a molecular link between budding with the position of the cell in G1. The microarray technique has become a standard tool to understand the mechanism of disease incidence and progression and has been applied successfully in cancer.

9.6.4 Mapping Transcription Network

In the early 1960s, bacterial genetics laid the concept of regulatory circuits controlling the expression of genes. In its simplest form, circuits are turned "on" or "off" by the binding of transcription factors or repressors to the upstream regulatory sequences of genes respectively. Expression analysis by microarray described in Section 9.6.3 revealed the co-regulated expression and repression of sometimes hundreds of genes during specific cellular events. Richard A. Young used the microarray technique to identify regulatory sequences bound by all transcription factors encoded by the yeast genome (Lee *et al.*, 2002). Based on the expectation that the promoter elements are located within the noncoding regions upstream of every gene, the group amplified the noncoding sequences between genes (intergenic regions) and spotted them on slides to create microarray of all promoter regions of yeast. Next, they identified 141 genes in the yeast genome that are predicted to code for transcription factors and tagged them with an epitope (myc-tag) at the c-terminus. Each epitope-tagged transcription factor was inserted into its native genomic locus by homologous recombination and analyzed for the level of expression by Western blot using an antibody specific to the myc-tag. To identify the DNA-binding sequences of the transcription factors, a genome-wide location analysis called ChIP (crosslinking chromatin immunoprecipitation) was performed (Ren *et al.*, 2000). The strategy is shown in Figure 9.15. Each yeast strain expressing a unique myc-tagged transcription factor was grown in rich medium, and the DNA was cross-linked to bound proteins *in vivo*. The protein-DNA complex was digested with the enzyme DNAse to cleave DNA into smaller fragments. The transcription factor-bound DNA fragments were enriched by immunoprecipitation using an antibody against the tag. After isolation of bound DNA from the protein-DNA complex, it was labeled and hybridized to a microarray slide spotted with the intergenic DNA. A positive hybridization signal identified specific intergenic regions to which the transcription factor was bound. Sequence comparison among the intergenic regions that lighted up with each transcription factor identified conserved motifs recognized by the transcription factor. A genome-wide search for the presence of motifs in upstream sequences of genes resulted in the discovery of genes that are co-regulated under the control of the transcription factor. Using this approach it was revealed that ~200 genes are regulated by SBF (SCB binding factor) and MBF [*MluI* cell cycle box (MCB) binding factor] transcription factors during G1–S phase transition of the cell cycle (Iyer *et al.*, 2001; Simon *et al.*, 2001). SBF and MBY are

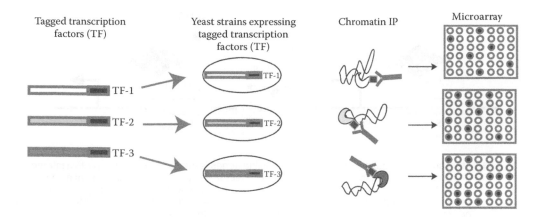

Figure 9.15   Genome-wide analyses of yeast transcription regulator binding sites. Tagged transcription regulators (TF-1, 2, and 3) are expressed in yeast. Bound regulators are crosslinked to DNA and DNA-protein complex is digested with an enzyme DNase to cleave DNA into fragments complexed with regulators. DNA bound to the transcription regulators is immunoprecipitated using antibodies specific to the tag. The DNA present in complex with the regulator proteins is isolated and hybridized to microarray spotted with intron sequences present in the yeast genome. The hybridization signal shown as dark spots in the figure indicates binding sites of each of the transcription regulators. (Modified from Lee, T.I., Rinaldi, N.J., Robert, F., *et al.*, *Science* 298, 799–804, 2002.)

sequence-specific transcription factors that activate gene expression during the G1/S transition of the cell cycle in yeast.

9.7 PROTEOMICS AND SYSTEM BIOLOGY MODELING

A cell continuously senses environmental stimuli and relays them inside across membranes to trigger a response. The process of signal transduction occurs by sequential transfer of information via protein–protein interaction. Take the example of a yeast cell growing in a glucose-rich environment suddenly encountering galactose. It senses the change in composition of the carbon source, shuts off the expression of genes required for glucose metabolism, and turns on the genes required for galactose utilization. What is the mechanism by which the cell makes necessary adjustments? It is known that galactose itself can induce changes in gene expression within the yeast cells. The mechanism of regulation of gene expression by galactose is mediated by altering the interaction between Gal4p and Gal80p. As shown in Figure 9.16, in the absence of galactose Gal4p is in complex with Gal80p and is functionally inactive as a transcription factor. However, the presence of galactose induces recruitment of Gal3p to Gal4–Gal180 complex. This binding revives the transcriptional competence of Gal4p by relieving the inhibition caused by Gal80p. As a consequence, Gal4 induces transcription of genes required for galactose metabolism.

Identifying the interacting partners of every protein in a cell is paramount to the understanding of how cells initiate and coordinate myriad functions to maintain homeostasis. Stanley Fields devised a yeast two-hybrid method of discovering protein–protein interaction (Fields and Song, 1989). Briefly, the two-hybrid method detects protein–protein interaction by transcriptional induction of reporter genes. Two separate reporter systems are used: A nutritional marker that allows growth of yeast strains on a selective medium and an enzyme (beta-galactosidase), whose activity can be measured by colorimetric assay. The transcriptional induction of the reporter genes is dependent on the availability of a functional Gal4 protein. Gal4p is a modular transcription factor with a distinct DNA-binding and transcription activation domain. Each domain retains its function in the absence of the other. In the two-hybrid method, the DNA-encoding the test proteins are fused *in vivo* either to the DNA-binding

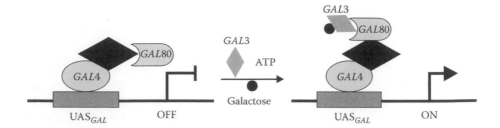

Figure 9.16 Transcriptional regulation of galactose-inducible genes is mediated by protein-protein interaction. In the absence of galactose, Ga14 transcription factor is kept transcriptionally silent as a result of interaction with its negative regulator Ga180 protein. This inhibition is relieved by Ga13 protein in the presence of galactose. Ga13 binds Ga180 and displaces the latter from the Ga14 transcription activation domain allowing transcriptional activation of the target genes.

domain or to the activation domain of the Ga14 transcription factor by homologous recombination in yeast. By themselves, the fusion proteins are incapable of inducing transcription. However, if the two test proteins interact with each other, the two domains of the transcription factor will be brought close together to reconstitute a functional Ga14 protein. As a result, the reporter gene under the control of the Ga14 promoter will be transcriptionally induced. Positive interactors are selected by monitoring the expression of beta-galactosidase using a synthetic substrate that turns blue following the action of beta-galactosidase. Yeast cells containing positive interactors grow into colonies that appear bluish when grown on selective plates. The two-hybrid method is shown in Figure 9.17. Yeast two-hybrid libraries are constructed by fusing cDNAs (obtained from mRNA) with the *activation domain* of Ga14 protein (AD-fusion) and cloned in the prey plasmid. The genes of interest are fused to the DNA *binding domain* of Ga14 protein (BD-fusion) and cloned in a bait plasmid.

The two-hybrid technology was employed for the first time to initiate a genome-wide interaction screen in *S. cerevisiae*. The complete sequence information of the yeast genome permitted cloning of all the ~6200 open reading frames (ORFs). Each gene (ORF) was screened against a yeast two-hybrid library and its interactors were identified. Since the first high-throughput screen (Uetz *et al.*, 2000) that detected ~1000 interactions, the yeast interaction database today holds ~5000 unique interactions. A recent analysis of the interactions has revealed many biologically relevant protein complexes associated with distinct cellular processes (Bader *et al.*, 2004).

A second approach of studying protein–protein interaction is by mass spectrometry. The major impetus of using a yeast system to analyze protein complexes by this method was the fact that the data generated using this developing technology can be cross validated easily with the vast amount of genetic, biochemical, and molecular biology information already available in this model organism. Two drug discovery companies have tagged ~1900 yeast proteins with an epitope tag and expressed them in yeast under the control of their native promoter. Protein complexes were purified from yeast lysates by affinity chromatography using an antibody against the tag. After separating individual proteins in the complex by denaturing gel electrophoresis, each protein band was excised from the gel, digested with trypsin, and identified by MALDI-MS (matrix-assisted laser desorption ionization-mass spectrometry). Together, the two studies identified 3018 interacting proteins (~50% of all yeast proteins) distributed in a variety of biologically relevant complexes (Gavin *et al.*, 2002; Ho *et al.*, 2002). These initial studies were followed by the generation of numerous protein interaction maps in both yeast and mammalian systems. A recent study reported interactors for 130 protein kinases, 38 protein phosphatases, 24 lipid and metabolic kinases, and 84 kinase and phosphatase regulators (Breitkreutz *et al.*, 2010). Interaction networks underscore the potential for kinases in regulating diverse biochemical pathways.

High-throughput protein–protein interaction analysis generates a huge amount of data that are not easy to tease apart to extract biologically meaningful interactions from the noise. A straightforward

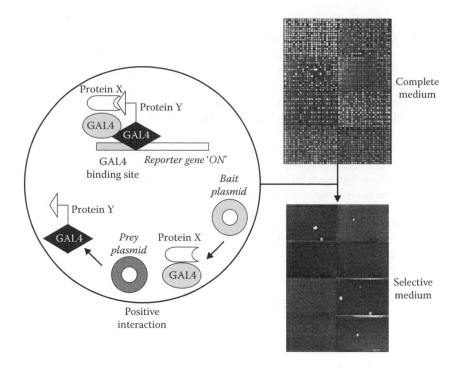

Figure 9.17 Yeast two-hybrid method of analyzing protein–protein interaction. Yeast cells are transformed with plasmids expressing two separate fusion proteins, one containing the Ga14 DNA-binding domain (bait plasmid) and the other containing the Ga14 activation domain (prey plasmid). The DNA-binding and the activation domains are brought closer to each other into a stable complex if the fusion proteins interact with each other. The interaction is detected by the expression of the reporter gene. Positive interactors show growth on selective plates.

approach to assess the quality of any interaction is to determine whether the interacting proteins are expressed together at any given time and also whether they localize within the same cellular compartment. *S. cerevisiae* was used to examine the expression and localization of the yeast proteome.

Using a GFP (green fluorescent protein) tag, a global analysis of protein localization in yeast was carried out to localize proteins to specific cellular compartments and to quantitatively determine the steady state level of all proteins in a yeast cell by Western blotting (Ghaemmaghami *et al.*, 2003; Huh *et al.*, 2003). Two different strains of yeast in which 6109 of the 6400 predicted yeast ORFs were fused either with a myc-tag or with a GFP-tag. Under normal growth conditions, expression of yeast cellular proteome by Western blotting and by protein fluorescence revealed surprisingly that ~80% of the proteome is expressed. The abundance of individual proteins ranged from 50 molecules to 100,000 molecules under normal growth conditions. The protein expression was complemented by examining protein localization using GFP fluorescence at high resolution and sensitivity. Two thirds of the previously unlocalized proteins were assigned to 12 subcellular compartments. The future goal is to understand how protein localization within a cell changes as a result of cellular signaling.

9.8 CONCLUDING REMARKS

For more than half a century yeast has contributed to our understanding of biological processes such as metabolism and enzyme regulation, cell recognition, the structure of chromosomes, mechanisms in meiotic recombination, epigenetic effects through mating type switching, cell cycle, the compartmental character of eukaryotic cells, protein targeting, etc. An area where yeast has provided valuable information is the emerging area of system biology. System biology aims to capture and combine information from multiple high-throughput platforms such as synthetic genetic interaction arrays (SGA), gene expression arrays (microarrays), and protein interaction maps to investigate complex cellular behavior such as a cell's perception of signals in a continuously changing microenvironment, or how cell division is linked to generation of cell polarity.

We conclude this chapter with an observation by Davis (2003):

> By a process of natural selection, as it were, yeast has attracted many post-1970 investigators with strong training in all three of vital disciplines: genetics, biochemistry, and molecular biology. Their ability to integrate these disciplines in their research and in their training of newcomers to yeast has led to connectivity of the many levels of organization that defines a model organism. This ability has also led to the explosive growth of the yeast community, which has been so self-sufficient in defining life's fundamental attributes that its members are no longer obliged to read the literature on other fungi. A serious asymmetry prevails in this matter: *Neurospora* and *Aspergillus* investigators ignore the yeast literature at their peril.

REFERENCES

Bader, J. S., Chaudhuri, A., Rothberg, J. M., *et al.* (2004). Gaining confidence in high-throughput protein interaction networks. *Nature Biotechnol.* 22:78–85.

Breitkreutz, A., Choi, H., Sharon, J. R., *et al.* (2010). A global protein kinase and phosphatase interaction network in yeast. *Science* 328:1043–1046.

Breslow, D. K., Cameron, D. M., Collins. S. R., *et al.* (2008). A comprehensive strategy enabling high-resolution functional analysis of the yeast genome. *Nature Methods* 5:711–718.

Chant, J., Corrado, K., Pringle, J. R., *et al.* (1991). Yeast BUD5, encoding a putative GDP-GTP exchange factor, is necessary for bud site selection and interacts with bud formation gene BEM1. *Cell* 65:1213–1224.

Chant, J. and Herskowitz, I. (1991). Genetic control of bud site selection in yeast by a set of gene products that constitute a morphogenetic pathway. *Cell* 65:1203–1212.

Collins, S. R., Miller, K. M., Maas, N. L., *et al.* (2007). Functional dissection of protein complexes involved in yeast chromosome biology using a genetic interaction map. *Nature* 446:806–810.

Costanzo, M., Baryshnikova, A., Bellay, J., *et al.* (2010). The genetic landscape of a cell. *Science* 327:425–431.

Davis, R. H. (2003). *The Microbial Models of Molecular Biology: From Genes to Genomes.* Oxford University Press, New York.

Dixon, S. J., Costanzo, M., Baryshnikova, A., *et al.* (2009). Systematic mapping of genetic interaction networks. *Annu. Rev. Genet.* 43:601–625.

Dohlman, H. G. and Thorner, J. W. (2001). Regulation of G protein-initiated signal transduction in yeast: Paradigms and principles. *Annu. Rev. Biochem.* 70:703–754.

Dujon, B. (2006). Yeasts illustrate the molecular mechanisms of eukaryotic genome evolution. *Trends Genet.* 22:375–387.

Fields, S. and Song, O. (1989). A novel genetic system to detect protein-protein interactions. *Nature* 340:245–246.

Foury, F. and Kucej, M. (2002). Yeast mitochondrial biogenesis: A model system for humans? *Curr. Opin. Chem. Biol.* 6:106–111.

Freifelder, D. (1960). Bud position in *Saccharomyces cerevisiae. J. Bacteriol.* 80:567–568.

Gavin, A. C., Bosche, M., Krause, R., *et al.* (2002). Functional organization of the yeast proteome by systematic analysis of protein complexes. *Nature* 415:141–147.

Ghaemmaghami, S., Huh, W. K., Connelly, C., *et al.* (2003). Global analysis of protein expression in yeast. *Nature* 425:737–741.

Giaever, G., Chu, A. M., Connelly, C., *et al.* (2002). Functional profiling of the *Saccharomyces cerevisiae* genome. *Nature* 418:387–391.

Giot, L., Bader, J. S., Bouwer, C., *et al.* (2003). A protein interaction map of *Drosophila melanogaster*. *Science* 302:1727–1736.

Goffeau, A., Barrell, B. G., Bussey, H., *et al.* (1996). Life with 6000 genes. *Science* 274:546, 563–567.

Good, M., Singleton, J., Tang, G., *et al.* (2009). The Ste5 scaffold directs mating signaling by catalytically unlocking the Fus3 MAP kinase for activation. *Cell* 136:1085–1097.

Hartwell, L. H. (2002). Nobel lecture: Yeast and cancer. *Biosci. Rep.* 22:373–394.

Hartwell, L. H. (2004). Yeast and cancer. *Biosci. Rep.* 24:523–544.

Hartwell, L. H., Culotti, J., Pringle, J. R., *et al.* (1974). Genetic control of the cell division cycle in yeast. *Science* 183:46–51.

Heitman, J., Movva, N. R., Hall, M. N., *et al.* (1991). Targets for cell cycle arrest by the immunosuppressant rapamycin in yeast. *Science* 253:905–909.

Ho, Y., Gruhler, A., Heilbut, A., *et al.* (2002). Systematic identification of protein complexes in *Saccharomyces cerevisiae* by mass spectrometry. *Nature* 415:180–183.

Huh, W. K., Falvo, J. V., Gerke, L. C., *et al.* (2003). Global analysis of protein localization in budding yeast. *Nature* 425:686–691.

Iyer, V. R., Horak, C. E., Scabe, C. S., *et al.* (2001). Genomic binding sites of the yeast cell-cycle transcription factors SBF and MBF. *Nature* 409:533–458.

Kanemaki, M., Sanchez-Diaz, A., Gambus, A., *et al.* (2003). Functional proteomic identification of DNA replication proteins by induced proteolysis *in vivo*. *Nature* 423:720–724.

Keaton, M. A. and Lew, D. J. (2006). Eavesdropping on the cytoskeleton: Progress and controversy in the yeast morphogenesis checkpoint. *Curr. Opin. Microbiol.* 9:540–546.

Koutnikova, H., Campuzano, V., Foury, F., *et al.* (1997). Studies of human, mouse and yeast homologues indicate a mitochondrial function for frataxin. *Nature Genet.* 16:345–351.

Lee, M. G. and Nurse, P. (1987). Complementation used to clone a human homologue of the fission yeast cell cycle control gene cdc2. *Nature* 327:31–35.

Lee, T. I., Rinaldi, N. J., Robert, F., *et al.* (2002). Transcriptional regulatory networks in *Saccharomyces cerevisiae*. *Science* 298:799–804.

Lodish, H., *et al.* (1995). *Molecular Cell Biology*. Scientific American Books, New York.

Mnaimneh, S., Davierwala, A. P., Hayne, J., *et al.* (2004). Exploration of essential gene functions via titratable promoter alleles. *Cell* 118:31–44.

Moseley, J. B. and Nurse, P. (2010). Cell division intersects with cell geometry. *Cell* 142:184–188.

Nelson, W. J. (2003). Adaptation of core mechanisms to generate cell polarity. *Nature* 422:766–774.

Novick, P., Field, C., and Schekman, R. (1980). Identification of 23 complementation groups required for post-translational events in the yeast secretory pathway. *Cell* 21:205–215.

Novick, P. and Schekman, R. (1979). Secretion and cell-surface growth are blocked in a temperature-sensitive mutant of *Saccharomyces cerevisiae*. *Proc. Natl. Acad. Sci. USA* 76:1858–1862.

Ren, B., Robert, F., Wyrick, J. J., *et al.* (2000). Genome-wide location and function of DNA binding proteins. *Science* 290:2306–2309.

Ross, E. and Schatz, G. (1976). Cytochrome c1 of bakers' yeast. II. Synthesis on cytoplasmic ribosomes and influence of oxygen and heme on accumulation of the apoprotein. *J. Biol. Chem.* 251:1997–2004.

Schatz, G. (2001). What mitochondria have told me? *Mol. Biol. Cell* 12:777–778.

Schena, M., Shalon, D., Davis, R. W., *et al.* (1995). Quantitative monitoring of gene expression patterns with a complementary DNA microarray. *Science* 270:467–470.

Schuldiner, M., Collins, S. R., Thompson, N. J., *et al.* (2005). Exploration of the function and organization of the yeast early secretory pathway through an epistatic miniarray profile. *Cell* 123:507–519.

Siller, K. H. and Doe, C. Q. (2009). Spindle orientation during asymmetric cell division. *Nat. Cell Biol.* 11:365–374.

Simon, I., Barnett, J., Hannett, N., *et al.* (2001). Serial regulation of transcriptional regulators in the yeast cell cycle. *Cell* 106:697–708.

Sollner, T., Griffiths, G., Pfanner, R., *et al.* (1989). MOM19, an import receptor for mitochondrial precursor proteins. *Cell* 59:1061–1070.

Spellman, P. T., Sherlock, G., Zhang, M. Q., *et al.* (1998). Comprehensive identification of cell cycle-regulated genes of the yeast *Saccharomyces cerevisiae* by microarray hybridization. *Mol. Biol. Cell* 9:3273–3297.

Steinmetz, L. M., Sinha, H., Richards, D. R., *et al.* (2002). Dissecting the architecture of a quantitative trait locus in yeast. *Nature* 416:326–330.

Tzagoloff, A. and. Myers, A. M. (1986). Genetics of mitochondrial biogenesis. *Annu. Rev. Biochem.* 55:249–285.

Uetz, P., Giot, L., Cagney, G., *et al.* (2000). A comprehensive analysis of protein-protein interactions in *Saccharomyces cerevisiae. Nature* 403:623–627.

Watson, J. D., Gilman, M., Witkowski, J., and Zoller, M. (1992). *Recombinant DNA*, Scientific American Books, New York.

Wiedemann, N., Frazier, A. E., Pfanner, N., *et al.* (2004). The protein import machinery of mitochondria. *J. Biol. Chem.* 279:14473–14476.

CHAPTER 10

Neurospora
A Gateway to Biology

Model organism ... a species that undergoes genetic domestication and intense genetic analysis at all levels of organization ... eventually becomes known so thoroughly through genetic dissection that biologists take it as the key representative of the family, order, or phylum ... in some cases of life itself.

—**Rowland H. Davis (2003)**

In 1843, a luxuriant growth of a pink-orange fungus was observed on bread in the bakeries of Paris. Subsequently this fungus was recognized as a common contaminant of bakeries and it came to be known as the pink bread mold. On September 1, 1923, an earthquake followed by fire struck Tokyo. The strange sight of a pink-orange growth that developed on almost all burnt trees and vegetation amazed the residents. Examination showed that the orange-colored growth was because of the profuse production of conidia by a fungus. The fungus was named *Oidium aurantiacus*. The name was later changed to *Monilia sitophila*. In 1927 Shear and Dodge discovered the sexual phase of the fungus in cultures grown in the laboratory and renamed the genus as *Neurospora* because it produced ascospores with neuron-like striations. Soon after its discovery by B. O. Dodge, who was then working at the New York Botanical Gardens, the ease of manually dissecting out its sexually formed spores under a binocular microscope and growing these into progeny colonies for analysis of phenotypic classes for genetic analysis was recognized. In a letter dated February 27, 1941, George W. Beadle, who was then at Stanford University, wrote to Dodge:

> Dr. Tatum and I are interested in doing some work on the mutation of Neurospora with the eventual aim of determining whether their requirements might be dependent on genetic constitution. I have written to Dr. Lindegren to see if he can supply us with some stocks. If preliminary experiments prove to be encouraging we will be interested in trying out the available species and also various collections of "wild type" if we can get them. Do you have a collection of material or could you give us some information as to who aside from Lindegren might have? We'd rather wait until we've made some trials before actually obtaining material, but we would be glad to have at hand information as to what is available.

Neurospora is the best studied of all fungi and is regarded as a model of all microbes (Davis and Perkins, 2002). This chapter describes some aspects of its history, lifestyle, attributes, and contributions to biology.

10.1 HABITAT AND LIFE CYCLE

Because in the past *Neurospora* was observed growing on bread made by bakeries in Europe, it came to be called the pink or the pink-orange bread mold. However, it is rarely observed in

Figure 10.1   Neurosopora. (A) On burnt sugarcane stump. (B) On sugarcane factory-waste dump. The genes controlling pigmentation have been used as a visual reporter system in gene-silencing experiments. (Photo by (late) P. Maruthi Mohan.)

modern-day bakeries. In Asia, it is commonly seen growing on stubbles of sugar cane after the canes have been harvested for milling when the agricultural field is burnt to clear the trash of the cut leaves (Figure 10.1) or on sugar cane factory waste (Figure 10.2). In addition to the pink-orange colored form, a yellow-colored *N. intermedia* is found in Asia almost exclusively on maize cobs after the roasted kernels have been eaten by human beings and the cobs discarded on the ground. Another species, *N. discreta*—thought to be infrequent and limited mainly to Ivory Coast and Papua New Guinea—was found growing beneath the bark of trees damaged by wildfires in western North America, including Alaska, and from southern Portugal to Switzerland and Sweden. This finding caught mycologists by surprise because hitherto *Neurospora* had been thought to be confined to warm and humid regions in the tropics. It is perhaps one of the most common globally distributed species (Jacobson *et al.*, 2006). In the laboratory the vegetative growth of all species of *Neurospora* occurs satisfactorily on a simple nutrient medium between < 20°C and 40°C. Hence the reason for substrate preferences or geographic distribution or its hiding places in nature is a mystery. Perhaps certain types of substrates are required for sexual reproduction to occur? For example, in the laboratory *N. crassa* and *N. intermedia* show good fertility in media containing sucrose, whereas the *N. discreta* crosses are infertile on sucrose media but satisfactory on media containing filter paper cellulose. The yellow strains of *N. intermedia*, unlike the orange strains, reproduce poorly in a medium containing sucrose though they are very fertile on corncobs that had been roasted. A general observation is that conditions that favor sexual reproduction are different from those that favor asexual reproduction. The finding of *Neurospora* on trees following wildfires indicates the importance of fire to the biology of this fungus. Based on the legendary phoenix that arises from the ashes, the term *phoenicoid* has been proposed for pioneer fungi on burned substrates. *Neurospora* is therefore an example of a phoenicoid fungus.

The linear growth rate of *Neurospora*—up to 5 mm per hour—is one of the fastest among the fungi, one of the reasons for its popularity among researchers. Another point in its favor is that it can be grown in completely defined media and is nonpathogenic. In Asia, *Neurospora* is used to

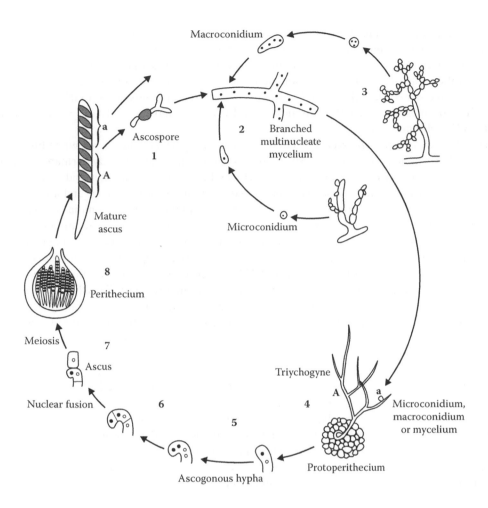

Figure 10.2 Life cycle of *Neurospora crassa*. The ascospore (1) germinates after chemical or heat activation and forms mycelium. Asexual reproduction occurs by formation of mitotically formed inconspicuous, uninucleate microconidia (2) or two- to three-nucleate, orange-colored macroconidia (3), which are dispersed. Sexual cycle begins with coiling of hyphae around ascogonium from which a trichogyne projects out forming protoperithecium (4). The trichogyne picks a nucleus of opposite mating type from microconidia, macroconidia, or hypha (stages 5 and 6). Nuclear fusion (stage 7) leads to formation of asci containing eight haploid ascospores of the two mating types, *A* and *a*. (From Perkins, D.D., Radford, A., and Sachs, M., *The Neurospora Compendium: Chromosomal Loci*, Academic Press, San Diego, 2000, with permission from Elsevier.)

prepare certain fermented products; hence this fungus is regarded as safe. Its growth rate is easily measured in a "race tube" in which mycelium growth on an agar surface is confined to one dimension, allowing the advancing mycelium to be marked at regular intervals. Morphologically similar strains can be physiologically different, as already hinted by occurrence of different species of *Neurospora* on different substrates. The heterothallic species, *N. crassa* and *N. sitophila*, comprised of strains of the opposite mating types, referred to as *A* and *a*, can be crossed (mated) to form asci (sing., ascus) bearing meiotically formed ascospores and are enclosed inside fruiting bodies called perithecia. In general, the species produce eight ascospores of which four each are of the two mating types. Barbara McClintock, who discovered mobile genetic elements (transposable elements) in maize (a discovery for which she received the Nobel Prize in Physiology or Medicine), also studied *Neurospora*. In 1945, using aceto-orcein staining and light microscopy, McClintock showed that *Neurospora* has seven chromosomes (haploid number n = 7, meaning it has one chromosome of each type and therefore one copy of every gene). The chromosomes are numbered based on their length, and this numbering based on their size does not correspond to the numbering of the linkage groups. The term *linkage group* refers to an order of linked genes whose linkage has been determined on the basis of their recombination frequency. The numbering of linkage groups is in the chronological order of their discovery. For *Neurospora*, LGI = chromosome 1, LGII = chromosome 6, LGIII = chromosome 3, LGIV = chromosome 4, LGV = chromosome 2, LGVI = chromosome 6, and LGVII = chromosome 7.

The life cycle of *Neurospora crassa* based on laboratory-grown cultures is shown in Figure 10.2. The sexually (meiotically) formed ascospores of *Neurospora* are constitutively dormant propagules that can survive in soil for several years. The activated ascospore germinates and forms a mycelium that produces two types of asexual spores. Under conditions of high sugar and nitrogen availability, the fungus produces powdery pink-orange macroconidia or simply conidia; whereas under low-nutrient conditions the fungus mainly produces inconspicuous, uninucleate microconidia (*see* Chapter 3, Table 3.1). The microconidia function as male cells. These cells are formed simultaneously with structures called protoperithecia that are formed by coiling of hyphae around ascogonia, one cell of which acts as the female gamete. DNA sequence analysis showed that the two alleles of the mating type locus, *mat A* and *mat a*, or *A* and *a*, have highly dissimilar DNA sequences and for this reason they were termed *idiomorphs* (Metzenberg and Glass, 1990). Mating occurs only between strains of *A* and *a* mating types, that is, between individuals that are sexually compatible. A slender hypha called a trichogyne projects out from the protoperithecia and is attracted toward macro- or microconidia of the opposite mating type, indicating that recognition of mating partner is based on the pheromone and pheromone receptor. The nucleus, picked up by this trichogyne (Figure 10.3), migrates to the ascogonium where fertilization occurs. In the laboratory, crosses are made by adding macroconidia (because these are obtained easily) from one culture to a culture of the opposite mating type that has been pre-grown in a medium with low nitrogen and carbon to induce the formation of protoperithecia. The haploid nuclei of the opposite mating type fuse in a hook-shaped structure called a crozier. The diploid zygote nucleus does not undergo divisions; rather, it immediately undergoes meiosis. The not-readily noticeable ascospores are formed at a different time from the striking conidial phase and had therefore been missed for a long time.

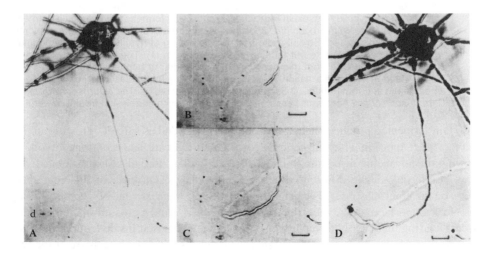

Figure 10.3   Chemotropic growth of trichogyne of Neurospora crassa. (A) Microconidia (male cells) were placed at lower left side. (B)–(D) Curvature of trichogyne is a visible evidence of production of a pheromone by conidia. (Reprinted from G.N. Bistis, *Mycologia* 63(5), 1981, with permission from the Mycological Society of America.)

In *Neurospora*, the four haploid nuclei, produced from a single meiotic division, divide mitotically to produce eight nuclei that are individually sequestered into eight oval-shaped cells called the ascospores and are enclosed in a single elongated cell, called the ascus. As this development is going on, the protoperithecia darken and form a flask-shaped structure called a perithecia with an opening, called an ostiole. The mature, dormant ascospores are shot out through the ostiole. The ascospores (haploid cells) germinate to produce a mycelium that contains haploid nuclei that have only one chromosome of each type, and therefore one copy of each gene. The fungus displays the alternation of haploid and diploid phases but the haploid phase persists for a long time and is the dominant phase. The haploid mycelium produces the characteristic pink-orange-colored conidia by which the fungus is easily recognized in nature and collected.

10.2 LIFE HISTORY

Based on fertility in crosses with known strains, approximately 65% of all *Neurospora* that have been collected globally is comprised of one species called *N. intermedia*. Its life history in nature was determined based on observations of the fungus on burnt sugarcane in agricultural fields and from simulated experiments in the laboratory with canes burned in the laboratory (Pandit and Maheshwari, 1996). Seasonal samplings of air spora in sugarcane fields did not reveal ascospores, discounting their aerial dissemination. However, virtually all samples of soil from the sugarcane field, after it had been heated to 60°C for 30–45 min—a treatment that kills conidia—yielded *Neurospora*. In the field-cultivated sugar cane, *Neurospora* colonies on the burnt stumps or the stubbles are initially in the part in contact with the soil, suggesting that ascospores present in soil infect the burnt cane. To confirm this, cane segments were planted in soil into which reciprocal mixtures of ascospores and of conidia of wild type (al^+, orange) and a color mutant (al, albino/white) had been mixed. The cane segments were burnt in the laboratory, simulating the conditions in the field. The phenotype of *Neurospora* colonies that developed on the cane segments was that of the genotype of the ascospores that had been mixed into the soil, confirming that the heat-resistant ascospores in soil initiate infection of burnt cane. In nature, the activation of dormant ascospores present in soil is brought about by furfural—a compound produced on heating xylose that occurs as a constituent of the polysaccharide xylan in the plant cell walls. This explains why Neurospora is commonly found on burnt plant tissues. This knowledge also accounts that the natural populations of Neurospora are largely sexual (Perkins and Turner, 1988).

The life history of *Neurospora* illustrates that most fungi produce spores both asexually and sexually, i.e., fungi are pleomorphic. Environmental conditions determine when, where, and how reproduction will occur. On the burnt sugarcane, conidia were formed in the sugar-rich cane tissue and their production continued until sugars had become depleted. The production of countless numbers of macroconidia externally and their constant dissemination by wind is seemingly wasteful, but it is the fungus's strategy of removing sugars from plant tissue and creating a nutritional environment for the mycelium inside the tissue to switch to sexual reproduction (Pandit and Maheshwari, 1966). The nutrient-depleted sugarcane stubble favors the development of protoperithecia and microconidia, i.e., for sexual reproduction to occur. Further, so long as the carbon source in juicy sugarcane stubble is available, the absorptive mycelium ramifies inside the tissue and forms closely packed lateral aggregates of conidiophores, termed sporodochium, beneath the epidermal tissue. The growth pressure of sporodochium causes the epidermis to separate from the ground tissue, thereby creating tissue pockets in which microconidiophores and protoperithecia, a biological solution for the need to ensure humid conditions and attract microfauna for effecting fertilization by transmitting microconidia to the trichogyne. The sexual phase is relatively inconspicuous, develops at a different location, and may appear unrelated. Indeed, it was unnoticed for a long time. The occurrence of morphologically distinct asexual and sexual phases at different times due to their differing environmental requirements illustrates that in fungi the conditions for asexual and sexual reproduction are often quite different. Media differing in composition have been devised for optimal production of asexual conidia or sexual ascospores.

10.3 ONE GENE—ONE ENZYME

In 1941, George Beadle and Edward Tatum, respectively a geneticist and a biochemist, wished to determine whether mutations can lead to nutritional requirements. For this they required an organism whose life cycle and genetics was worked out and that could be grown on a simple medium of known chemical composition so that any mutation-induced nutritional deficiency could be identified simply by supplying the fungus with a known metabolite. The organism of choice was *Neurospora crassa* as it requires a simple minimal salt and sugar medium with only a trace of biotin. Beadle and Tatum obtained cultures of *Neurospora* from B. O. Dodge for their experiments. They irradiated *Neurospora* with x-rays to induce mutations. The irradiated conidia were used to cross a culture of the opposite mating type, and the cultures derived from progeny ascospores were tested for mutant phenotypes (Figure 10.4). Strains were selected on the basis of differential growth response in media lacking or containing a single chemical compound. In each case, the auxotrophic mutation (a mutation that results in the requirement of any of the metabolic end products) was inherited as a

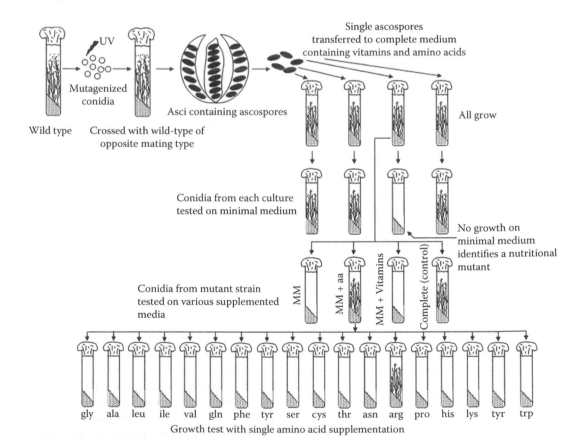

Figure 10.4   The procedure used by Beadle and Tatum for production and identification of auxotrophic mutants in *Neurospora crassa*. MM, minimal medium.

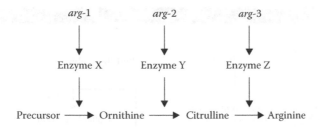

Figure 10.5   A biochemical pathway illustrating the one gene–one enzyme concept. Metabolic sequences are identified by the intermediates formed and purifying the enzymes that catalyze their conversion.

single-gene mutation, i.e., when crossed to a wild-type (original standard) strain it gave a 1:1 ratio of the mutant and the wild-type progeny.

As an example, a set of mutant strains required arginine to grow on minimal medium. By a gene mapping procedure, Beadle and Tatum found these mutants mapped into three different locations on separate chromosomes, even though all could grow if the same supplement (arginine) was added to the medium. However, the three mutants—*arg-1*, *arg-2*, and *arg-3*—differed in their response to related compounds, ornithine, citrulline, or arginine. The *arg-1* mutants grew with supplementation of ornithine, citrulline, or arginine. The *arg-2* mutants grew on either arginine or citrulline but not on ornithine. The *arg-3* grew only when arginine was supplied. The results demonstrated that synthesis of arginine proceeds from ornithine through citrulline and that each mutant strain lacks only one enzyme. Based on the properties of the *arg* mutants of *Neurospora*, Beadle and Tatum proposed a sequence of reactions in which each reaction is controlled by a specific enzyme, identified by a specific gene (Figure 10.5).

This model, known as the one gene–one enzyme hypothesis, was inferred from the properties of mutant classes. It provided the first insight into the functions of genes: That each gene controls one specific enzyme. Nearly all mutants responded to single growth supplement, i.e., each mutant was blocked in a single biosynthetic pathway. The position of the mutational block in the biochemical pathway could be determined by growth tests, whether the accumulation of the last intermediate preceded the block or occurred after the block. Although there are enzymes encoded by more than one gene, the one gene–one enzyme hypothesis fits most genes. This work demonstrated that the mutant approach can be generally exploited to determine individual steps in metabolic pathways. In 1958, George Beadle and Edward Tatum were awarded the Nobel Prize in Physiology or Medicine (shared with Joshua Lederberg) for their discovery that genes act by specifying enzymes that catalyze definite chemical reactions. Their discovery greatly promoted researches on fungi and laid the foundation of biochemical genetics. In 1966 Beadle recalled:

> During the course of this work, Tatum's late father, Arthur Lawrie Tatum, then professor of pharmacology at the University of Wisconsin, visited Stanford. On this occasion he called me aside and expressed concern about his son's future. "Here he is," he said, "not clearly either biochemist or geneticist. What is his future?" I attempted to reassure him—and perhaps myself as well—by emphasizing that biochemical genetics was a coming field with glowing future and there was no slightest need to worry.

10.4 MUTATIONAL ANALYSIS OF CONIDIA DEVELOPMENT

With the objective of identifying the genes that control development and if the related genes occur as clusters and if there is any order in their expression to give gene products, Charles Yanofsky initiated a research program on developmental genetics of *Neurospora*. Berlin and Yanofsky (1985) selected and grew mycelia in liquid, and then filtered and exposed the mycelial mat to nutrient-

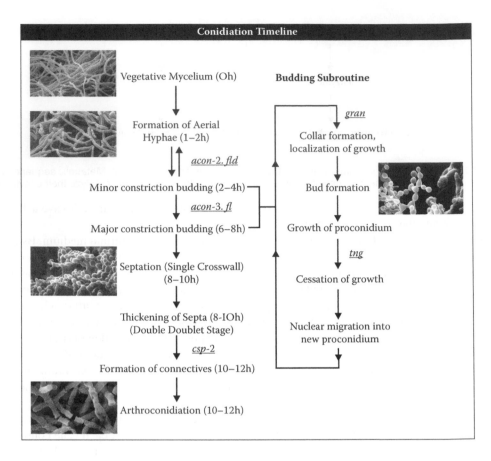

Figure 10.6   Developmental timeline of conidiation. (From Springer, M. L. and Yanofsky, C. *Genes Dev.* 3, 559-571, 1989, with permission.)

deprivation conditions under light, whence aerial mycelium grew and synchronously formed chains of conidia by apical budding, showing that light triggers differentiation of pink-orange conidia. Springer and Yanofsky analyzed the process of conidiation in wild-type and morphological mutants using scanning electron microscopy (SEM) and specific fluorescent probes. The first discernible morphological step of conidiation is the transition from growth by hyphal tip elongation to growth by repeated apical budding, resulting in the formation of chains of proconidia that resemble beads on a string. The initial proconidial chains are morphologically distinct from those that form later and are capable of reverting to hyphal growth, whereas the later chains are committed to conidiation (Figure 10.6). As the proconidial chains are formed, nuclei migrate into the conidiophore, and cross-walls arise between adjoining proconidia in a series of steps that have been defined by staining with calcofluor, a fluorescent chitin-binding probe. The chains ultimately disarticulate in several discrete stages into free, morphologically mature conidia. Conidiation-defective mutants were found blocked at distinct morphological stages in conidiation, allowing an opportunity to identify the genes controlling conidiation (named *con* genes) in the context of development blocks, and investigate the cues that induce *con* genes. Sequences homologous to Poly(A)+ RNA expressed in mycelia at discrete stages of conidiation were subtracted from the cDNA. Two transcripts complementary to *con-10a*, present at low levels in 8-h cultures, were present at high levels in 12-h cultures (Figure 10.7). Using mRNA, polypeptides synthesized in an *in vitro* system were analyzed by two-dimensional gel chromatography. Major changes were detected in profiles of protein *con-7*, *con-8*, *con-9*, *con-10a*, and *con-lla* hybridized with transcripts found exclusively in conidiating cultures.

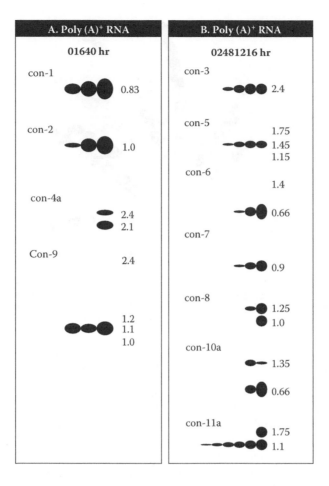

Figure 10.7 Expression of *con* genes. (Redrawn from Berlin, V. and Yanofsky, C., *Mol. Cell Biol.* 5, 849–855, 1985.)

10.5 MEIOTIC EVENTS

B. O. Dodge and C. C. Lindegren (see Davis, 2000) perceived *Neurospora* as a marvelous organism for genetic analysis. The four products of a single meiosis are packaged as ascospores in a single row in the order of their formation in ascus (usually called a tetrad, although in *Neurospora* the meiotic division is followed by a mitotic division; the four meiotically produced ascospores are duplicated as sister ascospores, hence an octad). The ascospores are large enough, approximately 29 × 15 μm, to be dissected out manually under a dissecting microscope, and their phenotypes and genotypes to be determined. This procedure, called tetrad analysis, allowed tracing the order of the events in meiosis. For example, by crossing a mutant strain that affects ascospore color, it could be determined simply by visual examination of the asci in opened perithecia (Figure 10.8) whether the separation of parental alleles occurred in the first meiotic division without crossing over in the first-division segregation or in the second division as a result of crossover between the gene pair and centromere (Figure 10.9). As an example, the 4:4 ratio of mating types (*A* or *a*) or of ascospore color marker (*cys-3*) in the octad demonstrates that alleles separate during meiosis, validating Mendel's first law of segregation. (In genetic nomenclature for *Neurospora*, the name of a gene is generally abbreviated to the first three letters. For example, the mutant allele *cys* refers to cysteine while the superscript "+" is used to distinguish the wild allele from the mutant allele.) In a cross of a wild *N. crassa* with pink-orange carotenoid pigment,

Figure 10.8 Photomicrograph of an opened perithecium of *Neurospora crassa* showing a rosette of maturing asci from a cross wild-type (+) x *cys-3* mutant (m) allele that affects cysteine biosynthesis and results in unpigmented (white) ascospores. Because the final division is mitotic, contiguous sister ascospores have the same genetic constitution. One ascus at top center and two asci at upper left show 4 black:4 white first-division segregation. The remaining mature asci show second division segregation patterns (2+, 2m, 2+, 2m) resulting from crossing over between *cys-3* and centromere. (From Raju, N.B., *Eur. J. Cell Biol.* 23, 208–223, 1980, with permission.)

i.e., *al⁺* and aerial mycelium (*col⁺*), to a double mutant albino (white, impaired in carotenoid biosynthesis) and colonial (restricted growth in the form of a colony), four types of progeny were produced in equal proportion: *al⁺ col⁺* (wild type with orange, aerial mycelium), *al col* (albino colonial), *al⁺ col* (pink-orange colonial), and *al col⁺* (albino, spreading mycelium), in conformity with Mendel's second law of independent assortment. From a *Neurospora* cross, a significant number of sexual progeny can be grown in small tubes and analyzed in about four weeks, providing considerable savings both in space and in time than would be possible with either a pea plant or a fruit fly. *Neurospora* has been the main source of information about basic recombination mechanisms in eukaryotes. It is an ideal classroom material for teaching microbiological and genetic methods.

Tetrad analysis demonstrated that meiotic recombination by crossing over between linked loci on the same chromosome occurs after chromosome replication between chromatids and not whole chromosomes. In a cross (*AB* × *ab*) where genes *A* and *B* are linked with alleles *a* and *b*, the appearance of two parental and two recombinant products (a tetratype) led to the conclusion that crossing over occurs after chromosome replication between nonsister homologous chromatids at the four-strand stage (Figure 10.10). Had it occurred before replication, no tetratypes would have been found. *Neurospora* began to be used for understanding recombination mechanisms. Infrequently, a small proportion of aberrant 6⁺:2 mutant and 2⁺:6 mutant asci instead of normal 4⁺:4 mutant asci were detected (Mitchell, 1955). Such aberrant segregation ratios were explained by gene conversion according to which allele at a locus converts a corresponding allele at the same locus to becoming its own type. This observation provided clues on molecular models of crossing over that could account for gene conversion (Holliday, 1964).

10.6 GENE MAPS

In *Neurospora*, the ascospores are ejected successively—the asci elongate, one at a time protrude through the opening in the perithecium, shoot ascospores in groups of eight, and then retract.

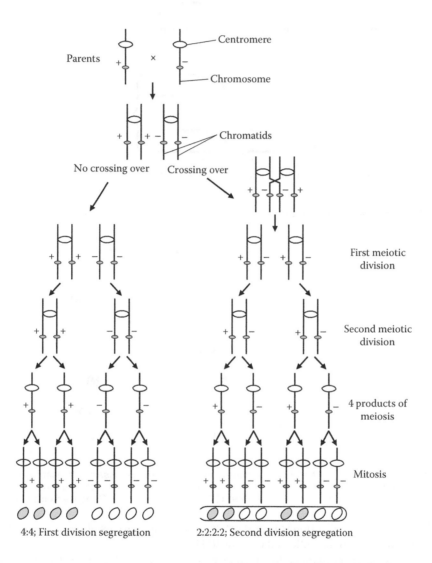

Figure 10.9   Scheme of first- and second-division segregation in *Neurospora*. Segregation of two + and − alleles can be deferred to the second division of meiosis, a phenomenon known as "second-division segregation."

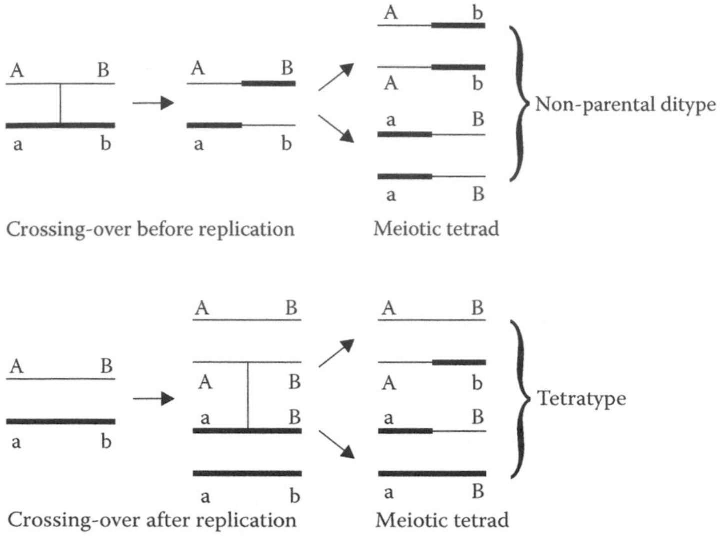

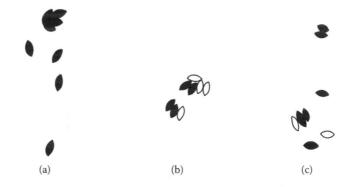

Figure 10.10   Meiotic recombination between linked loci. Crossing over (shown as a vertical line between homologous chromosomes) between genes *AB* and alleles *ab* occurs at the four-strand stage, that is, after homologous chromosomes had each duplicated into two chromatids. Products of meiotic division will divide by mitosis and the mature ascus will have eight ascospores.

Figure 10.11   Spontaneously shot ascospore octads from a *Neurospora crassa* cross insertional translocation × normal. (a) All normal (black). (b) 4 Normal:4 defective (white). (c) 6 normal:2 defective. (From Perkins, D.D., *Genetics* 77, 459–489, 1974, with permission of Genetics Society of America.)

Thus, if the time of their collection is kept short, well-separated groups of tetrads—i.e., products of single meiotic events—can be collected (Figure 10.11).

Neurospora is the first organism in which techniques were developed for detecting and analyzing chromosomal rearrangements in which segments between chromosomes become exchanged, deleted, or inverted (Figure 10.12), either spontaneously, by UV irradiation, or by DNA transformation. These are types of mutations. The rearrangement of chromosome structure is manifested as visually defective ascospores in shot octads (Perkins, 1974). The defective ascospores are white while nondeficiency ascospores are black. To identify the nature of chromosomal rearrangements, the strain to be tested is crossed to a normal strain in a Petri dish. After 10 days the perithecia are briefly inverted over a slab of water agar to collect spontaneously shot groups of eight ascospores (unordered tetrads) and examined under a binocular microscope to score defective octad (black:white) classes as 8:0, 6:2, 4:4, 2:6, and 0:8 The frequency of 4:4 type is diagnostic of the type

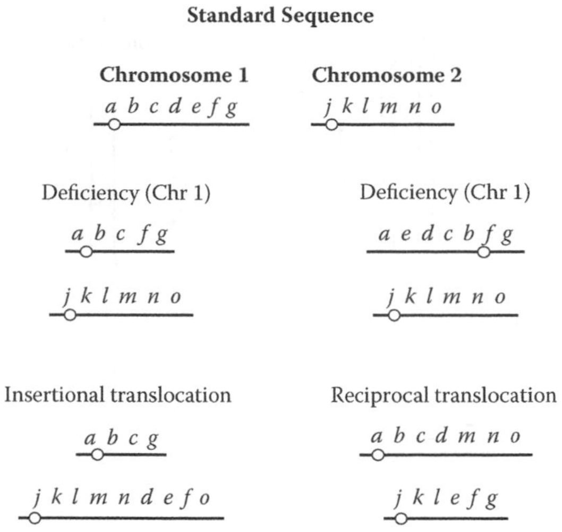

Standard Sequence

Chromosome 1   **Chromosome 2**
a b c d e f g   *j k l m n o*

Deficiency (Chr 1)   Deficiency (Chr 1)
a b c f g   *a e d c b f g*

j k l m n o   *j k l m n o*

Insertional translocation   Reciprocal translocation
a b c g   *a b c d m n o*

j k l m n d e f o   *j k l e f g*

Figure 10.12 Common types of chromosomal rearrangements found in *N. crassa*. (From Davis, R.H., *Neurospora: Contributions of a Model Organism*, Oxford University Press, New York, 2000, with permission from Oxford University Press.)

of translocation, whether insertional or reciprocal (Figure 10.13). Octads are used in the measurement of crossover frequencies based on the proportion of asci of the following classes: Parental ditypes, tetratypes, and nonparental ditypes. From the data, the distance between a gene and centromere can be determined and gene maps constructed.

To facilitate the mapping of a newly discovered gene, a linkage tester strain, *alcoy* (having three reciprocal translocations), was developed in which six linkage groups are represented as T(IR; IIR) *al-1*, T(IVR; VR) *cot-1*, and T(IIIR; VIL) *ylo*. The symbol T refers to translocation, the numbers refer to linkage groups, and R and L refer to the right and left arms of the linkage groups. The compound chromosomes have visible genetic markers near the junction point: *al*bino-1 (white color), *co*lonial *t*emperature sensitive 1 (button-shaped colony at 34°C), and *y*e*llo*w. Segregation of these markers among progeny helps in the detection of the linkage of an unknown gene to any of these markers. The use of this linkage tester facilitates the rapid mapping of new mutations on a particular chromosome, thereby greatly shortening the time required in finding linkage of an unknown gene than if it is to be found by crossing the unknown to each of the seven linkage-group tester strains. Here, we give the example of a wild-collected strain, senescent (*sen*), that resembles the mutant natural *d*eath (*nd*) in that both cease to grow in four to five subcultures. The question arose whether *sen* is the same as *nd*. A cross of a wild-type strain with conidia from a *sen* strain produced half wild type (⁺) and half senescent progeny, showing segregation of a single gene pair according to Mendel's law, identifying the "death" phenotype to be due to a single gene, called *senescent*, abbreviated as *sen* (Navaraj *et al.* 2000). A single cross of the mutant senescent to *alcoy* gave an essentially random recombination of *sen* with *al-1* (41%), eliminating linkage groups (LG) I and II from consideration. On the other hand, *sen* showed 22% recombination with *cot-1*, indicating that *sen* is either on LG IV or V. A follow-up cross of *sen* to a strain having *cot-1* marker on the normal chromosome produced parental and recombinant classes in nearly equal frequency (48 and 52%), showing that *sen* is not on LG IV. From a cross of *sen* to a strain containing two mutant markers, *cyh-2* and *al-3* in LGV, the frequency of recombination between *cyh-2* and *sen* (37%) and between *sen* and *al-3* (11%), *sen* gene was mapped between *cyh-2* and *al-3* (Figure 10.14). Whereas *nd* is in IR (r = right of centromere), *sen* is in VR; showing that *nd* and *sen* are different genes that control lifespan.

When a new gene is found, it needs to be shown that it is different from the previously discovered genes. Nearly 1000 genes have been mapped in *N. crassa* (Perkins *et al.*, 2000). It has the most

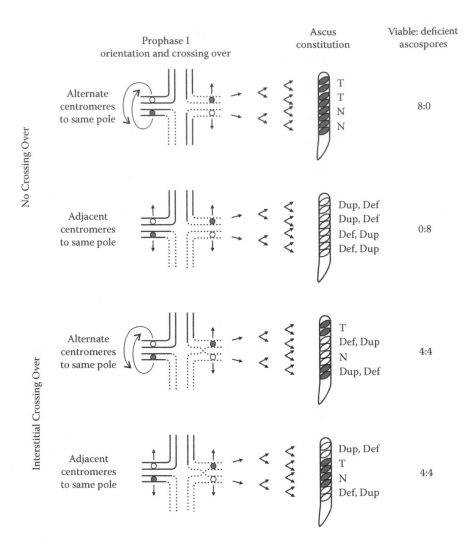

Figure 10.13 Segregation pattern for a cross-reciprocal translocation (black centromeres) × normal sequence (white centromeres). Def, deficient. (From Perkins, D.D., *Genetics* 77, 459–489, 1974, with permission from Genetics Society of America.)

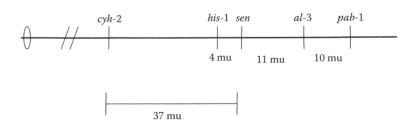

Figure 10.14 A partial gene map of linkage group VR. Abbreviations of genes: *cyh-2, cycloheximide resistant-2*; *his-1, histidine-1*; *sen, senescent*; *al-3, albino-3*; *pab-1, para-aminobenzoic acid-1*. 1 map unit (mu) = 1% recombination between gene loci. (From Navaraj, Pandit, and Maheshwari, *Fung. Genet. Biol.* 29, 165–1273, 2000.)

saturated genetic map of all fungi. The gene maps show that functionally related genes in the same biosynthetic pathway are not generally linked or clustered. Another thing learned from the mapping work is that crossover frequencies, while reproducible in crosses with the same two parents, vary widely in crosses with parents of different genetic backgrounds. The gene maps showed also that most genes related to the sexual cycle are not linked to the mating type locus (idiomorphs).

10.7 CHROMOSOME MORPHOLOGY AND GENE EXPRESSION

The development and improvements of cytological methods was instrumental in the adoption of *Neurospora* as a model organism for fungal cell biology. In 1945, McClintock used aceto-orcein to stain chromosomes of *Neurospora crassa* and showed that n = 7. Some half a century later, this finding of a small number of chromosomes was important for its genome analysis. The dormant ascospores of *Neurospora* have a four- to five-layered melanized cell wall, making it difficult to ascertain the chromosome number in mature ascospores that are used in sexual genetics. A *per-1* mutant has a block in formation of melanin and when used as female parent it produces unpigmented (hyaline) ascospores, allowing the number of nuclei in mature ascospores to be determined. It could be demonstrated that the original single nucleus in young ascospores divides, and after three to four successive divisions, the mature ascospores contain 16 to 32 nuclei, providing critical information that multinuclear condition is a basic feature that extends to nearly all types of fungal cells. Treatment with ethylene glycol greatly swells conidia, and staining showed that even though the nuclei share a common cytoplasm, their division is not synchronous.

An *N. crassa* mutant named *Banana* forms a single giant ascospore because the whole ascus is converted to one single ascospore instead of the usual eight ascospores (Raju, 2008). Following karyogamy and meiosis, the nuclei in the giant ascospore divide mitotically. Using the *Banana* mutant, it was shown that in a cross of *Ban* to *H1-GFP* strain, the *hH1-GFP* transgene segregated at the first division of meiosis; that is, the nuclei at one end of the ascospore were GFP-tagged, and the nuclei at the other end of the giant ascus were not (Raju, 2009; see Figure 8 G), demonstrating that GFP is not nucleus limited. A gradient of fluorescing protein suggested that transcripts are processed by ribosomes in the vicinity of *hH1-GFP* nuclei and the fluorescing transgene protein diffused into the cytoplasm and interacted with the coexisting heterozygous nuclei [Raju 2008, Figure 8(G)], demonstrating that GFP is not nucleus limited.

10.8 ASCUS DEVELOPMENT BIOLOGY

In *N. tetrasperma* the asci are four-spored instead of the eight-spore asci found in other species of the genus (Figure 10.15). Each ascospore contains two nuclei of opposite mating types, rather than heterothallic. A 180° rotation of nonoverlapping second-division spindles ensures that nuclei of opposite mating types are placed close together for inclusion in the ascospores. A single-gene mutation in the eight-spored heterothallic species could not in one step have given rise to the four-spored pseudohomothallic condition. Other changes affecting ascus cytoskeleton, spindle orientation, association of nuclei in pairs, and orderly sequestration of pairs into the four ascospores must occur.

In Chapter 3, we learned that an ascomycete fungus *Coniochaeta tetraspora* has four spores. It would have been natural to assume that this condition arises as in *Neurospora tetrasperma*. However, in the latter the immature asci are eight-spored, whereas the mature asci are four-spored. This is because four ascospores in each ascus abort, leaving only four mature ascospores, which showed either the first (4 viable:4 inviable) or the second-division segregation patterns (2:2:2:2 or 2:4:2) for ascospore death. In the related ascomycete *Cochliobolus heterostrophus*, a mature ascus has a single multinucleate, multisegmented ascospore; the remaining seven ascospores abort early

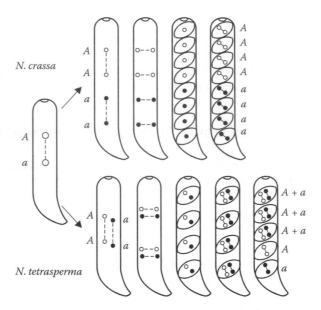

Figure 10.15   Schematic diagram of ascus development in the heterothallic species *N. crassa*, and in the pseudohomothallic species *N. tetrasperma*. Mating types (*mat A* and *mat a*) segregate at the first division of meiosis in both species. In *N. crassa*, the second-division spindles are aligned in tandem, and the four spindles at the third division (mitosis) are aligned equidistant and across the ascus. Ascospores are uninucleate at inception, four *mat A* and four *mat a*. In *N. tetrasperma*, the second- and third-division spindles overlap, and each of the four ascospores encloses two nuclei of opposite mating type. (From Jacobson, D.J., Raju, N.B., and Freitag, M., *Fung. Genet. Biol.* 45, 351–362, 2008, with permission.)

in development. Though nuclear fusion and meiosis follow very closely, some unknown mechanisms determine which meiotic product will survive. One of the interesting areas of study for the future will be the underlying evolutionary and developmental reasons behind variation in the shapes and number of ascospores surviving in different ascomycete genera.

10.9 MOLECULAR REVOLUTION

From the one gene–one enzyme hypothesis arose the concept of colinearity of gene and protein, i.e., the linear array of DNA nucleotide base sequence specifies the linear array of amino acids in the protein. Since the amino acid sequence of a protein was not yet known, Beadle and Tatum did not know the way in which genes determined enzymes. However, the biochemical genetic work initiated by *Neurospora* inspired workers using *E. coli* and yeast and was a starting point of a molecular revolution in biology. It was established that allelic forms of a gene specify structurally different forms of the same enzyme. Further, it was observed that certain pairs of mutants that were deficient in the same enzyme cooperated in diploid or in a heterokaryon to produce significant amounts of enzyme activity; i.e., the polypeptide chains from different mutants corrected each other's defects (allelic complementation). It was shown that many enzymes are multimeric, i.e., proteins containing more than one polypeptide chain, each specified by a different gene. The formation of active enzyme molecules was explained as molecular hybrids in which polypeptide chains from different mutants corrected each other's defects through effects on polypeptide conformation. These observations led to the modification of the one gene–one enzyme hypothesis to one gene–one polypeptide hypothesis. The basic relationships between genes, proteins, and phenotypes thus understood, attention turned to how the activities of enzymes are controlled. This led to the formulation of the operon model of control of gene activity from work done with *E. coli*. After the genetic basis of the enzyme structure was established, the work with *E. coli* led to investigations of the regulation of gene activity, stimulated by the lactose operon model of Jacob and Monod. The growth tests with the auxotrophic mutants of *Neurospora* started a molecular revolution in biology. "It became the preferred way to dissect complex biological systems such as embryonic development, cell division, the nature of sensory systems, and aging. Today, mutational analysis is the preferred way into a complex biological problem, especially as it provides access to the genes and protein players" (Horowitz *et al.*, 2004).

10.10 GENOME SEQUENCE

To understand how filamentous fungi work and to understand their evolution means that one must have a complete genomic sequence whereby one can monitor gene expression and correlate it with metabolic and developmental phenotype. Because in the model eukaryotic microbes (the yeast and in *Neurospora*) most genes have been characterized based on mutant phenotypes, these were chosen for genome-sequencing efforts in fungi The genome sequence of *Neurospora* was released in 2003 (Galagan *et al.*, 2003; Borkovich *et al.*, 2004). The features of sequence are given in Table 10.1. Although the *Neurospora* genome is, respectively, about 4 and 100 times shorter than those of fruit fly and human, the number of genes is high. The 43-Mbp genome of *N. crassa* encodes about 10,000 genes, nearly twice as many genes as in the yeast *Saccharomyces cerevisiae*. Many *Neurospora* genes lack homologues in yeast and may be specific for filamentous lifestyle. Approximately 14% of *Neurospora* genes match proteins in either plants or animals, suggesting a closer relation of filamentous fungi than of yeast to higher forms. A surprise revelation was that *N. crassa* has a homologue of phytochrome required for sensing red–far red in plants. The fungus shares genes with complex organisms that measure time (biological clock). The fungus has methylated sequences, which are relics of transposons. Another revelation is that though a saprophyte, the *Neurospora* possesses genes for virulence factors and enzymes for plant cell-wall digestion required for fungal pathogenesis, raising the possibility that this fungus, hitherto considered to be a saprophyte, could be an endophyte or an opportunistic plant parasite. Further similarity in the two types of fungi is suggested by genes for secondary metabolite production, for example, for biosynthesis of gibberellins and of polyketide. As we learned in Chapter 8, the fungus has mechanisms that inactivate duplicated (repeated) sequences in the sexual cycle. Many methylated DNA sequences, generally associated with duplicated genes, are present, indicating relics of transposons.

Table 10.1 Features of *Neurospora crassa* Genome

Size	38,639,769 bp
Chromosomes	7
% G+C	50
Protein-coding genes	10,082
Protein-coding genes (> 100 amino acids)	4200
tRNA genes	424
5S rRNA genes	74
Percent coding genes	44
Average gene size	1673 bp (481 amino acids)
% Protein-coding sequences similar to known sequences	13
Conserved hypothetical proteins	46
Predicted proteins (no similarity to known sequences)	40

Source: Adapted from Galagan, J.E., *et al.*, *Nature* 422, 859–868, 2003.

10.11 CONCLUDING REMARKS

A technical requirement of an organism that could be cultured in a defined medium led to the adoption of *Neurospora crassa* as an experimental organism with which to address the question as to what controls metabolic reactions. The resolution of this question by using induced mutants attracted the imagination of several workers on a genetic approach to biological problems. *N. crassa* played a role in connecting genetics and biochemistry, heralding biochemical genetics. More than a half century since its introduction in research, *Neurospora* continues to be used in studies on the hyphal mode of growth (Chapter 1), nuclear interactions in heterokaryons (Chapter 2), mutational and epi-genetic silencing mechanisms (Chapter 8), biological clocks (Chapter 13), speciation (Chapter 15), nuclear and plasmid-based senescence phenomenon (Chapter 16), and in many others, such as the meiotic drive elements that distort genetic ratios (Raju, 1996; Perkins, 2003). In-depth classical and molecular researches on *Neurospora* have made it a reference for eukaryotic organisms.

REFERENCES

Beadle, G. W. (1966). Biochemical genetics. Some recollections. In J. Cairns, G. S. Stent, and Watson, J. D., eds., *Phage and the Origins of Molecular Biology*. Cold Spring Harbor Laboratory, New York.

Berlin, V. and Yanofsky, C. (1985). Isolation and characterization of genes differentially expressed during conidiation in *Neurospora crassa*. *Mol. Cell Biol.* 5:849–855.

Borkovich, K. A., Alex, L. A., Yarden, O., Freitag, M., *et al.* (2004). Lessons from the genome sequence of *Neurospora crassa*: Tracing the path from genomic blueprint to multicellular organism. *Microbiol. Mol. Biol. Revs.* 68:1–108.

Davis, R. H. (2000). *Neurospora: Contributions of a Model Organism*. Oxford University Press, New York.

Davis, R. H. (2003). *The Microbial Models of Molecular Biology*. Oxford University Press, New York.

Davis, R. H. and Perkins, D. D. (2002). *Neurospora*: A model of model microbes. *Nature Revs. Genetics* 3:7–13.

Fincham, J. R. S. (1985). From auxotrophic mutants to DNA sequences. In J. W. Bennett and L. Lasure, eds., *Gene Manipulations in Fungi*. Academic Press, New York.

Galagan, J. E., *et al.* (2003). The genome sequence of the filamentous fungus *Neurospora crassa*. *Nature* 422:859–868.

Holliday, R. (1964). A mechanism for gene conversion in fungi. *Genet. Res.* 6:282–304.

Horowitz, N. H., Berg, P., Singer, M., Lederberg, J., Susman, M., Doebley J., and Crow, J. F. (2004). A Centennial: George W. Beadle, 1903–1989. *Genetics* 166:1–10.

Jacobson, D. J., Dettman, J. R., and Adams, R. I. (2006). New findings of *Neurospora* in Europe and comparisons of diversity in temperate climates on continental scales. *Mycologia* 98:550–559.

Maheshwari, R. (1999). Microconidia of *Neurospora crassa*. *Fung. Genet. Biol.* 26:1–18.

Metzenberg, R. L. and Glass, N. L. (1990). Mating type and mating strategies in *Neurospora*. *Bioessays* 12:53–59.

Mitchell, M. B. (1955). Aberrant recombination of pyridoxine mutants of *Neurospora*. *Proc. Natl. Acad. Sci. USA* 41:215–220.

Pandit, A. and Maheshwari, R. (1996). Life-history of *Neurospora intermedia* in a sugar cane field. *J. Biosci.* 21:57–79.

Perkins, D. D. (1974). The manifestation of chromosome rearrangements in unordered asci of *Neurospora*. *Genetics* 77:459–489.

Perkins, D. D. (1992). *Neurospora*: The organism behind the molecular revolution. *Genetics* 130:687–701.

Perkins, D. D., Radford, A., and Sachs, M. (2001). *The Neurospora Compendium: Chromosomal Loci*. Academic Press, San Diego.

Perkins, D. D. and Turner, B. C. (1988). *Neurospora* from natural populations: Toward the population biology of a haploid eukaryote. *Exp. Mycol.* 12:91–131.

Raju, N. B. (1996). Meiotic drive in fungi: Chromosomal elements that cause fratricide and distort genetic ratios. *J. Genet.* 75:287–296.

Raju, N. B. (2009). *Neurospora* as a model fungus for studies in cytogenetics and sexual biology at Stanford. *J. Biosci.* 34:139–159.

Springer, M. L. and Yanofsky, C. (1989). A morphological and genetic analysis of conidiophore development in *Neurospora crassa. Genes Dev.* 3:559–571.

Springer, M. L. and Yanofsky, C. (1992). Expression of *con* genes along the three sporulation pathways of *Neurospora crassa. Genes Dev.* 6:1052–1057.

Turner, B. C. (2001). Geographical distribution of *Spore killer* meiotic drive strains and of strains resistant to killing in *Neurospora. Fung Genet. Biol.* 32:93–104.

Turner, B. C., Perkins, D. D., and Fairfield, A. (2002). *Neurospora* from natural populations: A global study. *Fung. Genet. Biol.* 32:67–92.

CHAPTER **11**

Aspergillus nidulans
A Model for Study of Form and Asexual Reproduction

The contemplation of a single spore germinating to form a colony, perhaps anastomosing *en route* with a similar germ mycelium, provides a major challenge to the general mycologist. To him, the clear-cut, apparently simple, sequence of events provides intense aesthetic satisfaction and provokes unlimited enquiry and investigations into the hidden complexities which underly them.

—**J. H. Burnett (1976)**

Several fungi produce elaborate structures called conidiophores that bear asexual spores called conidia. These can be quite complex structures resembling a bunch of grapes, a spike of wheat, or a capitulum of dandelion. The conidiophore is a multicellular structure, but the cells are all attached transversely, not laterally. This is because in fungi cell wall is laid down transversely, not vertically. The conidiophore of *Aspergillus nidulans* (Ascomycotina) is a morphological device that loosely packs a large number of conidia in a small space in a short time and for the continuous dissemination of short-lived conidia by whiff of the wind, splash of rain, or insects. The principles elucidated from this relatively simple system apply to embryonic development, such as in how the embryo marks and measures space and time so that organs and tissues develop on schedule and in the right locations. A special advantage of studying development and differentiation in *Aspergillus* is that each of its conidia is uninucleate, providing a convenient source of identical haploid cells for UV mutagenesis for the selection of developmental mutants. *A. nidulans* is expected to prove more useful as a model for the study of fungal development as its genome sequence has been determined.

11.1 CONIDIOPHORE STRUCTURE

The development of *Aspergillus* conidiophore can be divided into steps that occur in temporal sequence. It begins with the differentiation of a hyphal cell into a thick-walled foot cell that extends into an aerial stalk of approximately 100 µm (Figure 11.1) and allows the asexual reproductive cells to be pushed out of the substratum for dissemination. Second, the stalk swells at the tip to form a globose, multinucleate vesicle that has a diameter of about 10 µm. Third, the vesicle buds out about 60 cigar-shaped uninucleate cells called metulae. Fourth, each metula in turn buds out two phialide cells. Fifth, a single nucleus enters into the phialide and by successive mitotic divisions each phialide buds out a vertical chain of conidia with the oldest conidium at the top of the chain. Coupling of budding to cytokinesis results in each conidium being uninucleate. Each conidium is approximately 3 µm in diameter. By this form of development, each conidiophore can produce over 1000 conidia with economy of space and in a short time. We are beginning to understand how genes specify the development of a conidiophore stalk of a definite height that swells into a vesicle forming a number of metulae and phialides—the latter budding out airborne uninucleate conidia.

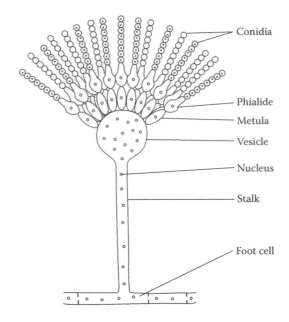

Figure 11.1   Diagrammatic structure of a conidiophore of *Aspergillus nidulans*.

The time line of conidiophore development in *A. nidulans* is as follows (Boylan *et al.*, 1987): Undifferentiated hyphae (0 h) → aerial stalk (5 h) → vesicle (10 h) → metula and phialide (15 h) → immature conidia (20 h) → mature dark green conidia (25 h).

11.2 CONIDIATION GENES

Mutants of *A. nidulans* were obtained that were normal in hyphal growth and sexual reproduction but were defective in conidiophore development (Clutterbuck, 1969). Since conidia are not formed, the conidiation mutants are detected by the lack of green color and inability to be replica plated—a simple technique that allows colonies producing loose conidia in a Petri dish to be sampled by pressing a velvet cloth secured over the top of a cylinder and then pressed onto the surface of new medium in a Petri dish. The procedure transfers each conidiating colony in a way that the pattern of colonies is maintained on the replica plate. The colonies that are missing because of nonconidiation are identified in the master plate. In one study the number of conidiation-specific genes was estimated by comparing the frequency of conidiation mutants with the frequency of auxotrophic mutants:

$$\left(\frac{\text{frequency of conidiation mutants}}{\text{frequency of auxotrophic mutants}} \right) \times \text{number of auxotrophic genes}$$

It was estimated that approximately 100 genes specifically affect conidiation (Martinelli and Clutterbuck, 1971). A larger number of genes (about 1000) was estimated based on the diverse messenger RNA that accumulated specifically during conidiation (Timberlake, 1980). Based on the mutant phenotypes and the comparison of gene expression in wild-type and conidiation mutants, three genes were identified as essential regulators of conidiophore development: The bristle (*brl*) gene controls the swelling of the conidiophore stalks into vesicles; the abacus (*abaA*) gene controls the budding of conidia by the conidiophore stalk; and the wet-white (*wetA*) gene controls the hydrophobicity that is essential for the aerial dissemination of conidia. These three genes are themselves not specific to conidiation; rather, they integrate the expression of other genes that determine the structure of the conidiophore. The genes were therefore considered to be regulatory genes that play a key role in conidiophore development and act in the sequence *brlA* → *abaA* → *wet A*. Using a genomic library, the wild *brlA*, *abaA*, and *wetA* were isolated by complementation of the mutant strains. The nucleotide sequence of *brlA* showed that it has a sort open reading frame (ORF) that regulates transcription. The deduced amino acid sequence of *brlA* protein product can form a secondary structure with a loop of the polypeptide chain, to which a Zn atom can bind. Since this type of structure is common in DNA-binding proteins, the *brlA* gene is thought to encode a transcription regulator that activates genes required for conidiophore formation (Adams *et al.*, 1990).

To determine whether conidiation-specific genes occur in a close physical order that reflects their order of action or are distributed randomly in the genome, 30 cDNA clones containing approximately 1.5-kb-long DNA inserts were hybridized to poly(A⁺) RNA (Orr and Timberlake, 1982). On the basis of the random distribution of conidiation genes, the majority of clones were expected to have only one spore-specific transcript. However, the experimental finding was that many clones hybridized to several spore-specific transcripts of different molecular weights, indicating that spore-specific genes in *A. nidulans* may be clustered and constitute a functional unit of developmental gene regulation.

The genome sequence analysis of *A. nidulans* was completed in 2005 and showed that it has 9541 protein-coding genes (Galagan *et al.*, 2005). A clone called SpoC1 was isolated from a genomic library by hybridization with a cDNA probe complementary to poly(A⁺) RNAs that accumulated preferentially during conidiation. Restriction fragment analysis revealed five overlapping genes. It is not known whether each of the SpoC1 genes responds independently to a specific regulatory molecule. Evidence from a related species, *A. fumigatus*, suggests that the genes controlling pigmentation in spores are also clustered.

11.3 CONIDIATION TRIGGER

In 1900, Klebs, a German botanist, postulated that the conditions for sporulation are different from vegetative mycelium, with growth being favored by starvation. This is consistent with the common observation that sporulation is first observed in the center of the colony growing on a solid medium from which the mycelium has absorbed the nutrients. A molecular approach was taken to test whether conidiation can be induced in the presence of an excess of nutrients (Adams and Timberlake, 1990). Adopting the recombinant DNA methodologies, they constructed strains of *A. nidulans* in which the promoter gene (*p*) from the alcohol dehydrogenase gene (*alcA*) was fused to conidiation regulatory genes (*brlA* or *abaA*) and critically tested the nutrient exhaustion hypothesis of conidiation. The genotype of constructed gene fusion strains is denoted as *alc(p)::brlA abaA* and *alcA(p)::abaA*. The fusion strain germlings were grown as homogeneous mycelium in submerged cultures containing a carbon source. After 3 h, the cultures were shifted to medium containing threonine or ethanol to force the induction of *brlA* or *abaA* gene. The forced expression of the conidiation-specific regulatory genes *brlA* or *abaA* led to a generalized metabolic shutdown as indicated by the reduction in total protein and RNA and to a loss in the ability to take up nutrients from the medium. Vegetative growth ceased and conidia were formed at the ends of hyphae, although without the formation of vesicle, metula, and phialide (Adams *et al.*, 1988; Adams and Timberlake, 1990). The ability to induce conidiation by direct activation of the *brlA* gene suggested that *brlA* mediates the development switch from polarized growth of stalk cell to budding growth leading to conidium formation. It was inferred that reproduction is a genetically programmed event in development that is triggered by an internal signal that shuts down the genes involved in nutrient uptake and activates the conidiation pathway by activating the expression of *brlA* alone. The fungus must sense the external environment and transduce the stimulus (signal) across the cell membrane into the intracellular environment.

11.4 DEVELOPMENTAL COMPETENCE

A. nidulans does not conidiate in submerged liquid but will do so in culture if the mycelium is simply filtered off to expose the mycelium to air, whence it produces green (viable) conidia, i.e., induction. The time for such a liquid-grown mycelium of normal (wild-type) strain to produce the green, viable conidia is 22.7 h (competence time), when there is a very rapid increase (doubling time 10–15 min) in the number of conidia in the colony. To understand events that occur during this time, conidia were UV mutagenized and a set of strains isolated. These sporulated normally at 28°C (permissive temperature) but not at 42°C (restrictive temperature). Colonies of these mutants were subjected to temperature upshifts and after a fixed time of 72 h the total number of conidia harvested from each mutant colony was determined (Figure 11.2). For example, conidia yields for *aco A-49* upshifted after 30 h at 28°C was determined after 42 h at 42°C. The conidia yield for *aco-75* upshifted after 50 h at 28°C was determined after incubation of mycelium for 22 h at 42°C. If the mutant's sensitive event was pre-induction, the event would have been passed over at the permissive temperature so it would conidiate at the restrictive temperature. But if the temperature-sensitive event was post-induction, conidiation could not occur and would be inhibited. The conidial yields

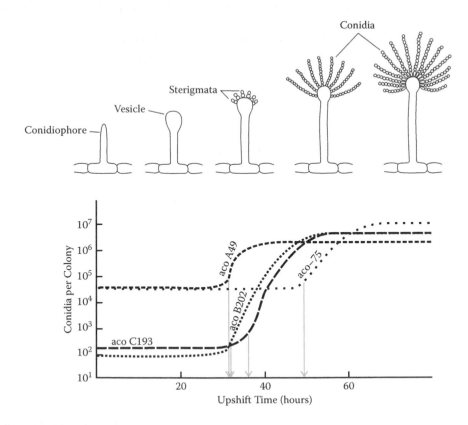

Figure 11.2 Temperature shift analysis of critical time between induction and onset of conidiation. Above, morphological stages in conidiation of *Aspergillus nidulans*. Below, identification of a pre-induction and post-induction mutant. (Based on data of Yager, L. N., Kurtz, M. B., Champe, S. P. 1982. Temperature-shift analysis of conidial development in *Aspergillus nidulans*. *Dev Biol.* 93:92–103.)

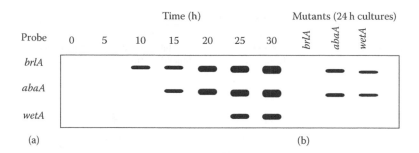

Figure 11.3 Temporal pattern of expression of regulatory genes in *Aspergillus nidulans*. (Adapted from Boylan, M.T., Mirabito, P.M., Willett, C., Zimmerman, C.R., and Timberlake, W.E., *Mol. Cell. Biol.* 7, 3113–3118, 1987.)

of all mutants was similar, from 10^7–10^8 per plate, demonstrating that these did not suffer from sickness. The experimental results indicated that there are a number of steps leading to conidiation, and each biochemical step is controlled by specific genes whose expression occurs in linear order. The competence time will be affected not only by products of several genes but also by physical and chemical conditions of culture.

11.5 REGULATORY PATHWAY

Having identified the genes that regulate conidiation, the time of their expression could be studied. Total RNA was isolated from cultures and analyzed by Northern blots using *brlA*, *abaA*, and *wetA* probes (Figure 11.3). The experiment detected transcripts of *brlA* at 10 h, of *abaA* at 15 h, and of *wetA* at 25 h. The *brlA* mutation blocked the accumulation of all three RNAs. The *abaA* and *wetA* mutations reduced the accumulations of the *brlA* and *abaA* RNAs and blocked accumulation of the *wetA* RNA. The *brlA* and *abaA* mutations affected their own expression and the expression of one another. The *wetA* transcript was absent in *wetA* temperature-sensitive mutant grown at restrictive temperature, implying that the *wetA* gene is autoregulatory (Mirabito *et al.*, 1989). Subsequently mutations in six genes were isolated that affect conidiophore development and result in cotton-like colonies with "fluffy" morphology. These were designated *flbA–E and fluG*. The expression of *brlA* was reduced in these mutants. The nonregulatory, developmentally activated genes were divided into early, middle, and late depending on the timing of their expression. The three genes—*brlA*, *abaA*, and *wetA*—define a linear dependent pathway in which the activation of *brlA* is sufficient to initiate a cascade of events that involve other genes (Adams, 1995). By examining patterns of RNA accumulation in mutant strains, the developmentally activated genes were divided into four categories (Timberlake and Marshall, 1988; Mirabito *et al.*, 1989): Class A genes are activated by either *brlA* or *abaA* or both, independently of *wetA* (Figure 11.4); *wetA* activates Class B genes, independent of *brlA* and *abaA*; the *brlA*, *abaA*, and *wetA* together activate Class C and Class D genes. The accumulation of *wetA* mRNA requires *wetA*+ activity, suggesting that *wetA* is autogenously regulated (Marshall and Timberlake, 1991).

11.6 MASTER REGULATORY GENES

In *A. nidulans* a single-gene mutant initiates conidiophore development earlier than the wild type from which it was derived in the absence of an environmental signal. The mutant supports the idea that one or a few genes are master regulators, switching on the conidiation pathway (Figure 11.4). A homologue is the *FluG* involved in the production of conidiophore that controls six genes: *fluG*,

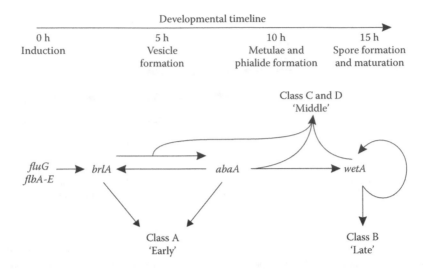

Figure 11.4   Interactions among the regulatory genes during conidiophore development in *Aspergillus nidulans*. (Based on Adams, T.H., Asexual sporulation in higher fungi, in N.A.R. Gow and G.M. Gadd, eds., *The Growing Fungus*, Chapman and Hall, London, 1995.)

flbA, *flbB*, *flbC*, *flbD*, and *flbE* (Seo *et al.*, 2006). Interestingly the precocious sporulation that occurs without development of a conspicuous mycelium in the absence of environmental signals (nitrogen starvation) in this mutant triggers the switch to conidiophore development as a germ tube in microcycle conidiation, bypassing the normal vegetative phase (mycelium development). The *Fluffy* (*fl*) gene of *Neurospora crassa* is required for asexual sporulation and encodes an 88-kDa polypeptide containing a typical DNA-binding motif, suggesting that it is a major regulator gene. Increased expression of *fl* from a heterologous promoter is sufficient to induce expression of hydrophobin protein typically present in the cell walls of reproductive structures of fungi in the absence of ACON2 or nitrogen starvation. This suggests that a *fluffy* or a homologue is a master regulator gene that controls a subset of genes in the sequence of development of mature conidiophore.

11.7 MICROCYCLE CONIDIATION

A type of conidiation in which a spore directly produces a conidiophore almost bypassing the vegetative mycelium is indicative of a small set of master elements that control several genes required for the differentiation of functional conidiophore. In *Aspergillus niger* and some other species of fungi, microcycle conidiation can be induced under certain regimes of nutrition and temperature (Figure 11.5). Apparently, in this type of development, the conidiation genes are activated and expressed precociously. The idea that a small number of genes act as master switches of conidiophore development is strengthened by the discovery of a phenomenon called *microcycle conidiation*, *precocious conidiation*, or *paedogenetic conidiation* in some species of fungi (Figure 11.6). Microcycle conidiation could be a common feature of fungi growing in nature, allowing the formation of asexual propagule in the shortest possible time.

11.8 CONCLUDING REMARKS

The conidiophore is a device for continuous production and dissemination of asexual spores for aerial dissemination. The production of conidiophore occurs after a prior period of vegetative

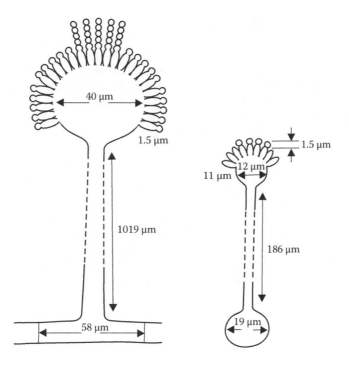

Figure 11. 5   Diagram of microcycle conidiation in *Aspergillus niger*. The mycelial phase was bypassed when conidia were subjected to a regime of heat shock. (After Anderson, J.G. and Smith, J.E., *J. Gen. Microbiol.* 69:185–197, 1971.)

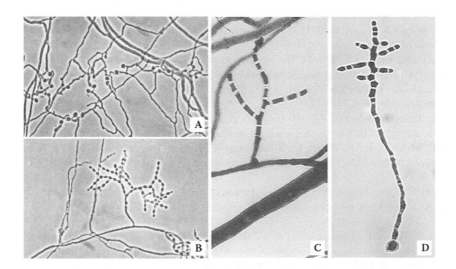

Figure 11.6   Microcyclic condiation in a strain of *Neurospora crassa*. (A) Fusion of conidial germ tubes and formation of mycelium. (B) Conidiophore in surface-grown mycelium bearing macroconidia formed by budding. (From Maheshwari, R., *J. Gen. Microbiol.* 137, 2103–2116, 1991, with permission from Society of General Microbiology.) (C) A conidiophore in surface-grown mycelium producing arthroconidia by septation. (D) Microcyclic conidiation in the same strain grown in submerged shake culture.

growth that prepares the mycelium until it is ready to respond to external stimuli and form conidia. The amount of vegetative growth required before a colony can produce conidiophores is relatively short—about 20 h, or even less—during which time an erstwhile dormant conidia directly germinates to form a conidiophore. It appears that the period just prior to production of conidiophore is most critical, during which the endogenous macromolecular substrates RNA and protein in mycelium are broken down and, simultaneously, new precursors (notably for spore wall polysaccharides and hydrophobin protein) are synthesized, as well as spore-specific carbohydrates, lipids, and compounds that will give protection and serve as an energy source during dissemination and until it can germinate to produce its absorptive mycelium. The numbers of conidiation-specific genes are in the hundreds, not thousands as previously supposed. Genes are being identified that determine how the development of a hyphal cell in the hyphae growing on the surface is switched into a stalk of determined length that grows into air, and how an orderly expression of cells occurs for maximizing the number of conidia, packing these into a small space for dissemination by insects, wind, or rain. Conidiogenesis offers an opportunity to probe how nuclear division and cell wall formation are coupled such that each conidium is uninucleate. Remarkably, only a few regulatory genes control the production of conidiophore in time and space. Although nutrient starvation had long been thought to be the signal for asexual spore formation, the molecular data suggest that this alone is insufficient.

REFERENCES

Adams, T. H., Boylan, M. T., and Timberlake, W. E. (1990). *brlA* is necessary and sufficient to direct conidiophore development in *Aspergillus nidulans*. *Cell* 54:353–362.

Adams, T. H. and Timberlake, W. E. (1990). Developmental repression of growth and gene expression in *Aspergillus*. *Proc. Natl. Acad. Sci. USA* 87:5405–5409.

Adams, T. H. and Timberlake, W. E. (1990). Upstream elements repress premature expression of an *Aspergillus* developmental regulatory gene. *Mol. Cell. Biol.* 10: 4912–4919.

Adams, T. H. (1995). Asexual sporulation in higher fungi. In N. A. R. Gow and G. M. Gadd, eds., *The Growing Fungus*. Chapman and Hall, London.

Adams, T. H., Wieser, K. K., and Yu, J.-H. (1998). Asexual sporulation in *Aspergillus nidulans*. *Microbiol. Mol. Biol. Revs.* 62:35–54.

Anderson, J. G. and Smith, J. E. (1971). The production of conidiophores and conidia by newly germinated conidia of *Aspergillus niger* (microcyclic conidiation). *J. Gen. Microbiol.* 69:185–197.

Axelrod, D. E., Gealt, M., and Pastushok, M. (1973). Gene control of developmental competence in *Aspergillus nidulans*. *Develop. Biol.* 34:9–15.

Bailey-Shrode, L. and Ebbole, D. J. (2004). The *fluffy* gene of *Neurospora crassa* is necessary and sufficient to induce conidiophore development. *Genetics* 166:1741–1749.

Boylan, M. T., Mirabito, P. M., Willett, C., Zimmerman, C. R., and Timberlake, W. E. (1987). Isolation and physical characterization of three essential conidiation genes from *Aspergillus nidulans*. *Mol. Cell. Biol.* 7:3113–3118.

Burnett, J. H. (1976). *Fundamentals of Mycology*. Edward Arnold.

Clutterbuck, A. J. (1969). A mutational analysis of conidial development in *Aspergillus nidulans*. *Genetics* 63:317–327.

Galagan, J. E., Calvo, S. E., Cuomo, C., *et al.* (2005). Sequencing of *Aspergillus nidulans* and comparative analysis with *A. fumigatus* and *A. oryzae*. *Nature* 438:1105–1115.

Gwynne, D. I., Miller, B. L., Miller, K. Y., and Timberlake, W. E. (1984). Structure and regulated expression of the SpoCl gene cluster in *Aspergillus nidulans*. *J. Mol. Biol.* 180:91–109.

Maheshwari, R. (1991). Microcycle conidiation and its genetic basis in *Neurospora crassa*. *J. Gen. Microbiol.* 137:2103–2116.

Marshall, S. D. and Timberlake, W. E. (1991). *Aspergillus nidulans wetA* activates spore-specific gene expression. *Mol. Cell Biol.* 11:55–62.

Martinelli, S. D. and Clutterbuck, A. J. (1971). A quantitative survey of conidiation mutants in *Aspergillus nidulans*. *J. Gen. Microbiol.* 69:261–268.

Mirabito, P. M., Adams, T. H., and Timberlake, W. E. (1989). Interactions of three sequentially expressed genes control temporal and spatial specificity in *Aspergillus* development. *Cell* 57:859–868.

Mooney, J. L. and Yager, L. N. (1990). Light is required for conidiation in *Aspergillus nidulans*. *Genes Develop.* 4:1473–1480.

Moore, D. (1998). *Fungal Morphogenesis*. Cambridge University Press, Cambridge.

Orr, W. C. and Timberlake, W. E. (1982). Clustering of spore-specific genes in *Aspergillus nidulans*. *Proc. Natl. Acad. Sci. USA* 79:5976–5980.

Rerngsamran, P., Murphy, M. B., Doyle, S. A., and Ebbole, D. J. (2005). *Fluffy*, the major regulator of conidiation in *Neurospora crassa*, directly activates a developmentally regulated hydrophobin gene. *Mol. Microbiol.* 56:282–297.

Seo, J.-A., Guan, Y., and Yu, J.-H., (2006). FluG-dependent asexual development in *Aspergillus nidulans* occurs via derepression. *Genetics* 172:1535–1544.

Timberlake, W.E. (1980). Developmental gene regulation in *Aspergillus* nidulans. *Dev. Biol.* 78:497–510.

Timberlake, W. E. and Marshall, M. A. (1988). Genetic regulation of development of *Aspergillus nidulans*. *Trends Genet.* 4:162–169.

Yager, L. N., Kurtz, M. B., Champe, S. P. (1982).Temperature-shift analysis of conidial development in *Aspergillus nidulans*. *Dev Biol.* 93:92–103.

CHAPTER 12

Ustilago maydis and Other Fungi as Models of Sexual Reproduction

The fungi are remarkable for the diversity of their sexual processes and involve the essential features of nuclear fusion and meiosis which follow each other closely, usually in that order. Superimposed upon these basic events are a variety of regulating mechanisms and range of morphological developments, the effects of which are to determine with more or less precision the kinds of nuclei that will fuse at fertilization.

—J. H. Burnett (1976)

What is mating good for? After the obligatory round of facetious answers, we are still left with a dilemma: Some organisms mate regularly as part of their life cycles, but others seem to be able to propagate themselves without limit by vegetative growth. Obviously, since both forms of reproduction exist, there must be advantages to each of them.

—Robert L. Metzenberg and N. Louise Glass (1990)

12.1 HETEROTHALLISM VS. HOMOTHALLISM

One of the fungi more commonly encountered on damp bread or <u>rotten</u> vegetable matter or freshly deposited dung belongs to the genus *Mucor*. Its colonies are white and fast growing but become grey to brown or black due to the development of erect hyphal sporangiophore in which the tip swells to form a globose sporangium containing uninucleate, haploid sporangiospores. During sexual reproduction, two compatible strains form short, specialized hyphae called gametangia. At the point where two complementary gametangia contact, they fuse to form thick-walled, black zygospores. Nuclear <u>karyogamy</u> and <u>meiosis</u> (sexual recombination) occur within the zygospores.

Imagine Albert F. Blakeslee who, while collecting cultures of zygospore-forming fungi for his PhD research for developing a method of their identification and classification, noted that *single* individual cultures sent to him had black zygospores in them. He wrote (1907):

Inoculations were at once made into sterilized nutrient and a large number of the young zygospores resulting were isolated and laid on cleared nutrient agar in Petri dishes. Paired cultures X & XX and Y & YY were obtained by making mycelial transfers from the outgrowths from the suspensors of two zygospores both of whose suspensors had germinated. On the assumption that the species is heterothallic, each pair should contain the two opposite strains, and each strain when grown alone should be incapable of zygospore formation. In addition, cultures Z & ZZ were obtained from suspensors that could not be traced with certainty to the same zygospore, and cultures A–D were obtained from the germination of single isolated sporangiospores. By opposing inoculations from these cultures against one another and against standard (+) and (–) test strains, it was seen that X and Y are (–) and XX, YY,

Z, ZZ, and A–D are (+). The contrasts numbered over 40 and were made for the most part on sterilized flour paste in slender dishes, but were controlled by cultures on nutrient agar in test tubes. None
of the strains produced zygospores on substrata favorable to zygospore formation when sown alone in
pure cultures, nor when sown together with the same strain from a different source. They have always
formed them, however, whenever the opposite strains are grown.

Whether or not it may ever be found that a homothallic race may occur in a species normally heterothallic (perhaps not an impossible condition in view of the writer's having obtained a homothallic
mycelium of the heterothallic Phycomyces), the evidence at hand leads one to the conclusion that the
large majority of the Mucorineae are heterothallic.

Figure 12.1 illustrates heterothallism. The + and the − cultures are morphologically indistinguishable; the term *heterothallic* is used to denote the physiological difference between the + and
the − strains.

Conversely, the term *homothallic* is used when a single individual that originates from a single spore forms morphologically similar male and female gametangia and is able to complete a
sexual cycle, i.e., it is self-fertile and sexually self-sufficient. For example, in the homothallic fungus *Rhizopus sexualis* (Zygomycotina) neighboring hyphal branches of the same individual form
gametangia that fuse (Figure 12.2). Here, sexual fusion occurs between cells containing genetically
identical (sister) nuclei. Although they do not seek a mate, the homothallic fungi still engage in
sexual reproduction, retaining karyogamy and meiosis. From Blakelee's observations, the concept
arose that genes control the mating types. To indicate that the differences between the strains are due
to two alleles of a single gene, the mating types of strains are designated as a/α in *Saccharomyces
cerevisiae* (Chapter 9) or *A/a* as in *Neurospora crassa* (Chapter 10).

The mating type of a strain is specified by the unique DNA sequence in the mating-type region
or the MAT locus. Classical mutation technique yielded a number of strains of *Neurospora crassa*
that failed to participate in sexual cycle events. From libraries of *N. crassa*, DNA sequences were
isolated that restored the ability of these sterile mutants to form perithecia (Vollmer and Yanofsky,
1986). The two mating-type loci *MAT1-1* and *MAT1-2* occupy the same chromosomal location but
lack sequence similarity and are thus termed *A* or *a* idiomorphs rather than alleles (Metzenberg
and Glass, 1990). The analyses of the DNA sequence of the MAT locus showed that these encode
homeodomain protein. Originally identified in ascomycete fungus, homologues are present in zygomycetes. In a cross of (+) and (−) strains of *Phycomyces blakesleeanus*, 33 progeny were (+) and 33

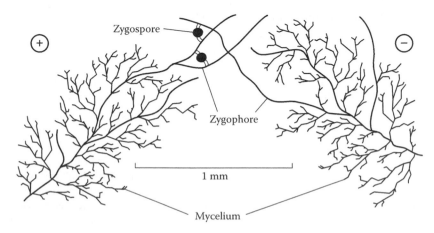

Figure 12.1   Sexual reproduction in heterothallic *Mucor hiemalis* showing mating of morphologically similar
cells from two individuals (designated plus and minus) to form zygospore. (From Ingold, C.T., *The
Biology of Fungi,* Chapman and Hall, London, 1961, with kind permission of Springer Science
and Business Media.)

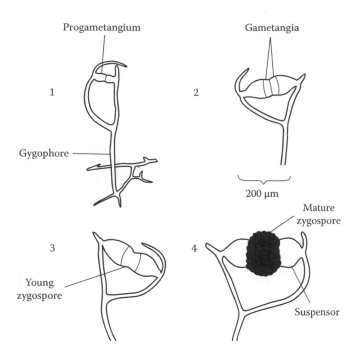

Figure 12.2   Sexual reproduction in homothallic *Rhizopus sexualis*. (1) Mating of a pair of morphologically similar cells or gametangia formed by same mycelium. (2) Delimited gametangia. (3) Fusing gametangia. (4) Mature zygospore.

were (–). The 1:1 segregation showed that sex difference is due to a single gene with + and – alleles (bipolar mating system). The sex gene + (or, sex P for parental) or sex gene – (or, sex M for maternal) co-segregate with HMG (high-mobility group) domain protein (Figure 12.3).

12.2   CELL–CELL RECOGNITION

12.2.1   Mating Types

The fungi grouped in Basidiomycotina are remarkable for having thousands of potential mating partners or mating types. How do these fungi select a mating partner in the absence of any morphological differentiation? They are remarkable too in having an extended phase of life cycle in which the nuclei of opposite mating types divide synchronously and remain in close proximity as $(n + n)$ pair by a special type of hyphal growth called *clamp connection* in which a short hyphal branch grows backwards as a hook that fuses with the penultimate cell, with the nucleus passing through it such that each binucleate compartment in the hypha acquires a nucleus of the opposite mating type (see Chapter 2, Figure 2.7). The nuclei of opposite mating types in each hyphal compartment remain associated for an extended period of time without fusing; a diploid (2n) phase is postponed in favor of an extended dikaryophase $(n + n)$, the significance of which is a mystery. Coming together of compatible mating types results in the formation of a dikaryophase, not a 2n zygote.

Among the fungi in this group that have been used as models to investigate the mate recognition process are the mushrooms *Coprinus cinereus* and *Schizophyllum commune* and the corn smut fungus, scientifically known as *Ustilago maydis* (Casselton, 2002). Here, we shall highlight the efforts aimed at understanding sexual reproduction in *U. maydis*. This fungus is of interest for other

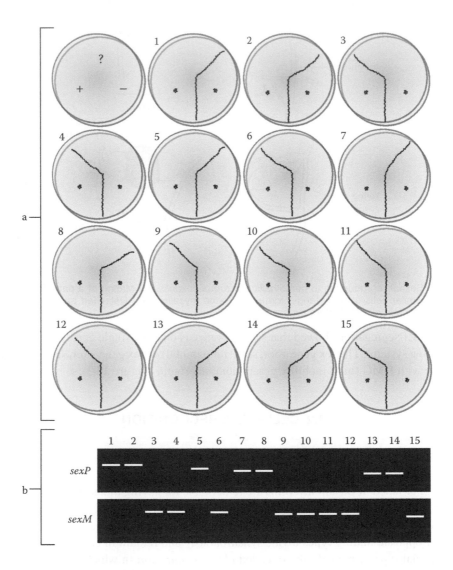

Figure 12.3 Diagram of method used for identification of sex genes in *Phycomyces blakesleeanus* (Zygomycotina). (a) Interaction of 15 zygospore progeny from (+) × (−) cross co-inoculated on medium against (+) and (−) tester strains. Black lines of zygospores indicate mating. (b) Genomic DNA from progeny probed with PCR amplified DNA from sex M and sex P genes encoding HMG-domain proteins. (Data from Idnurm, A., Walton, F.J., Floyd, A., and Heitman, J., *Nature* 451, 193–197, 2007.)

reasons as well: In this fungus, dikaryophase (sexual) development and pathogenicity are interconnected (Brachmann *et al.*, 2003). Moreover, the fungus exhibits the phenomenon of dimorphism: It switches from a yeast-like saprophytic form to a filamentous, pathogenic growth. Regine Kahmann's group in Germany is attempting to explain mating and fusion of conjugant cells (sporidia) in biochemical terms and the switch of the dikaryotic mycelium into a pathogenic mycelium.

U. maydis causes abnormal growth of kernels or leaves into galls or tumors that are filled with countless numbers of brownish-black spores. These spores are formed by the separation of binucleate hyphal cells and are called *chlamydospores*. The chlamydospores (Figure 12.4)—functionally analogous to teliospores of rust fungi in which the two haploid nuclei fuse (see the Appendix)—are thick-walled dormant structures that can survive in crop refuse and in soil for many years. The spore, called a chlamydospore but appropriately a teliospore, germinates to form a short tube that becomes divided by septa, a distinguishing structure of fungi belonging to the order Ustilaginales in phylum Basidiomycotina of the kingdom Eumycota. This structure is called a *basidium* in which the diploid nucleus undergoes meiosis and produces four haploid nuclei that are abstricted as basidiospores (also called sporidia). The sporidia can be grown on an artificial medium on which they multiply by yeast-like budding. The characteristics of the colonies formed from the sporidia from a single basidium differ in color, topography (smooth or rough colonies), and mating type, demonstrating that they are recombinant progeny.

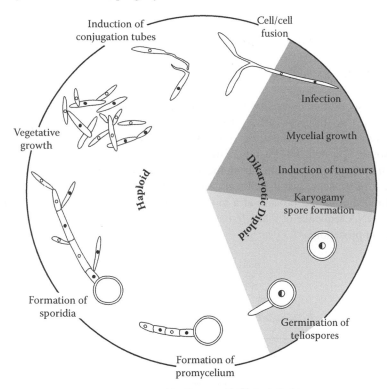

Figure 12.4 Diagram of life cycle of *Ustilago maydis*. The haploid (n) sporidia multiply to form secondary sporidia by budding. Sporidia of compatible mating type fuse to form a dikaryotic mycelium (n + n) that infects a corn plant, the kernels of which are filled with a powdery mass of spores (chlamydospores). Nuclear fusion occurs in chlamydospores (teliospores), which are 7–10 μm in diameter. These germinate to form a tube-like basidium in which the diploid (2n) nucleus divides by meiosis. The four haploid nuclei are abstricted in four cells (sporidia) that can grow saprophytically on refuse and multiply by budding. (From Kämper, J., Bölker, M., and Kahmann, R., *The Mycota*, vol. 1, Springer-Verlag, Berlin, 1994, pp. 323–332, with kind permission of Springer Science and Business Media.)

In the first step in sexual reproduction, the cells must distinguish self from non-self and choose a potential mating partner in the environment. The partners must coordinate their choices by making contact, but is the contact based on chance or physiological mechanism? The sporidia secrete small polypeptides called *pheromones* that are sensed by sporidia of the opposite mating type in the vicinity, reorient their growth, and fuse via conjugation tube. An example of visible cell–cell signaling in a fungus was given in Figure 10.3.

12.2.2 Assay for Mating Compatibility

A simple test of mating compatibility in *U. maydis* is to co-spot basidiospores sporidia on nutrient agar medium containing charcoal where the fused sporidia show filamentous growth (the "fuz" reaction). However, the sustained growth of the hypha can occur only as a parasite in the corn plant (*Zea mays*). Only the haploid cells (sporidia) that differ at two mating type loci can mate, resulting in hypha (dikaryon) in which the two nuclei divide several times but maintain their nuclear identity. Fertilization is postponed in favor of dikaryosis. This mycelium is a special type of heterokaryon in which each cellular compartment contains two nuclei of opposite mating types in close proximity without fusion. The dikaryotic hypha recognizes and enters into the corn plant, which responds to infection by forming galls (tumors) that are filled with chlamydospores (teliospores).

12.3 EXTRACELLULAR RECOGNITION

12.3.1 The *a* Locus

The fusion of sporidia is controlled by two unlinked mating type loci, designated as *a* and *b*. Compatible mates have different alleles of genes at both loci. The *a* mating type locus has two "alleles," *a1* and *a2*, which are structurally quite dissimilar even though they are at the same position in the chromosome. This unusual situation of dissimilar mating type genes being present at the same locus in a chromosome is distinguished from the term *allele* by the term *idiomorph*. The idiomorphs are responsible for extracellular cell–cell recognition and the fusion of sporidia (Figure 12.5).

To understand the mating process at a molecular level, Regine Kahmann's group cloned and determined the DNA sequence of *a* and *b* mating type loci and performed functional assays based on transformation procedure.

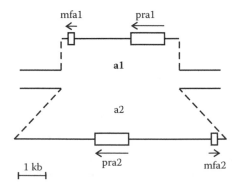

Figure 12.5 Schematic representation of the *a* idiomorphs. The *a* alleles have little sequence homology but occupy the same chromosomal location. *mfa*, pheromone precursor gene; *pra*, receptor gene. Arrows indicate the direction of transcription. (From Kämper, J., Bölker, M., and Kahmann, R., *The Mycota*, vol. 1, Springer-Verlag, Berlin, 1994, pp. 323–332, with kind permission of Springer Science and Business Media.)

12.3.2 Pheromone and Receptor

The cloning strategy of the *a* locus was based on the observation that when both *a1* and *a2* alleles are introduced by transformation in the same cell, *U. maydis* behaves as a "double mater," i.e., it can mate with strains of either *a* mating type (Figure 12.5 and Figure 12.6). The nucleotide sequence of *a* loci revealed that they encode small polypeptides with a carboxy-terminus sequence of Cys-*AAX*, where *A* is an aliphatic amino acid and *X* any amino acid with a farnesyl group attached to cysteine—a motif that is characteristic of pheromones (Figure 12.7). Moreover, the nucleotide sequencing data disclosed two genes, called *pra1* and *pra2*, that encode proteins having seven hydrophobic, potential membrane-spanning regions, which is a characteristic feature of the receptors found in membrane. From this it is inferred that the *a* locus encodes a pheromone polypeptide and a receptor; it is the primary determinant of cell–cell recognition. Mating of *a1* and *a2* compatible cells occurs if the *mfa1* receptor recognizes the *a2* pheromone and the *mfa2* receptor recognizes the *a1* pheromone. Haploid strains of the same *a1* mating type fused if both were transformed with the receptor gene (*pra2*), provided they contained a different *b* allele.

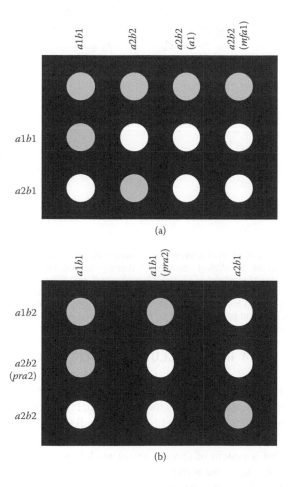

Figure 12.6 Diagram of test for mating reaction in *Ustilago maydis* on charcoal-containing medium. Cells were co-spotted as indicated on the top and on the left. Genes introduced by transformation are in parentheses. (a) Double mater phenotype of *a2b2* transformed with *a1*. (b) *a1b1* and *a1b2* were transformed with receptor gene *pra2*. Compatible cells produce white-filament (Fuz⁺ phenotype). (Based on Bölker, M., Urban, M., and Kahmann, R., *Cell* 68, 441–450, 1992, with permission from Elsevier.)

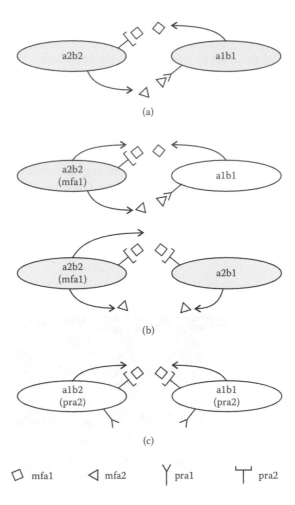

(a)

(b)

(c)

□ mfa1   ◁ mfa2   Y pra1   ┬ pra2

Figure 12.7 Schematic representation of interactions between pheromones and receptor coded by the *a* locus of *Ustilago maydis*. (a) Mating interaction between compatible cells differing at both *a* and *b* loci. (b) Double mater phenotype of *a2* strain transformed with the pheromone gene from the opposite mating type; mating occurs between haploid strains of both *a* mating types. (c) Strains of the same *a1* mating types can fuse if both are transformed with the receptor gene (*pra2*) from the opposite mating type. *mfa* and *pra* introduced by transformation are thickened. (Reprinted from Bölker, M., Urban, M., and Kahmann, R., *Cell* 68, 441–450, 1992, with permission from Elsevier.)

12.4 INTRACELLULAR RECOGNITION

After recognition of mating partners through a system based on pheromones, further sexual development occurs through a multiallelic *b* locus which determines whether a filamentous, i.e., a pathogenic, dikaryon is formed or not. For fusion of sporidia to occur, the nuclei must have two different *b* alleles (Table 12.1). Thirty-three different *b* alleles (*bE1* to 33 and *bW1* to 33) have been identified. This means that a system of recognition of nuclei exists by which any combination of $33^2–33$ or 1056 different *b* alleles are active, and 33 combinations of the same allele are inactive (Kämper *et al.*, 1995). The clue to this system emerged from the cloning of the *b* locus. The cloning strategy for the *b* locus is based on the observation that the diploid *a1a2 b1b2* strain develops fuzziness (limited aerial hyphae) on artificial medium containing charcoal. The introduction by transformation of *b* allele in a diploid cell that is heterozygous for *a* but homozygous for *b* gives the fuz, indicative of a mating reaction. The fuz will induce tumors if introduced in corn plant; therefore

Table 12.1   Mating Reaction of Sporidia of *Ustilago maydis*

Genotype	a1b1	a1b2	a1b3	a2b1	a2b2	a2b3
a1b1	–	–	–	–	+	+
a1b2	–	–	–	+	–	+
a1b3	–	–	–	+	+	–
a2b1	–	–	–	–	–	–
a2b2	+	–	+	–	–	–
a2b3	+	+	–	–	–	–

Note: + = Successful mating reaction results in formation of a dikaryotic mycelium (fuz reaction). The *a* locus is biallelic. The *b* locus is multiallelic; however, only three are shown.

mfa1 MLSIFAQTTQTSASEPQQSPTAPQGRDNGSPIGYSS<u>CVVA</u>
mfa2 MLSIFETVAAAAPVTAETQQASNNENRGOPGYY<u>CLIA</u>

Figure 12.8   *Ustilago maydis.* Polypeptide sequences encoded by *mfa1* and *mfa2*. Amino acids are shown by one-letter symbols. The Cys-*AAX* motif at the C-terminal end is underlined. (After Bölker, M., Urban, M., and Kahmann, R., *Cell* 68, 441–450, 1992.)

the reaction is a valid test of successful mating. The fuz reaction allows the cloning of the *b* locus and determination of its nucleotide sequence.

The *b* locus contains a pair of divergently transcribed genes, *bE* (east) and *bW* (west), separated by a spacer region (Kämper *et al.*, 1995). From the DNA sequence information, *b* polypeptides were predicted to have a constant and a variable domain and a homeodomain motif. Homeodomain proteins, first identified as products of *Drosophila* genes with an important role in embryonic development, bind to DNA and function as transcription factors regulating development. A yeast two-hybrid system (Chapter 9) was used to demonstrate that bE and bW proteins can associate into a dimer, but only if they are derived from different alleles. It is postulated that only the pairwise combinations of specific *bE* and *bW* that originate from different alleles of the locus interact through complementary hydrophobic sequences to form heterodimers that bind to promoter regions acting as transcription factors (Kämper *et al.*, 1995; Kahmann and Bölker, 1996). There is no interaction between proteins if they are encoded by the same mating type: Self/non-self recognition is at the level of protein dimerization (Figure 12.9). Thus, no single strain is self-fertile. It is expected that use of microarray will allow identification of the regulatory DNA sequences to which one *bE* and one *bW* heterodimer bind, what these transcripts code for, and how the cascades of signal transduction lead to pathogenic growth in the corn plant leading to the fusion of the nuclei in the dikaryon and the completion of the life cycle.

Sequencing the genome of *U. maydis* (20.5 Mb) revealed 12 clusters of genes in small gene families that encode small secreted proteins (Kämper *et al.*, 2006). Expression analysis showed that the clusters are regulated together and induced in infected tissue. Deletion of individual clusters altered the virulence of *U. maydis* in five cases, ranging from a complete lack of symptoms to hypervirulence; however, no "true" virulence factors were found. Secreted novel protein effectors are essential for fungal proliferation inside the plant host. It is thought that these novel effector proteins are translocated into plant cells for biotrophic mode of life. The minimal set of hydrolytic enzymes found in *U. maydis* is in line with its biotrophic lifestyle, in which damage to the host is minimized.

12.5 GENETIC ENGINEERING FOR SEX

The fungus *Trichoderma* is common in soils in both temperate and tropical regions and has been in the limelight because of its potentialities in biotechnology, especially the species *T. reesei* QM6a,

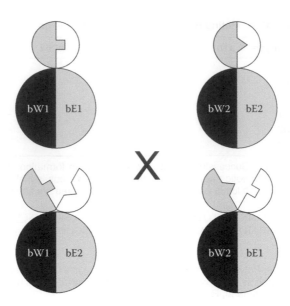

Figure 12.9 *Ustilago maydis.* Schematic model proposed for allele-specific dimerization. The interaction of bW and bE polypeptides from the same b allele produces inactive heterodimers because the DNA-binding domains are buried. Upon mating, two additional new types of active heterodimers are formed in which DNA-binding motifs are exposed. (After Gillissen, B., Bergemann, J., Sandmann, C., Schroeer, B., Bölker, M., and Kahmann, R., *Cell* 68, 647–657, 1992.)

which was derived from a strain that was causing rot of army tents and clothing in the Solomon Islands during World War II. The strain closely resembles the ascomycete fungus *Hypocrea jecorina* (teleomorph), which has an asexual stage (anamorph). This strain had not produced a sexual stage ever since it was discovered. The genome sequence revealed that QM6a has a *MAT1-2* mating type locus that encodes an HMG-domain protein (Seidl *et al.*, 2009). Could a partner with opposite *MAT-1* mating type be found? A worldwide collection of *H. jecorina* was searched for strains with *MAT1-1* and mating experiments set up, taking into account that the fungus has a requirement for light for sexual reproduction. Fruiting bodies containing ascospores were formed between the oppositely paired cultures (Figure 12.10). To confirm that sexual recombination had occurred, a *T. reesei* mutant strain containing a single copy of hygromycin-phosphotransferase gene (*hph*) was used in

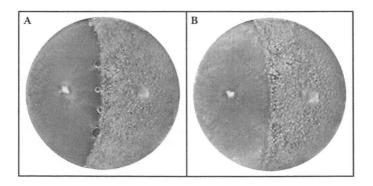

Figure 12.10 Mating in *Trichoderma* (A) *H. jecorina* MAT1-1 was mated to *Trichoderma viride* QM6a on agar plates, and the formation of fruiting body containing ascospores was observed at the interaction zone of the fungal colonies. (B) No stromata were formed upon mating of QM6a with MAT1-2 strains. (From Schmoll, M., *PNAS* 106, 13909–13914, 2009, with permission.)

crossing experiments. Approximately 50% of progeny were resistant to hygromycin B, confirming that mating between two strains had indeed occurred.

12.6 CONCLUDING REMARKS

In the early 1900s, sex in animals was thought to exist only between morphologically distinguishable individuals; A. F. Blakeslee's discovery of mating between morphologically similar individuals was therefore stupefying. In fungi, sexual interaction between two partners is governed by bipolar or tetrapolar mating-type systems. In bipolar mating a *MAT* chromosomal locus has two quite dissimilar mating type "alleles," instead called *idiotypes*. They encode proteins with homeodomains that dimerize into a heterodimeric transcriptional factor. In tetrapolar sex, two different chromosomal loci govern mating between two individual thalli; the necessary condition for a successful mating is that the two conjugant partners must be complementary at both loci. Currently, the structure of mating-type loci in different fungi is receiving attention to trace the steps in the evolution from sex-specific mating-type loci to sex-specific chromosomes of animals and some plants. The *U. maydis* system has provided some insights into the first step in sexual reproduction, i.e., the detection of a potential mating partner in the environment (mostly soil). It is hoped that fungal genome sequencing projects will provide evolutionary insights into structures and functions of mating-type locus. Using primers based on similarities to unique sex (mating type) locus encoding HMG-domain proteins, cryptic sex genes have been discovered in some asexual fungi and have even led to experimental induction of sexual reproduction in fungi where this was not known to occur previously, opening up the possibilities of strain improvement by breeding. *Ustilago maydis* reveals that a function of sexual reproduction is to reprogram development from a saprophytic to a parasitic mode of life.

REFERENCES

Blakeslee, A. F. (1907). Heterothallism in bread mold, *Rhizopus nigricans*. *Bot. Gaz.* 43:415–418.

Bölker, M., Urban, M., and Kahmann, R. (1992). The *a* mating type locus of *U. maydis* specifies cell signaling components. *Cell* 68:441–450.

Brachmann, A., Schirawski, J., Müller, P., and Kahmann, R. (2003). An unusual MAP kinase is required for efficient penetration of the plant surface by *Ustilago maydis*. *EMBO J.* 22:2199–2210.

Casselton, L. A. (2002). Mate recognition in fungi. *Heredity* 88:142–147.

Gillissen, B., Bergemann, J., Sandmann, C., Schroeer, B., Bölker, M., and Kahmann, R. (1992). A two-component regulatory system for self/non-self recognition in *Ustilago maydis*. *Cell* 68:647–657.

Glass, N. L., Vollmer, S. J., Staben, C., Grotelueschen, J. N., Metzenberg, R. L., and Yanofsky, C. (1988). DNAs of the two mating-type alleles of *Neurospora crassa* are highly dissimilar. *Science* 241:570–573.

Heitman, J., Kronstad, J. W., Taylor, J. W., and Casselton, L., eds. (2007). *Sex in Fungi: Molecular Determination and Evolutionary Implications*. ASM Press, Washington.

Idnurm, A., Walton, F. J., Floyd, A., and Heitman, J. (2008). Identification of the *sex* genes in an early diverged fungus. *Nature* 451:193–197.

Ingold, C. T. and Hudson, H. J. (1993). *The Biology of Fungi*. Chapman and Hall, London.

Kahmann, R. and Bölker, M. (1996). Self/nonself recognition in fungi: Old mysteries and simple solutions. *Cell* 85:145–148.

Kämper, J., Bölker, M., and Kahmann, R. (1994). Mating-type genes in heterobasidiomycetes. In *The Mycota*, vol. 1. Springer-Verlag, Berlin, pp. 323–332.

Kämper, J., Kahmann, R., Bölker, M., *et al.* (2006). Insights from the genome of the biotrophic fungal plant pathogen *Ustilago maydis*. *Nature* 444:97–101.

Kämper, J., Reichmann, M., Romeis, T., Bölker, M., and Kahmann, R. (1995). Multiallelic recognition: Nonself-dependent dimerization of the bE and bW homeodomain proteins in *Ustilago maydis*. *Cell* 81:73–83.

Kronstad, J. W. and Staben, C. (1997). Mating type in filamentous fungi. *Annu. Rev. Genet.* 31:245–276.

Metzenberg, R. L. and Glass, N. L. (2000). Mating type and mating strategies in *Neurospora*. *Bioessays* 12: 53–59.

Seidl, V., Seibel, C., Kubicek, C. P., and Schmoll, M. (2009). Sexual development in the industrial workhorse *Trichoderma reesei*. *Proc. Natl. Acad. Sci. USA* 106:13909–13914.

Vollmer, S. J. and Yanofsky, C. (1986). Efficient cloning of genes of *Neurospora crassa*. *Proc. Natl. Acad. Sci. USA* 83:4869–4873.

Wahl, R., Zahiri, A., and Kämper, J. (2009). The *Ustilago maydis* b mating type locus controls hyphal proliferation and expression of secreted virulence factors in planta. *Mol. Microbiol.* 75:208–220.

Zheng, Y., Kief, J., Auffarth, K., Farfsing, J. W., Mahlert, M., Nieto, F., and Basse, C. W. (2008). The *Ustilago maydis* Cys2His2-type zinc finger transcription factor Mzr1 regulates fungal gene expression during the biotrophic growth stage. *Mol. Microbiol.* 68:1450–1470.

PART V

Adaptations

CHAPTER 13

Photoresponses and Circadian Rhythm

The environment is highly periodic with respect to many of its geophysical variables, such as light–dark or temperature cycles, and so there may be implications for Darwinian fitness for the organism to anticipate and respond to these monotonously regular environmental changes.

—Rachel Ben-Shlomo and Charalambos P. Kyriacou (2002)

My fungus shows no response to light.

—Alexander Idnurm, Surbhi Verma, Luis M. Corrochano (2010)

Because the mycelium penetrate into the substratum for absorbing nutrients, where little light penetrates, there has been a tendency to think that—unlike in plants—light is not essential for the growth and development of fungi. Indeed, even when fungi and green plants share a platform (as in university botany departments), it is understandable if in the past the fungi were discriminated against on this basis. The data in Table 13.1 show that just a brief flash of light is sufficient to evoke a development response. Analysis of genome sequences of *Neurospora crassa* and *Aspergillus nidulans* led to unexpected identification of homologues of plant photoresponse proteins—the phytochrome and cryptochrome—which sparked an interest in the role of light in the development and behavior of fungi. The increasing awareness that light regulates metabolic pathways in fungi (Tisch and Schmoll, 2010) should encourage investigators to report environmental conditions for greater reproducibility of results. Perhaps the lack of realization that light influences fungal biochemistry and physiology led to ignoring the need for controlling illumination, resulting in irreproducible results, and to the fungi being dubbed as "a mutable and treacherous tribe." Biosynthesis of penicillin in certain fungi is inhibited by light. Cellulase gene expression in *Trichoderma* is reported to be stimulated by light.

13.1 PHOTORESPONSE PHENOMENA

What should we study to know that light affects growth processes in the fungi? As a guide, some experimental findings are reviewed.

13.1.1 Pigmentation

One of the first experimental works on the effect of light on fungi was that of F. A. F. C. Went (father of plant physiologist F. W. Went, discoverer of the plant growth hormone auxin). A century ago, Went was working at the Bogor Botanic Gardens on the island of Java in Indonesia where he observed that *Oidium aurantiacus* growing on bread developed an orange color in the daylight but remained white when grown in darkness. This fungus was later renamed *Neurospora sitophila*. To determine the active wavelength of light that induced pigmentation in this fungus, Went used

Table 13.1 Effects of Light on Fungi

Fungus	Phenomenon	Threshold
Aspergillus nidulans	Induction of conidiation	8×10^{-4} mol m^{-2} 8×10^{-4} mol m^{-2}
Aspergillus ornatus	No conidia in continuous dark; cleistothecia and ascospores in continuous light	375–400 ft candle
Alternaria tomato	Inhibition of photoinduced conidiation	10^{-5} mol m^{-2}
Magnaporthe oryzae	Suppression of spore release	10^{-5} mol m^{-2}
Neurospora crassa	Induction of protoperithecia	4 J m^{-2}
	Circadian clock resetting	10^{-5} mol m^{-2}
Phycomyces blakesleeanus	Induction of sporangiophores	10^{-10} mol m^{-2} 10^{-8} W m^{-2} 10 J m^{-2}
Puccinia graminis var. tritici	Inhibition of urediospore germination	>200 ft-c
Trichoderma reesei	Cellulase gene expression	25 µmol m^{-2} s^{-1}

Note: Light energy is expressed as a fluence rate in W/m^{-2} for threshold obtained under continuous illumination or as a fluence in J/m^{-2} for threshold obtained after light exposure of a certain duration. The use of monochromatic light allows the measurement of the number of photons that have been applied and the determination of thresholds in mol m^{-2}.

Source: Data from Corrochano, L.M., *Photochem. Photobiol. Sci.* 6, 725–736, 2007, and others.

double-walled glass jars filled with different colored salt solutions that absorbed light of different wavelengths and acted as a light filter. He discovered that blue light was the active region of the spectrum that induced pigmentation. Other blue light (maximum at 440–450 nm) responses in *Neurospora* are the formation of conidia and protoperithecia, and phototropism of the perithecia neck.

13.1.2 Zonations

Cultures of several fungi on solid media show alternating zones of sporulation and nonsporulation (Figure 13.1). For example, cultures of *Sclerotinia fruticola* show zonations of decreasing intensity every 24 h after dark-grown cultures are exposed to light. In dark-grown cultures of *Fusarium discolor* var. *sulfureum*, *Trichothecium roseum*, and *Verticillium lateritium* (Fungi Anamorphici), daily illumination of 1000–3000 lux for a few seconds suffices to induce zonations. The rhythmic zonations can continue for some time after the fungus is transferred to complete darkness and is almost certainly due to the diurnal periodicity of alternating light and darkness. This rhythmic growth pattern is a manifestation of an endogenous clock—they have a period of approximately 24 h, and are independent of temperature. In some fungi, zonation is stimulated by supplementation of growth medium by factors present in yeast, malt, or potato extract (Jerebzoff, 1965), suggesting that they need excitation by certain chemical factors for the manifestation of rhythms. The endogenous rhythms in *Sclerotinia fruticola* (Ascomycotina) continues without attenuating for three weeks, for one month in *Aspergillus ochraceus* and *A. niger*, and for 70 days in *Alternaria tenuis* (Fungi Anamorphici).

13.1.3 Reproduction

More striking is the effect of light on differentiation of reproductive structures, demonstrating that light acts as a signal leading to production of spores for the survival of species. Colonies exposed to diurnal periodicity exhibit alternating zones of sporulating and nonsporulating hyphae. The brown-rot fungus *Monilia* spp. (Fungi Anamorphici) forms concentric rings of spores on fruits. *Phyllosticta* (Fungi Anamorphici) and other fungi form rings of flask-like fruiting body on leaves.

Figure 13.1 *Sclerotinia fruticola.* Growth rhythm in culture maintained in dark for two days were first exposed to 250 lux light for 12 h, and a second after five days. After each stimulus, four zonations of decreasing intensity occurred, one every 4 h. (From Jerebzoff, S., in G.C. Ainsworth and A.S. Sussman, eds., *The Fungi*, vol. 1, Academic Press, New York, chap. 27, pp. 625–645, 1965. With permission of Elsevier.)

The zones of spores are usually stimulated by light and may actually form during the subsequent dark periods. *Aspergillus ornatus* (Ascomycotina) reproduces asexually under light but switches to a sexual mode of reproduction under dark (Hill, 1976), hence nicknamed "the bashful fungus." Under illumination, the uptake of glucose by mycelium is markedly lowered, suggesting that conidiation may be triggered by starvation.

The soil fungus *Trichoderma* has emerged as a new model for investigation of mechanisms in photoresponses. In the dark the fungus grows as a nonsporulating mycelium. A brief pulse of light applied to the actively growing zone of the mycelium leads to the formation of dark green mature conidia as a ring at the edge of the colony when light was applied (Figure 13.2). The fungus is responsive to light (competent) only after 16 h of growth (Gressel and Galun, 1967). Three to seven hours after the induction, abundant branching of septated aerial hyphae is found, leading to conidiophore development (Galun and Gressel, 1966). Once triggered, this developmental program was suppressed using RNA synthesis inhibitors, such as 5-fluorouracil.

13.1.4 Periodicity of Spore Discharge

A frequently observed response to light or dark is the rhythmic discharge of spores into the environment. Spores of some fungi can be confidently identified by microscopy following impact on a sticky microscope slide. Sampling of air in agricultural fields in England at intervals showed maximum spores of *Phytophthora infestans* (Straminipila) in the early morning, of *Ustilago nuda* (Basidiomycotina) at forenoon, of *Cladosporium* (Fungi Anamorphici) around noon, and of *Sporobolomyces* (Fungi Anamorphici) during the night (Figure 13.3). In an infected apricot orchard in California, the maximum conidia of *Monilinia laxa* (Ascomycotina) occurred at night.

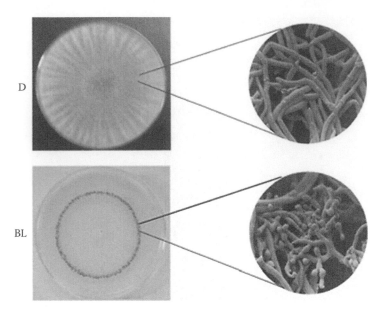

Figure 13.2   *Trichoderma atroviride.* Top, photographs of a colony growing in the dark. Bottom, a colony 36 h
after 5-min pulse of blue light, a ring of conidia developed at the colony perimeter at the time of
exposure. At the right are scanning electron micrographs of hyphae and conidiophores. (From
Schmoll, M., Esquivel-Naranjo, E.U., and Herrera-Estrella, A., *Fung. Genet. Biol.* 47, 909–916,
2010, with permission of Elsevier.)

Air sampling close to the ground of rust-infected wheat fields in Kansas showed a diurnal pattern
of liberation of urediospores of *Puccinia recondita* (Basidiomycotina). The ascospores of *Sordaria
fimicola* (Ascomycotina) are discharged mainly in day and of *Daldinia concentrica* (Ascomycotina)
during the night. These observations suggest that light influences maturation and discharge of
spores, although the influence of temperature, humidity, and wind velocity cannot be overlooked.
Fungi take advantage of the conditions of light, high humidity, and wind velocity to maximize the
dispersal and germination of their propagules. Other examples of seasonality and photoperiodism in
fungi are given by Ingold (1971) and by Roenneberg and Merrow (2003). Some examples considered
in Section 13.1.5 show that the intensity (brightness of illumination), the quality (wavelength of the
light), and the duration (relative lengths of the alternating periods of light and darkness) influence
fungal development.

13.1.5 Phototropic Curvature

13.1.5.1 Pilobolus spp.

A common fungus that grows on the dung of herbivorous animals is *Pilobolus* (Zygomycotina).
The dung contains nitrogen, vitamins, growth factors, and minerals, and satisfies the fungus's
unusual nutritional requirement of a chelated form of iron. The fungus can be grown in media
incorporating a decoction of dung or on a synthetic medium containing a complex iron-containing
compound called coprogen. Exposure to visible light (380 to 510 nm) triggers the formation of a
large bulbous cell called a *trophocyst* that is embedded in the substratum. The trophocyst elongates
into a stalk about 0.5–1 cm high, called a *sporangiophore*. The sporangiophore enlarges upward
into a crystal-clear subsporangial bulb, which is capped by a sporangium that contains asexual
spores. When dung is kept under a bell jar to provide damp atmosphere and light, in a few days it
becomes covered by turgid sporangiophores. The sporangiophore bends toward the light source

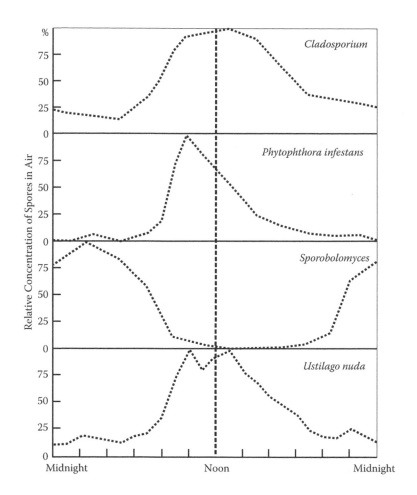

Figure 13.3 Diurnal variation in spore content in air in agricultural field in England. (Tracing from Ingold, C.T. and Hudson, H.J., *The Biology of Fungi,* Chapman and Hall, London, 1993.)

and the entire black sporangium containing the spores (Figure 13.4) is shot toward the light with a velocity of 5 to 10 m/s. The empty sporangiophore is thrown flat on the medium by the recoil (Page, 1962). Covering the bell jar with a black paper into which a small window has been cut demonstrates the precision with which the fungus can direct shooting of sporangium toward a light source. The released sporangia strike the inside of the bell jar in the area that receives illumination. This phototrophic response permits the sporangiophores to grow out of dung; the sporangia strike a blade of grass and adhere to it. When a grazing animal eats grass, the spores within each sporangium pass unharmed through its alimentary canal and are voided with the dung, wherein they germinate.

Klein (1948) subjected *Pilobolus* grown on dung-decoction agar to light–dark (LD) cycles. He found that although sporangium formation occurred predominantly at the end of a dark period, a dark period is essential to establish periodicity of growth and maturation of fruiting bodies. Periodicities other than those observed in nature could be established by artificial illumination. Among these were light–dark cycles (in hours) not only of 12–12 but 16–16, 15–9, and 9–15. When subjected to continuous darkness, the rhythm synchronized by the previous light cycle persists and cultures initiate a self-sustained rhythm under circadian control.

Canadian mycologist A. H. Reginald Buller suggested that the subsporangial vesicle acts as a lens, focusing the rays of light at the base of the orange-colored vesicle (Figure 12.3). The curvature results from an increase in growth in this region. A fresh crop of sporangiophores is formed daily

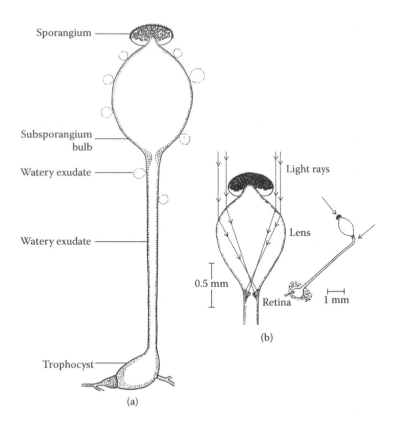

Figure 13.4 *Pilobolus kleinii.* (a) Upper part of sporangiophore acting as a simple eye. (b) Sporangiophore originally developed in light from direction 1, but 2 h later illumination was altered to direction 2. (From Ingold, C.T. and Hudson, H.J., *The Biology of Fungi,* Chapman and Hall, London, 1993, with kind permission of Springer.)

and the discharge occurs around noon. To determine the region of the sporangiophore that is sensitive to light, Robert Page (1962) made a series of photomicrographs at intervals following unilateral illumination. The tip (about 0.5 mm) of the sporangiophore curved sharply toward light in 10 min and reached its maximum curvature in about 1 h. To respond in this way, some chemical substance must perceive light. E. Bünning (1960) compared the phototropic response of various wavelengths (action spectrum) of light by means of glass and liquid filters with the absorption spectrum of extracted pigments. From the close resemblance between the action and absorption spectrum, he suggested that photosensitive pigment is a carotenoid. Page (1962) exposed sporangiophores to wavelengths of light dispersed by a prism and caught the shot sporangia on a glass plate (Figure 13.5). The distribution of the sporangia gave an action spectrum for phototropism. The response was strong, between 410 and 420 nm. Based on the similarity of the action and the absorption spectrum, the photoreceptor is most likely a carotenoid or a flavin. The adaptive response of *Pilobolus* to illumination suggests that light ensures the discharge of sporangium and dissemination, ensuring the survival of the fungus.

13.1.5.2 *Phycomyces blakesleeanus*

The fungus *Phycomyces blakesleeanus* (Zygomycotina) occurs in decaying organic matter but it is easily grown in laboratory medium that is supplemented with the vitamin thiamin. Its sporangiophore is a 10- to 15-cm-long single cell (http://www.es.embnet.org/~genus/phycomyces.html),

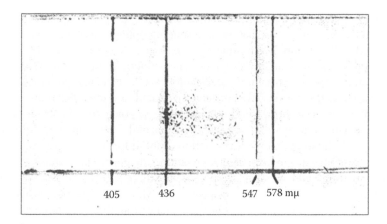

405   436   547 578 mμ

Figure 13.5 Diagram of distribution of *Pilobolus* sporangia adhering to a glass plate interposed between a culture and light dispersed by a prism. (Tracing from Page, 1962.)

suggesting that its cell wall must be very rigid. Nobel Laureate Max Delbrück was attracted to this fungus. He left his research on phage and began experiments that showed a sporangiophore is very sensitive to blue light (400 to 450 nm), with a threshold close to 10 photons per square micrometer, indicating that it has extremely sensitive photoreceptors. The observations suggest that the dosage of the light required by the fungi is very small, allowing sporangiophores to grow out of the decaying organic matter.

When viewed from above, the sporangiophore grows counterclockwise toward the light. The blue-light photoresponse indicates that the photoreceptor is a flavin or a carotenoid molecule. The bending response (Figure 13.6) is limited to the growing zone that extends from 0.1 to 2 mm below the bright yellow sporangium, darkening to nearly black (Bergman *et al.*, 1969). From multiple-exposure

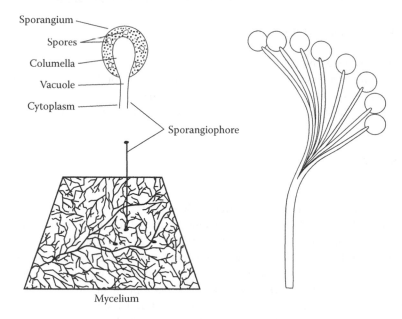

Figure 13.6 Bending of *Phycomyces* sporangiophores toward unilateral light source. Direction of illumination was from left. Tracings from photographs taken at 5-min intervals. (From Bergman, K. et al. (1969). *Phycomyces. Bacteriol. Revs.* 33: 99–157.)

photographs of a sporangiophore, Dennison (1958) discovered that sporangiophore elongates by spiral growth. To explain the phototropic oscillations of a sporangiophore about its stable equilibrium position, a three-layered wall structure was postulated. The photoreceptor is attached to a "receiving and adaptive shell" in the sporangium and twists but does not stretch. An inner wall (responsive structure) bends and grows, and a "reactive structure" passively follows the bend and growth of the "inner wall" (Shropshire, 1963). These three walls have not yet been related to visible structures.

A sporangiophore placed horizontally bends upward (negative geotropism). Mutants of *Phycomyces* were isolated that are phototropically abnormal. These mutants fall into seven complementation groups, indicating that the transduction of the information from the photoreceptor to the morphogenetic response mechanism involves a number of steps. The mutants, *madA* and *madB*, are defective in gravitropic and avoidance responses. It was hypothesized that the inputs from the gravity and chemical sensors feed into the tropism path. Many physiological processes have been shown to be affected by light, including inhibitory effect of blue light on upward bending of horizontally placed sporangiophores and growth away from solid barriers (Bergman *et al.*, 1969).

In 1976 (http://www.es.embnet.org/~genus/phycomyces.htm), Delbrück said,

> I feel that if I make a serious experimental research effort (necessarily a very strenuous exercise) it should be in *Phycomyces*. I am still convinced that *Phycomyces* is the most intelligent primitive eukaryote and as such capable of giving access to the problems that will be central in the biology of the next decades. If I drop it, it will die. If I push it, it may yet catch on as phage…caught on. Since I invested 25 years in this venture I might as well continue. I do not expect to make great discoveries, but if I continue to do the spade work my successors may do so.

The difficulties in genetic analysis of *Phycomyces*—for example, the failure of stably transforming it with exogenous DNA—have dampened interest in this fungus.

13.2 MORPHOGENESIS OF FRUITING BODIES

13.2.1 Development of Fruit Bodies

Light accelerates development of fruit bodies of the coprophilic fungus *Coprinus sterquilinus* (Basidiomycotina). Primordia appear in 8 to 11 days under continuous illumination, but none appear in darkness. Sequential light exposures are necessary to initiate fruit body morphogenesis in Basidiomycotina. Additional periods of illumination are required for the development of the cap; blue (400 to 520 nm) to near ultraviolet (320 to 400 nm) are the most effective.

13.2.2 Other Fungi

Barnett and Lily (1950) reported that a strain of the fungus *Choaneophora cucurbitum* (Zygomycotina) formed conidia only in alternating light and dark, whereas another strain formed conidia in complete darkness. There are other examples of striking differences in response to light between members of the same family (Hawker, 1966). In *Penicillium isariiforme* and *P. claviforme* (Fungi Anamorphici), light triggers development of aggregated conidiophores (coremium). In *Aspergillus nidulans*, conidiation is promoted by light of red wavelength and is reversed by far-red (Mooney and Yager, 1990). This is characteristic of the plant pigment phytochrome (a protein containing a chromophore, the light-absorbing portion), which undergoes a reversible interconversion between the biologically inactive red-light-absorbing form (P_r) and the active far-red light-absorbing form (P_{fr}) and controls behavior responses in plants such as seed germination, stem elongation, and flowering. These fungi should be attractive for further experimentation.

13.3 CIRCADIAN RHYTHM IN *NEUROSPORA*

The fungi are a richly diverse collection of sessile poikilotherms. They extract what they can from their immediate environment, and are subject to the extreme conditions of nature. Thus, there is an obvious selective advantage for possessing an endogenous timing system with which to anticipate the regular, daily changes in temperature, humidity, and light.

—**M. Merrow, T. Roenneberg, G. Macino, and L. Franchi (2001)**

How is an intracellular molecular cycle used to regulate the behavior of the cell?

—**Jay C. Dunlap (2008)**

Virtually all forms of life, from unicellular bacteria to multicellular organisms, including humans, exhibit the daily cycles known as *circadian* (from Latin: *circa*, about; *diem*, a day). A critical feature of circadian rhythms is their self-sustained nature; that is, under constant environmental conditions they continue to repeat the daily cycle. For example, when macroconidia of *Neurospora crassa* (refer to Chapters 3 and 10) are inoculated at one end of an agar growth medium in a race tube (Figure 13.7), the surface mycelium begins to grow toward the other end, and after 21.6 h the mycelium produces aerial hyphae once a day that form conidia. This cycle repeats in a regular manner every 21.6 h. Thus after a few days, the culture in the race tube exhibits conidiating bands alternating with undifferentiated surface growth. The time between bands (period length) is close to 24 h (circadian), which persists ("free runs") in the laboratory under constant illumination or darkness for several days. The rhythmic formation of conidia is a manifestation of an endogenous time-keeping system, a biological clock (Figure 13.8). This feature of *Neurospora* has made it an attractive model for investigating how living organisms keep time. In nature the rhythm is synchronized (entrained) to a 24-h light and dark cycle. How this happens is not understood, although the light–dark cycle is undoubtedly the major factor that continuously resets the internal time-keeping mechanism.

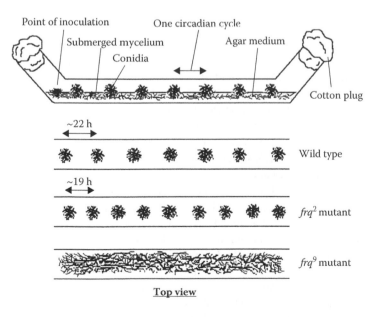

Figure 13.7 Circadian rhythm in macroconidia production in *Neurospora crassa*. Top, diagram of fungus growing in a race tube. Bottom, photograph of a race tube in top view.

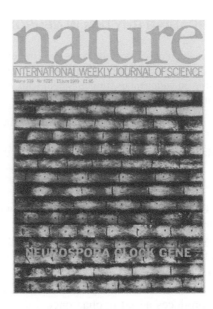

Figure 13.8   Race tube assays of the conidiation rhythms in various strains of *Neurospora* on cover of *Nature* magazine. The short black lines mark the variations in period length. (Reprinted from Macmillan Publishers Ltd. *Nature* 15 June 1989, with permission.)

13.3.1 Clock Gene

Using race tubes, it was shown that several mutants of *Neurospora* show altered period lengths of 16 to 29 h or are *arrhythmic* (Figure 13.8), suggesting that genes control the circadian clock. For example, one mutant has a period of about 19 h, another has a period of about 22 h, and yet another is arrhythmic. Genetic tests showed that a single gene named *frequency* (*frq*) with different alleles frq^2, frq^7, and frq^8, controls the period length (16 to 29 h) of the strains. The gene *frq* is in chromosome VIII between oligomycin resistance (*oli*) and formate (*for*). Dunlap's laboratory used the chromosome walk method to clone *frq*. By transformation experiments, it was found that a 7.7-kb DNA fragment from a wild-type strain could restore circadian rhythm. An arrhythmic mutant allowed cloning of the *frq* gene (McClung *et al.*, 1989). Using the *frq* probe, Northern and Western blot analyses were made to monitor *frq* transcription and translation in liquid-grown cultures growing in continuous light (LL), which were shifted to continuous dark (Figure 13.9). The relative levels of the *frq* mRNA and FRQ protein levels cycle with a 22-h period in the wild-type strain grown in constant darkness. It was inferred that *frq* is the oscillator determining the conidiation rhythm. The FRQ protein has helix-turn-helix DNA-binding domain and nuclear-localization motifs, which suggest that it is a transcription regulator.

The circadian day, about 22 h in length, is divided into 24 equal parts: The circadian hours. A molecular model for the *Neurospora* clock views the day in a linear fashion (Figure 13.10). By convention, CT0 corresponds to subjective dawn and CT12 to subjective dusk. The cultures maintain their rhythmicity in liquid media, making it possible to monitor changes in mRNA and protein product FRQ at different time intervals. At dawn of circadian time (CT0), both *frq* mRNA and FRQ proteins are low, but *frq* transcript starts to rise at CT4. The amount of frq mRNA and FRQ protein cycle in circadian manner, suggesting they control a cascade of clock-controlled output genes and thereby rhythmicity of the organism. The stability of FRQ is a major determining factor for the period length of the clock. Light causes a rapid increase in the levels of the *frq* transcript and resets the clock by resetting the rhythm of *frq* transcription. This autoregulatory loop involving expression of genes to proteins, which, in turn, inhibit their own expression, is thought essential for the

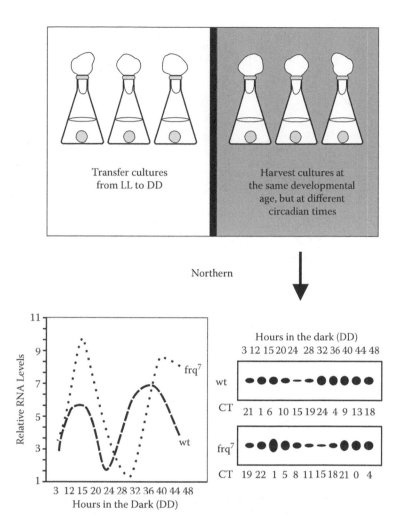

Figure 13.9 Rhythmic liquid culture assay. Assay of rhythm in static *Neurospora* cultures. Transfer of the cultures from LL to DD is staggered to allow harvesting at different circadian times but near the same developmental age. Rhythmic Northern blot is shown for morning peaking clock-controlled gene *ccg-2* from wild-type (wt) and long-period (29-h) *frq*[7] mutant strain. Note that at DD32, *ccg-2* mRNA cycles almost 180° out of phase in the mutant versus the *wt* strain, reflecting the long period of *frq*[7] allele. DD = constant dark; LL = constant light. (Redrawn from Vitalini, M.W., Dunlap, J. C., Heintzen, C., Liu, Y., Loros, J.J., and Bell-Pedersen, D. (2009). Circadian rhythms. In *Cellular and Molecular Biology of Filamentous Fungi*. Borkovich, K. A. and Ebbole, D. J. (Eds.) ASM Press.)

self-sustained circadian rhythmicity. FRQ can be considered to have a function similar to oscillator clock proteins Period (Per) and Timeless (Tim) of *Drosophila melanogaster*, and CRY1 (cryptochrome) and CRY2 of mammals. *Neurospora* clock gene frequency shares a sequence element with the *Drosophila* clock gene period (see McClung *et al.*, 1989).

13.3.2 Feedback Loops

In *Neurospora*, mutants called the *white collar* (*wc*) produce pigmented conidia on a collar of white mycelium and exhibit arrhythmic conidiation. They are "blind," being blocked in light-induced pigmentation (carotenogenesis) in mycelium.

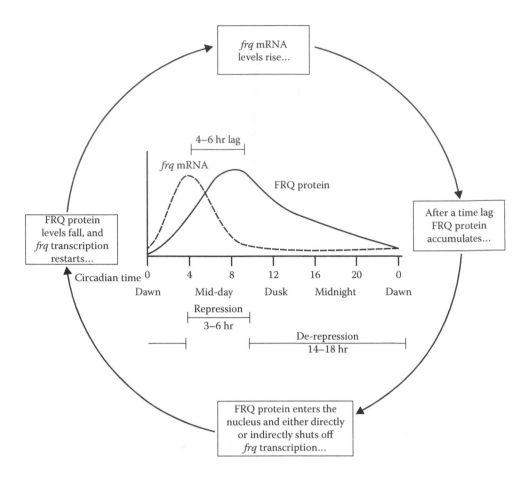

Figure 13.10 Oscillations in *Neurospora* clock components. The relative amounts of *frq* mRNA and FRQ protein are plotted over 24 h, one circadian cycle. (From Bell-Pedersen, D., Garceau, N., and Loros, J.J., *J. Genet.* 75, 387–401, 1966, with permission of Indian Academy of Science.)

The finding of *frq* mRNA in a *wc-2* mutant suggested that light signaling acted primarily through *wc-1*. Cloning of *wc-1* and *wc-2* revealed that their sequences are quite similar. Both genes encode zinc-finger transcription factors and have a "PAS" domain in common that serves as a protein–protein dimerization domain found in many transcription factors and signaling components. Purified WC-1 and WC-2 form a *white collar complex*, a protein complex. (In the *Neurospora* nomenclature for proteins, the protein is given the same name as the gene, but with letters in uppercase without italicization.) The association of a flavin chromophore with WC-1 suggested that complex is a blue-light photoreceptor. Its nucleotide sequence suggested that the complex of WC-1 and WC-2 also act as transcription factor; they bind to a *frequency* (*frq*) gene (pronounced "freak") and induce its expression (Crosthwaite *et al.*, 1997). Null mutants of either gene are arrhythmic.

A model of regulation of *frq* by white-collar genes proposed by Dunlap and co-workers (Liu *et al.*, 2000) is shown in Figure 13.11. The model envisages that FRQ plays a dual role: Acting to depress its own synthesis by allowing transcription only when the concentration falls below a certain threshold, and also to activate *wc-1* and *wc-2* genes, which encode the DNA-binding proteins that act as transcriptional factors (positive element) in the feedback loop for the photoinduction of *frq* transcript. Phosphorylation of FRQ triggers its turnover and is a major determinant of period length in the clock. The rate of degradation of FRQ is most likely the major determining factor for the period length of the clock. *Neurospora* has become a model for clock studies.

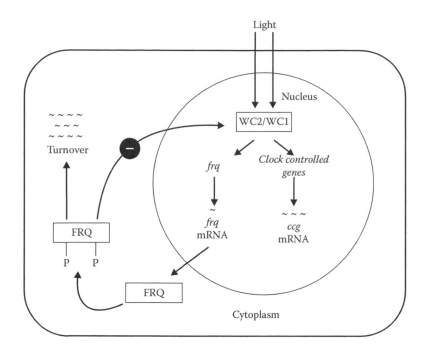

Figure 13.11 Genes and regulation of *Neurospora* clock.

13.4 LIGHT-RESPONSIVE GENES

Many genes have a role in conidiation (Chapter 8). In *Neurospora* screening for clock-regulated genes was carried out using a subtractive hybridization of subjective morning versus subjective evening mRNAs (Loros *et al.*, 1989). In general, the transcripts of these genes accumulate in late night to early morning. The best-characterized clock-controlled gene is *easily wettable*, abbreviated *eas*, which encodes a hydrophobic component of conidial wall and is important for spore dissemination (Lauter *et al.*, 1992). Conidia of *eas* are easily wettable because hydrophobin is missing. Genes associated with carotenoid biosynthesis and conidiation were also found to display a daily rhythm in mRNA accumulation (Lauter, 1996). Experiments using transcriptional profiling with DNA microarray will provide a means to determine the full extent of clock-regulation of genes.

In *Trichoderma atroviride*, a soil-inhabiting fungus, microarray expression analysis of 1438 genes—one fifth of its total genes—indicated that there are genes that are light responsive: About 2% induced and 0.8% repressed (Rosales-Saavedra *et al.*, 2006). A gene *bld-1*, encoding a putative polyketide synthase that may be involved in the first step in the biosynthesis of melanin, which protects cells from the harmful effect of UV irradiation and oxidative stress, is repressed rather than induced by light. The early physiological responses induced by light in *Trichoderma viride* include changes in protein phosphorylation patterns and membrane potential (Gresik *et al.*, 1988, 1989), increases in the activity of adenyl cyclase (Kolarova *et al.*, 1992) as well as in the levels of cAMP and ATP (Gresik *et al.*, 1988), and the accumulation of mRNA of the photolyase-encoding gene *phr-1* (Berrocal-Tito *et al.*, 2000). Photolyases are enzymes that repair damage to DNA caused by exposure to ultraviolet light. The promoter region of most of the light-responsive genes had sequences under control of blue light, elements similar to those defined for light-regulated genes in *N. crassa* (Berrocal-Tito *et al.*, 2000). Orthologues of the *N. crassa wc-1* and *wc-2*, essential for photoconidiation, were detected. In *N. crassa*, a similar number of genes (3%) are activated by blue light, with no evidence of repression at the transcriptional level associated with the stimulus (Lewis *et al.*, 2002). Interestingly, only two of the *T. atroviride* genes *(blu-1* and *blu-17)* induced by light have been found to be light inducible in *N. crassa*. In comparison, 6000 genes in the model plant *A. thaliana* (34% of the genes) are regulated by light (Ma *et al.*, 2001).

13.5 ENTRAINMENT

The time-keeping mechanism has the property of being reset or synchronized. This property is known as *entrainment*. That is, the rhythm can be advanced or delayed although the cycle has the same length, allowing the organism to coincide its clock to the day phase of the environment. It seems that both dawn and dusk are entraining cues. Entrainment is easily demonstrable in fungi. In *Neurospora* a single light pulse in the late night advances the clock, whereas a single light pulse in the early evening or early night delays the clock. In cultures of *Pilobolus* growing in a Petri dish, a single 1/2000-second high-intensity light flash was enough to rephase sporangiophore development (Bruce *et al.*, 1960). In *Sordaria fimicola* (Ascomycotina) grown in plate cultures under alternating conditions of 12 h light and 12 h dark, the discharge of ascospores occurred 6 h after the beginning of the light period (Austin, 1968). The phasing of the light period was then shifted through 12 h by imposing either a 24-h period of light or dark (Figure 13.12). Spore discharge was immediately entrained to the new light phase. That circadian rhythm can be entrained demonstrates it is an adaptive mechanism, although the biochemical mechanism of the process is not known.

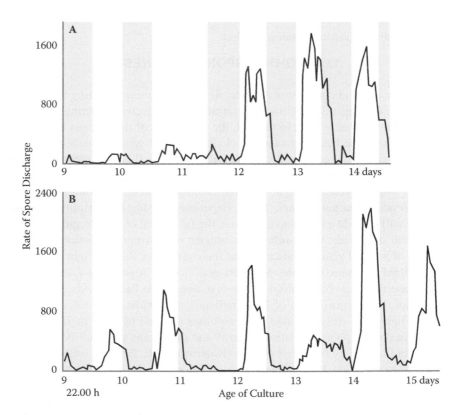

Figure 13.12   Rate of spore discharge from cultures of *Sordaria fimicola* under alternating conditions of light (unshaded area) and dark (shaded area). Cultures were grown in continuous light until day 12, and then light periods corresponding to natural period for three days. The phasing of the light periods was then shifted through 12 h by imposing either a 24-h light period (A) or darkness (B). (Redrawn from Austin, B., *Ann. Bot.* 32, 261–274, 1968.)

13.6 POSSIBLE ROLE OF CIRCADIAN RHYTHM IN *NEUROSPORA*

We will recall from Chapter 10 that orange- or pink-orange airborne macroconidia produced in prodigious quantities and most familiar to biologists do *not* have a direct role in propagation of species or as fertilizing male element in sexual reproduction. Their longevity in sunny daylight is limited; they are not likely to access protoperithecia, which are formed beneath the epidermis or bark of plant tissue. Their massive production in heterothallic species is a requisite for reproduction, but only indirectly for depleting the substrate of nutrients through their constant discharge. This physiological conditioning of the substrate—i.e., creation of a nutritional balance by the fungus itself—triggers formation of the protoperithecia and microconidia that are required for sexual reproduction (fertilization). The observed rhythm in macroconidia production may be a manifestation of a basic endogenous rhythmic phenomenon in mycelium as suggested by rhythm in enzyme activities of glyceraldehyde-3-phosphate dehydrogenase and geranylgeranyl pyrophosphate synthase and of cell wall protein hydrophobin. Continental sampling of single *Neurospora* colonies on scorched vegetation after natural fires have shown that in nature even adjacent colonies differ in mating type, strongly suggesting that if sexual reproduction is to occur, spatially separated cell types, *mat A* and *mat a*, must be together (Powell *et al.*, 2003; Jacobson *et al.*, 2006). Therefore, for sexual reproduction to succeed, the requirement arises for the male gamete to be transmitted to the trichogyne. Though capable of chemotropic curvature, the trichogyne is of limited length. Circadian rhythm presumably evolved as a natural mechanism to facilitate fertilization of protoperithecia (sexual fruiting body) by microconidia taking advantage of the foraging time of vectors (bees, ants, mites, and nematodes), their transmission of microconidia, such that new recombinant, dormant ascospores are produced for survival of species in nature. Both macroconidia and microconidia development may be controlled by shared genetic components without circadian rhythm being essential for macroconidiation (Maheshwari, 2007).

13.7 CONCLUDING REMARKS

To conclude, understanding photoresponses in fungi is of practical interest because light acts as a timing mechanism for initiation of fungal fruiting bodies, for example in mushroom production. Additionally, as noted, light affects spore discharge and their spread in nature. Other applications of studying this phenomenon are because of the link between light-sensing and possible virulence in pathogenic fungi, and photoregulation of enzyme production and secondary metabolite biosynthesis.

REFERENCES

Austin, B. (1968). An endogenous rhythm of spore discharge in *Sordaria fimicola. Ann. Bot.* 32:261–274.

Bell-Pedersen, D., Garceau, N., and Loros, J. J. (1996). Circadian rhythms in fungi. *J. Genet.* 75:387–401.

Ben-Shlomo, R. and Kyriacou, C. P. (2002). Circadian rhythm entrainment in flies and mammals. *Cell Biochem. Biophys.* 37:141–156.

Bergman, K., Burke, P. V., Cerdá-Olmedo, E., David, C. N., Delbrück, M., Foster, K. W., Goodell, E. W., Heisenberg, M., Meissner, G., Zalokar, M., Dennison, D. S., and Shropshire, W., Jr. (1969). *Phycomyces. Bacteriol. Revs.* 33:99–157.

Berrocal-Tito, G., Esquivel-Naranjo, E. U., Horwitz, B. A., and Herrera-Estrella, A. (2007). *Trichoderma atroviride* PHR1, a fungal photolyase responsible for DNA repair, autoregulates its own photoinduction. *Euk. Cell* 6:1682–1692.

Borkovich, K. A., Alex, L. A., Yarden, O., *et al.* (2004). Lessons from the genome sequence of *Neurospora crassa*: Tracing the path from genomic blueprint to multicellular organism. *Microbiol. Mol. Biol. Revs.* 68:1–108.

Bruce, V. C., Weight, F., and Pittendrigh, C. S. (1960). Resetting the sporulation rhythm in *Pilobolus* with short light flashes of high intensity. *Science* 131:728–730.

Bünning, E. (1960). Biological Clocks. *Cold Spring Harbor Symp. Quant. Biol.* 25:1–9.

Castellanos, F., Schmoll, M., Martínez, P., Tisch, D., Kubicek, C. P., Herrera-Estrella, A., and Esquivel-Naranjo, E. U. (2010). Crucial factors of the light perception machinery and their impact on growth and cellulase gene transcription in *Trichoderma reesei*. *Fung. Genet. Biol.* 47:468–476.

Corrochano, L. M. (2007). Fungal photoreceptors: Sensory molecules for fungal development and behavior. *Photochem. Photobiol. Sci.* 6:725–736.

Crosthwaite, S. K., Dunlap, J. C., and Loros, J. J. (1997). *Neurospora wc-1* and *wc-2:* Transcription, photoresponses, and the origins of circadian rhythmicity. *Science* 276:763–769.

Crosthwaite, S. K., Loros, J. J., and Dunlap, J. C. (1995). Light-induced resetting of a circadian clock is mediated by a rapid increase in *frequency* transcript. *Cell* 81:1003–1012.

Dennison, D. S. (1959). Phototropic equilibrium in the *Phycomyces*. *Science* 129:775.

Dunlap, J. C. (2008). Salad days in the rhythm trade. *Genetics* 178:1–13.

Galun, E. and Gressel, J. (1966). Morphogenesis in *Trichoderma*: Suppression of photoinduction by 5-fluorouracil. *Science* 151:696–698.

Grešík, M., Kolarova, N., Farkas, V. (1988). Membrane potential, ATP, and cyclic AMP changes induced by light in *Trichoderma viride*. *Exp. Mycol.* 12:295–301.

Grešík, M., Kolarova, N. and Farkaš, V. (1989) Light-stimulated phosphorylation of proteins in cell-free extracts from *Trichoderma viride*. *FEBS Lett.* 248:185–188.

Gressel, J. and Galun, E. (1967). Morphogenesis in *Trichoderma*: Photoinduction and RNA. *Dev. Biol.* 15:575–59.

Hawker, L. E.: *Physiology of Fungi* (1956). University of London Press, London.

Hill, E. P. (1976). Effect of light on growth and sporulation of *Aspergillus ornatus*. *J. Gen. Microbiol.* 95:39–44.

Idnurm, A., Verma, S. and Corrochano, L. M. (2010). A glimpse into the basis of vision in the kingdom *Mycota*. *Fung. Genet. Biol.* 47:881–892.

Ingold, C.T. and Hudson, H. J. (1984). *The Biology of Fungi*. Chapman & Hall.

Ingold, C. T. (1971). *Fungal Spore: Their Liberation and Dispersal*. Clarendon Press, Oxford, chap. 12.

Ingold, C. T. and Hudson, H. J. (1993). *The Biology of Fungi*. Chapman and Hall, London.

Jacobson, D. J., Dettman, J. R., Adams, R. I. (2006). New findings of *Neurospora* in Europe and comparisons of diversity in temperate climates on continental scales. *Mycologia* 98:550–559.

Jerebzoff, S. (1965). Growth rhythms. In G. C. Ainsworth and A. S. Sussman, eds., *The Fungi*, vol. 1. Academic Press, New York, chap. 27, pp. 625–645.

Klein, D. (1948). Influence of varying periods of light and dark on asexual reproduction of *Pilobolus kleinii*. *Bot. Gaz.* 110:139–147.

Kolarova, N., Haplová, J., and Grešík, M. (1992). Light-activated adenyl cyclase from *Trichoderma viride*. *FEMS Microbiol. Let.* .93:275–278.

Lauter F.-R. (1996). Molecular genetics of fungal photobiology. *J. Genet.* 75:375–386.

Lewis, Z. A., Correa, A., Schwerdtfeger, C., Link, K. L., *et al.* (2002) Overexpression of White Collar-1 (WC-1) activates circadian clock-associated genes, but is not sufficient to induce most light-regulated gene expression in *Neurospora crassa*. *Mol. Microbiol.* 45:917–931.

Lilly, V. G and Barnett, H. L. (1951). *Physiology of the Fungi*. McGraw-Hill, New York.

Liu, Y., Loros, J., and Dunlap, J. C. (2000). Phosphorylation of the *Neurospora* clock protein frequency determines its degradation rate and strongly influences the period length of the circadian clock. *Proc. Natl. Acad. Sci. USA* 97:234–239.

Loros, J. J., Denome, S. A. and Dunlap, J. C. (1989). Molecular cloning of genes under control of the circadian clock of *Neurospora*. *Science* 243:385–388.

Ma, L., Li, J., Qu, L. *et al.* (2001). Light control of *Arabidopsis* development entails coordinated regulation of genome expression and cellular pathways. *Plant Cell.* 13:2589–2608.

Maheshwari, R. (2007). Circadian rhythm in the pink-orange bread mould *Neurospora crassa*: For what? *J. Biosci.* 32:1053–1058.

McClung, C. R., Fox, B. A., and Dunlap, J. C. (1989). The *Neurospora* clock gene frequency shares a sequence element with the *Drosophila* clock gene *period*. *Nature* 339:558–562.

Merrow, M., Brunner, M., and Roenneberg, T. (1999). Assignment of circadian function for the *Neurospora* clock gene *frequency*. *Nature* 399:584–586.

Merrow, M., Roenneberg, T., Macino, G., and Franchi, L. (2001). A fungus among us: the *Neurospora crassa* circadian system. *Cell & Develop. Biol.* 12:279–285.

Mooney, J. L. and Yager, L. N. (1990). Light is required for conidiation in *Aspergillus nidulans. Genes Dev.* 4:1473–1482.

Page, R. M. (1962). Light and asexual reproduction of *Pilobolus. Science* 138:1238–1245.

Powell, A. J., Jacobson, D. J., Salter, L., and Natvig, D. O. (2003). Variation among natural isolates of *Neurospora* on small spatial scales. *Mycologia* 95:809–819.

Roenneberg, T. and Merrow, M. (2003). The network of time: understanding the molecular circadian system. *Curr. Biol.* 13:R198–R207.

Rosales-Saavedra, T., Esquivel-Naranjo, E. U., Casas-Flores, S., Martínez-Hernández, P., Ibarra-Laclette, E., Cortes-Penagos, C., and Herrera-Estrella, A. (2006). Novel light-regulated genes in *Trichoderma atroviride*: A dissection by cDNA microarrays. *Microbiology* 152:3305–3317.

Schmoll, M., Esquivel-Naranjo, E. U., and Herrera-Estrella, A. (2010). *Trichoderma* in the light of day: Physiology and development. *Fung. Genet. Biol.* 47:909–916.

Shropshire, W. W. (1963). Photoresponses of the fungus, *Phycomyces. Bacteriol. Rev.* 33:99–157.

Tisch, D. and Schmoll, M. (2010). Light regulation of metabolic pathways in fungi. *Appl. Microbiol. Biotechnol.* 85:1259–1277.

Vitalini, M. W., Dunlap, J. C., Heintzen, C., Liu, Y., Loros, J. J., and Bell-Pedersen, D. (2009). Circadian rhythms. In *Cellular and Molecular Biology of Filamentous Fungi.* Borkovich, K. A. and Ebbole, D. J. (Eds.) ASM Press.

CHAPTER 14

Thermophilic Fungi
Eukaryotic Life at High Temperature

It is of interest to study the systematics, distribution, and physiological adaptations of organisms which have been successful in colonizing high temperature environments in order to examine the limits to which evolution can be pushed. From an ecological point of view, high temperature environments usually have relatively simple species composition and short food chains, which make a study of productivity, trophodynamics, population fluctuation, and species interaction more simple. From the viewpoint of applied ecology, an understanding of the biology of high temperature habitats is essential if we are to predict and control the consequences of thermal pollution by various industrial sources.

—**Thomas D. Brock (1995)**

The majority of fungi grow best between 20 and 37°C. This range is called mesophilic. This temperature range overlaps the temperature range 18–22°C where the vast majority of plants and animals thrive. The vegetative mycelium of these fungi cannot survive prolonged exposure above 40°C. However, some 30 species show optimum growth between 40 and 50°C, with a few species capable of growth up to 60–62°C. These species are grouped as thermophilic fungi. Their minimum temperature of growth is around 20–25°C. The temperature range for growth of the thermophilic or the mesophilic fungi is about the same, approximately 32°C, which means that in thermophilic fungi their maximum temperature of growth is extended, while their minimum temperature of growth has also been raised.

While some species of bacteria and archaebacteria grow at 80–113°C in hot springs, solfataras, or hydrothermal vents, the thermophilic fungi are the only eukaryotes (organisms with a distinct nucleus) that thrive at 45–60°C, at which no plant or animal can survive (Figure 14.1). They are valuable systems for investigations of genetic and biochemical mechanisms that allow systems to adapt and deal with heat stress. The thermophilic fungi are arbitrarily distinguished from the thermotolerant fungi, which have a minimum temperature of growth between 12 and 17°C and maximum temperature of growth between 50 and 62°C. In their general morphology or ultrastructural features, the thermophilic fungi closely resemble the mesophilic or the thermotolerant fungi.

14.1 DISCOVERY

Thermophilic fungi were discovered when Hugo Miehe (1907) in Germany investigated the cause of self-heating and spontaneous combustion of damp haystacks. He was drawn to study the role of microbial flora in thermogenesis. From the self-heating haystacks, Miehe isolated several microorganisms, including four species of thermophilic fungi: *Mucor pusillus* (renamed *Rhizomucor pusillus*), *Thermomyces lanuginosus* (*Humicola lanuginosa*), *Thermoidium sulfureum* (renamed *Malbranchea cinnamomea*), and *Thermoascus aurantiacus*. To assess their role in self-heating

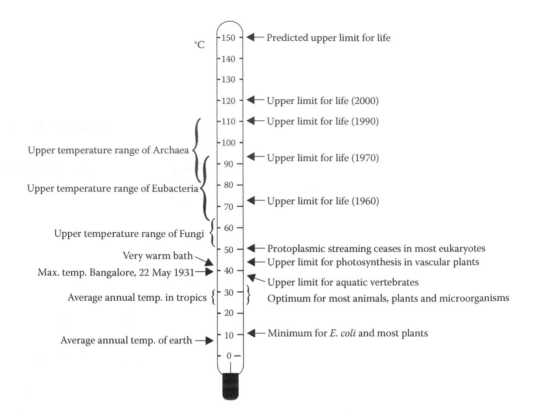

Figure 14.1   Thermometer of life (°C).

of agricultural residues, Miehe inoculated moist hay and other plant materials kept inside Dewar flasks with pure cultures of individual fungi. Whereas sterilized hay did not heat, that inoculated with the fungus did; the attained final temperature of the material showed a correlation with the maximum temperature of growth of the fungus used. Furthermore, by controlled experiments, he demonstrated that the naturally occurring microorganisms present in moist haystacks cause it and other plant materials to heat. Initially, the heat produced by metabolism of mesophilic fungi raises the temperature of the biomaterial to approximately 40°C, after which the warmed-up environment of the compact material favors the growth of the thermophilic fungi and actinomycetes. The latter raise the temperature to about 60°C or higher, corresponding to their upper temperature limit of growth. In 1964, D. G. Cooney and R. Emerson published a monograph that provided taxonomic descriptions of the 13 species known at that time, their sources, and their morphology and general biology. They explained the role of thermophilic fungi in spontaneous combustion of hay. Above 60°C the growth of thermophilic fungi declines and they survive as heat-resistant spores, whereas the actinomycetes become active and raise the temperature of the mass up to 70–75°C, which is their maxima. Beyond this temperature, autocatalytic chemical reactions are presumed to be triggered that ignite the haystack. Figure 14.2 shows a photograph of spontaneous combustion of a big pile of sugarcane bagasse stored near a sugar manufacturing factory. The discovery of thermophilic fungi provided a link in the puzzle of spontaneous combustion of stored agricultural products.

14.1.1 Guayule Rets

During the World War years (1914–1945), the need for finding alternate sources of rubber led to studies of the latex-bearing guayule plant *Parthenium argentatum* as a potential source of rubber.

Figure 14.2 Spontaneous combustion of a heap of sugarcane bagasse near a sugar factory in India.

The extractability and quality of rubber improved if, before milling, this shrub was chopped and stored in a mass ("rets") that self-heated to 65–70°C. From the guayule rets, Paul J. Allen and Ralph Emerson (1949) isolated some 10 species of thermophilic fungi, with temperature limits up to 60°C. All species were capable of decomposing resin. The improvement from retting resulted primarily from the reduction in the amount of contaminating resin in the latex by a thermophilic microflora. Allen and Emerson found that for the optimal development of thermophilic fungi in the guayule rets, its moisture content, its porosity, and its size were crucial factors. These parameters were important for microbial growth, diffusion of air for aerobic respiration, and reducing the dissipation of heat produced by the exothermic metabolic reactions.

14.1.2 Composts

Compost is prepared by gathering material such as garbage, plant residues, kitchen waste, herbivore dung, municipal waste, etc., in a heap that hastens the decay of the gathered material and results in the reduction of the bulk of the organic matter. The heat produced in the heaped mass kills unwanted pests and mesophilic microorganisms, and drives off toxic ammonia. The process of composting, resulting in the production of organic manure, is an unwitting exploitation of thermophilic microorganisms. For mushroom cultivation, a mixture of herbivore dung and straw is composted to give the material a physical texture that favors the growth of mycelium of the edible mushroom *Agaricus bisporus*. Thermophilic fungi—in particular *Scytalidium thermophilum* (= *Torula thermophila/Humicola grisea* var. *thermoidea/Humicola insolens*)—play a dominant role in the preparation of mushroom compost. The majority of the nearly 30 currently known species of thermophilic fungi were originally isolated from composts of various types. Since composts are man-made habitats, they are not considered as natural habitats where thermophilic fungi could have evolved. Cooney and Emerson suggested nests of incubator birds (mallee fowl) as a possible natural habitat of thermophilic fungi. These are large-sized birds that have existed for 50 to 60 million years and are found in Australia and the islands of the southwestern Pacific. These birds build large mounds (composts) out of the forest litter and incubate their eggs inside wherein the microbial activity warms the mounds up to 33–50°C for several months (http://www.abc.net.au/science/scribblygum/October2000/default.htm). Perhaps such warm, humid, and aerobic environments were the sites where thermophilic fungi evolved from related mesophilic forms. Their limited species diversity suggests their relatively recent origin.

14.1.3 Soil

Thermophilic fungi have been isolated from almost any soil, even in the temperate zone. Tendler *et al.* (1967) remarked, "The ubiquitous distribution of organisms, whose minimal temperature for growth exceeds the temperature obtainable in the natural environment from whence they were isolated, still stands as a 'perfect crime' story in the library of biological systems." A controversial question is whether their widespread presence in soil is because of dissemination of spores from the compost heaps occurring worldwide or because of their growth therein. This question has not been easy to resolve since the opaqueness of soil precludes the direct visualization of fungal growth in soil by light microscopy. Therefore indirect approaches have been taken to assess soil as a habitat of thermophilic fungi. Eggins *et al.* (1972) used an immersion tube (Figure 14.3) for discriminating

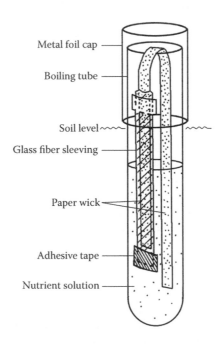

Metal foil cap

Boiling tube

Soil level

Glass fiber sleeving

Paper wick

Adhesive tape

Nutrient solution

Figure 14.3 Diagram of immersion perfusion tube. (Redrawn from Eggins, H.O.W., von Szilvinyi, A., and Allsopp, D., *Internat. Biodeterioration Bull.* 8, 53–58, 1972.)

the mycelial form from dormant spores. A cellulose paper (a source of carbon) was enclosed inside a screen of glass fiber and placed in soil so that only the active mycelium penetrated through the screen and colonized the substrate. The tubes were removed from soil at intervals and the paper strips plated on cellulose agar to determine if the paper strips were invaded by growth of mycelium from the soil. By this method the thermophilic fungi (*Humicola grisea* and *Sporotrichum thermophile*) were detected but only in the sun-heated soil. It would appear that the device used by Eggins *et al.* has the potential of picking out active mycelium, but Tansey and Jack (1977) pointed out that an incorrect impression of growth of fungi can occur if spores were passively carried onto the test substrate by soil arthropods or by capillary action. Therefore, these authors modified the protocol: They studied the development of spores of individual species of thermophilic fungi in pure cultures in Petri dishes that were buried in field soil in Indiana. All tested species germinated and formed sporulating colonies in plates that had been placed inside soil, leading the authors to conclude that the extent and duration of elevated temperature reached in the sun-heated soil are sufficient for thermophilic fungi to grow in soil. However, this extrapolation from pure cultures in Petri dishes that were buried in soil is also doubtful since the propagules of thermophilic fungi are present in soil with a quantitatively higher population of mesophilic fungi, which are potential competitors in fluctuating conditions of temperature.

Rajasekaran and Maheshwari (1993) attempted to forecast the potential of thermophilic fungi to grow in soil based on competitive growth in mixed cultures under a fluctuating temperature regime. Incubation of soil plates at temperature regimes from 24 to 48°C, from 32 to 48°C, and from 36 to 48°C in a programmed incubator (Figure 14.4) yielded either a mesophilic or a thermotolerant fungus (*Aspergillus fumigatus*). At the temperature regime from 36 to 48°C, the predominant fungi that developed in soil plates were still the mesophilic types, with occasional colonies of the thermophilic fungus *Humicola grisea* var. *thermoidea*. Thermophilic fungi developed only when the incubation temperature fluctuated to a small extent between 40 and 48°C, about the summed average optimum temperature of common species of thermophilic fungi (46°C). The results suggested that high temperatures are necessary for thermophilic fungi to compete with the numerically larger mesophilic

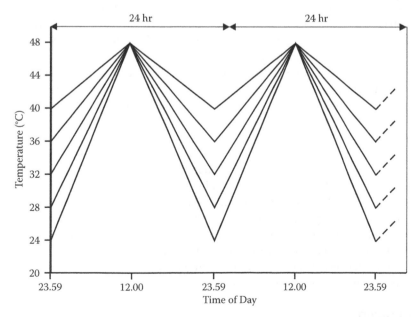

Figure 14.4 Diagram of five diurnally fluctuating temperature regimes used to study competitive growth of a mixed soil fungal flora on soil plates in nutrient medium. (Redrawn from Rajasekaran, A.K. and Maheshwari, R., *J. Biosci.* 18, 345–354, 1993.)

population in soil. Several workers have isolated thermophilic fungi from desert soils that heat up to 40–50°C or more with the implied conclusion that desert soil is a natural habitat of thermophilic fungi. However, since water is required for growth, it is unlikely that their presence in the dry soils is a consequence of active growth *in situ*.

In yet another attempt to resolve whether soil is a natural habitat of thermophilic fungi, Rajasekaran and Maheshwari (1993) used immunofluorescence microscopy to detect the growth of a thermophilic fungus in soil. Spores of *Thermomyces lanuginosus*—a ubiquitous thermophilic fungus—were adhered on glass cover slips as the support and buried in field soil where moisture levels seemed favorable for microbial growth and the soil temperature varied between 22 and 28°C during the time when the experiment was done. The cover slips bearing spores were retrieved after three weeks and germination of spores examined by staining with a fluorescein isothiocyanate (FITC) conjugated antibody preparation that could specifically identify the growth of the test fungus. However, the spores remained dormant, though spores that had been similarly placed under soil in Petri dishes in an incubator at 50°C in the laboratory had germinated. These results did not support the view that soil is a "natural habitat" of thermophilic fungi. Various air-sampling devices have captured spores of thermophilic fungi from air even in temperate countries. Their widespread occurrence in soil is most likely due to the aerial dissemination of spores elsewhere (composts) and fallout from air. The concentration of spores of thermophilic fungi per gram material is approximately 10^6 times higher in composts than in soils (see Maheshwari, 1997). Thermophilic fungi are primarily compost fungi.

14.2 PHYSIOLOGY

14.2.1 Nutrition

The early workers experienced difficulty in growing thermophilic fungi unless the medium incorporated a decoction of hay, casamino acids, peptone, yeast extract, or a tricarboxylic acid cycle intermediate (succinic acid), leading investigators to the view that thermal adaptation is associated with special nutritional requirements. However, subsequently some species were successfully grown in synthetic minimal medium (Gupta and Maheshwari, 1985) containing only glucose (a carbon source), ammonium phosphate, urea, or asparagine (a source of nitrogen), magnesium sulfate (a source of sulfur), potassium phosphate (buffering anions), and trace elements. The optimal growth of thermophilic fungi that have been studied occurs between pH 7.0 and 8.0, which is close to the pH of composts. Controlling the pH of the medium is an important factor in the cultivation of thermophilic organisms because of the unavailability of carbon dioxide at elevated temperatures. For example, if ammonium sulfate is used as a source of nitrogen, the absorption of ammonium ions and the counter transport of H^+ from the cells result in rapid acidification of the medium, and the growth stops. The acidic conditions that develop reduce the solubility of CO_2 required for anaplerotic reactions—for example, for the reaction catalyzed by the enzyme pyruvate carboxylase: Pyruvate + CO_2 + ATP $\rightarrow$ oxaloacetate + ADP + P_i. This enzyme functions to replenish the intermediates of the tricarboxylic acid cycle as they are used for generating energy and for biosynthesis. Alternately, the medium requires that a tricarboxylic acid cycle intermediate, such as succinic acid, be supplied in the medium. Increasing the buffering capacity of the medium, or automatic pH control by addition of alkali in an instrumented fermenter, are common methods used for pH control. A convenient practice is to replace the inorganic nitrogen source with an organic nitrogen source, such as L-asparagine.

14.2.2 Respiration

In 1920 Kurt Noack in Germany compared the metabolic rates of thermophilic and mesophilic fungi to examine their adaptive variations in different thermal regimes. Using the volume of carbon dioxide evolved over time as a measure of the metabolic rate, he compared a *Thermoascus aurantiacus* (thermophilic fungus) with *Penicillium glaucum* (mesophilic fungus), both grown in identical medium. He observed that the volume of carbon dioxide released by the mesophilic fungus in 24 h was equivalent to 67% of its dry weight at 15°C and 133% at 25°C. He argued that if this fungus could grow at 45°C, the extrapolated value of carbon dioxide according to the van't Hoff rule would be 532%. However, the actual value for the thermophilic fungus used at 45°C was 310%. From this data, Noack inferred that at a given temperature the metabolism of thermophilic fungi is actually *slower* than what would be predicted based on the van't Hoff rule.

Two parameters are commonly used in comparing metabolism of different fungi. One is exponential growth rate (μ), which is determined from the exponential portion of semi-logarithmic plots of growth curve:

$$\mu = [2.303 \, (\log x_2 - \log x_1)/t_2 - t_1] \, h^{-1}$$

where x_2 and x_1 are mg dry wt at time t_2 and t_1, respectively. The other parameter is the molar growth yield, which is determined as yield (dry wt of mycelium) per mole of glucose utilized. Interestingly, a feature of shaker-grown mycelia of some thermophilic fungi is the formation of homogeneous mycelial suspension that is amenable to pipeting, allowing reliable sampling and monitoring of growth. The exponential growth rates and molar growth yield of thermophilic and mesophilic fungi studied are similar. On average, both mesophilic and thermophilic fungi convert approximately 55% of sugar for the synthesis of biomass and 45% for metabolism. This situation is similar to that in thermophilic bacteria.

Subsequent measurements of oxygen uptake of mycelial suspensions have dispelled the assumption that thermophilic fungi have a faster rate of metabolism than their mesophilic counterparts: At their respective temperature optima, the two types of fungi have nearly comparable respiratory rates (Prasad *et al.*, 1979; Rajasekaran and Maheshwari, 1993). An unexpected disclosure from the experiments was that the respiratory rate of mesophilic fungi was higher than that of thermophilic fungi over a broad range of temperatures (Figure 14.5). All major chemical components—proteins, lipids, and nucleic acids—have their structural and functional properties altered by changes in temperature. How different organisms have adaptively modified their basic biochemical structures and functions to exploit the natural environments continues to be the focus of biochemical adaptation.

14.2.3 Utilization of Carbon Sources

In a compost heap, as the temperature begins to rise, the mesophilic microflora are succeeded by thermophilic microflora. Therefore, the availability of soluble carbon sources (sugars, amino acids, and organic acids) should lessen; that is, the carbon source available to thermophilic fungi would mainly be the polysaccharide constituents of the biomass, chiefly cellulose and hemicelluloses. Thermophilic fungi are therefore especially well adapted for polysaccharide utilization. The growth rate of the thermophilic fungus *Sporotrichum thermophile* on cellulose (paper) was similar to that on glucose (Bhat and Maheshwari, 1987). *Chaetomium thermophile* and *Humicola insolens* grew better on xylan than on simple sugars (Chang, 1967).

The extracellular hydrolysis of polysaccharide constituents of biomass by the secreted enzymes would be expected to result in a mixture of sugars in the growth environment of thermophilic fungi. One of the adaptive strategies for their growth could be the simultaneous utilization of a mixture of sugars. To test this, the thermophilic fungi *Thermomyces lanuginosus* and *Penicillium duponti* were grown in a mixture of glucose and sucrose in liquid media (Maheshwari and Balasubramanyam, 1988). Both fungi concurrently utilized glucose and sucrose at 50°C, with sucrose being utilized faster than glucose (Figure 14.6). This is quite the opposite of the phenomenon of diauxic growth observed in bacteria that utilize one carbon source at a time; for example, glucose is utilized before lactose in a mixture of the two sugars.

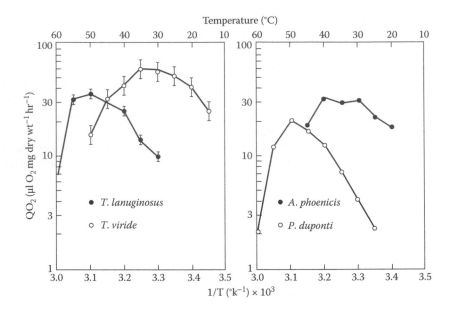

Figure 14.5 Respiration of thermophilic (*Thermomyces lanuginosus, Penicillium duponti*) and mesophilic (*Trichoderma viride, Aspergillus phoenicis*) fungi measured by Warburg manometry. The data are given as Arrhenius plots of the logarithm of respiratory rate (Q_{O_2}) of shaker-grown mycelia against reciprocal of absolute temperature. (Redrawn from Rajasekaran, A.K. and Maheshwari, R., *J. Biosci.* 18, 345–354, 1993.)

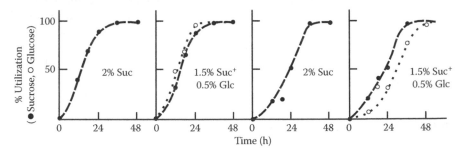

Figure 14.6 Simultaneous utilization of glucose and sucrose by *Thermomyces lanuginosus*. Panels 1 and 2, *Thermomyces lanuginosus*. Panels 3 and 4, *Penicillium duponti*. (Redrawn from Maheshwari, R. and Balasubramanyam, P.V., *J. Bacteriol.* 170, 3274–3280, 1988.)

14.2.4 Transport of Nutrients

Nutrients from the environment are transported inside the cell by specific carrier proteins in the plasma membrane. One reason thermophilic fungi fail to grow at ordinary temperatures could be that their transporter proteins are transformed into a rigid conformation affecting the binding and release of ions and nutrients, and therefore functioning very poorly. Future possibilities include the cloning of transporters from mesophilic and thermophilic fungi, their reconstitution in lipid vesicles (liposomes), and a comparative study of the kinetics of transport at a range of temperatures.

14.2.5 Protein Turnover

Thermophily in bacteria was discovered earlier than in the eukaryotes. An early hypothesis put forward to explain thermophily in bacteria proposed that growth at high temperatures occurs because the denatured cellular proteins are replaced quickly by resynthesis. Two different groups compared the rate of protein breakdown in thermophilic and mesophilic fungi by feeding mycelia with radioactively labeled amino acid and monitoring the radioactivity in mycelial proteins after transferring mycelia to nonradioactive medium (Miller *et al.*, 1974; Rajasekaran and Maheshwari, 1990). The results of the two studies were at variance. Although both groups measured only protein breakdown, very likely the results would have been similar had protein turnover been measured. Since at their respective temperature optima the mesophilic and thermophilic fungi form comparable biomass, it is unlikely that this hypothesis is generally valid because the energy expended in increased protein turnover in thermophilic fungi would be expected to affect their yield, which does not happen.

14.3 EXTRACELLULAR ENZYMES

There has been much interest in thermophilic fungi because of the prospects of finding enzyme variants with high-temperature optima and a long "shelf life," which are desirable properties for enzymes used in industries. Hence this topic has been recently reviewed (Maheshwari, 2003; Brienzo *et al.*, 2008). Chiefly, the extracellular enzymes have been studied as the mycelium can be removed by a simple filtration method and large amounts of culture filtrates obtained as starting material for enzyme purification. The extracellular enzymes of thermophilic fungi have a molecular mass ranging from 20 to 60 kDa and are thermostable. Maheshwari *et al.* (2000) may be consulted for original references cited therein. Examples of a few extracellular enzymes purified from cultures of thermophilic fungi are given below.

14.3.1 Protease

These enzymes act on proteins or polypeptides. Historically, attention on thermophilic fungi was catapulted by the finding of a strain of *Mucor* that produced milk-curdling activity that could substitute for chymosin (rennin), required in the manufacture of cheese. Hitherto rennin was obtained from the stomach of suckling calves. The *Mucor* acid protease, induced in medium containing wheat bran, had maximal activity at pH 3.7 and was stable at 55°C. The *Mucor* rennin gene was expressed in yeast that secreted the foreign (heterologous) protein at concentrations exceeding 150 mg/liter. A recombinant mesophilic fungus, *Aspergillus oryzae*, produced *Mucor* acid protease in excess of 3 g/liter of medium. Among the varied applications of proteases are tenderizing of meat, the manufacture of cheese, as a component of detergents for the removal of proteinaceous dirt from fabric, and for dehairing of leather in the leather industry.

14.3.2 Lipase

Lipases hydrolyze triacylglycerols. Thermostable lipases, stable at 60–65°C at pH 10–11, have application as ingredients of laundry detergents for removing oil stains. Strains of *Thermomyces lanuginosus* (= *Humicola lanuginosa*) secrete appreciable quantities of lipase when grown in medium containing animal or vegetable oils as inducer. The lipase gene from *T. lanuginosus* and *Rhizomucor miehei* (= *Mucor miehei*) were cloned and expressed in a heterologous mesophilic fungus, *Aspergillus oryzae*. The thermophilic *R. miehei* lipase was the first lipase whose three-dimensional structure was deduced by x-ray analysis. The catalytic site of the lipase protein is covered by a short α-helical loop that acts as a "lid" that moves when the enzyme is adsorbed onto the oil–water interface, allowing access of the substrate to the active site.

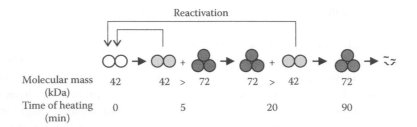

| Molecular mass (kDa) | 42 | 42 | > | 72 | 72 | > | 42 | 72 |
| Time of heating (min) | 0 | 5 | | | 20 | | | 90 |

Figure 14.7 A schematic representation of the observations on the effect of heat on α-amylase. The native dimeric enzyme is shown by open circles, the partially active enzyme by light gray circles, and the inactive trimeric species by dark gray circles.

14.3.3 Amylase

Amylases are a group of enzymes that hydrolyze starch, forming oligosaccharides, maltose, or glucose. Amylases have applications in the industrial-scale conversion of starch into glucose, for which heat-stable amylases are preferred since the saccharification reaction for manufacture of glucose syrups can be done at 70–80°C in large reactors with minimal threat of contamination by common microbes in the environment.

Starch is hydrolyzed by two main types of enzymes: An endo-acting α-amylase that produces maltooligosaccharides and an exo-acting glucoamylase that produces chiefly glucose. *T. lanuginosus* produces both enzymes that are fully stable at 50°C though inactivated at 70°C (Mishra and Maheshwari, 1996). In the presence of calcium, α-amylase was nearly eight times more stable at 65°C. Addition of calcium increased the melting temperature of α-amylase from 66°C to 73°C. The molecular changes in 42-kDa amylase upon heating were investigated by measurement of enzyme activity after heating protein to 94°C for different time periods, in parallel with electrophoretic analysis of structural changes (Figure 14.7). The data revealed that α-amylase of *T. lanuginosus* is a dimer, which upon heating to boiling temperature of water undergoes structural reorganization and is progressively converted to an inactive trimeric species by self-association of subunits (Figure 14.7). However, this conformational state was potentially capable of withstanding an otherwise lethal temperature and returned to the native state and regained enzymatic activity in a temperature- and time-dependent process. Thus results of this study suggest that aggregation and disaggregation of enzyme protein subunits may be one of the mechanisms allowing growth and survival of thermophilic fungi at fluctuating cycles of high and low temperatures. The starch-degrading enzymes of *T. lanuginosus* are the most thermostable enzymes among fungal sources. The high purity of products (glucose and maltose) and their thermostability suggest potential usefulness of *T. lanuginosus* glucoamylase and α-amylase in enzymatic saccharification of starch.

14.3.4 Cellulase

Cellulase refers to a group of enzymes (endoglucanase, exoglucanase, and β-glucosidase) that act together to solubilize cellulose. Cellulase enzymes are used in detergent formulations to remove unwanted pill-like balls of fuzz that form on clothes due to repeated washing and wearing and that collect dirt. The endoglucanases (mol wt ranging from 30 to 100 kDa) are thermostable, and active between 55 and 80°C at pH 5.0–5.5. The exoglucanases (mol wt 40–70 kDa) are thermostable glycoproteins that are optimally active at 55–75°C. The molecular characteristics of β-glycosidase are variable, with mol wt ranging from 45 to 250 kDa and carbohydrate content ranging from 9 to 50%. Except for their thermostability, the molecular characteristics of cellulase components of thermophilic fungi are quite similar to those from mesophilic fungi. *T. aurantiacus* grown on paper and *H. insolens* grown on wheat bran were exceptionally good producers of cellulase enzymes.

14.3.5 Xylanase

Several thermophilic fungi have been identified as exceptionally high producers of xylanase that have application in bio-bleaching of pulp for paper manufacture. They are generally single-chain glycoproteins, ranging from 6 to 80 kDa, active between pH 4.5 to 6.5, and at temperatures between 55 and 65°C. Xylanase is commonly induced with cellulase in response to xylan or natural substrates containing hemicelluloses, the monosaccharide constituent of xylan. *Thermoascus aurantiacus* and *Paecilomyces varioti* are exceptionally high producers of xylanase: The enzyme simply crystallizes out in a concentrated protein solution. The thermostability of xylanase from thermophilic fungi was postulated to be due to an extra disulfide bond and the preponderance of salt bridges holding secondary structures of protein.

14.4 INTRACELLULAR ENZYMES

Because of the general difficulty in disrupting mycelia and extracting cellular protein from mycelia, few attempts have been made to compare functional or physical-chemical properties of intracellular enzymes from mesophilic and thermophilic fungi. Here two examples of glycosidases are given that showed opposite patterns of activity during growth.

14.4.1 Trehalase

The substrate of this enzyme is a small sugar molecule, trehalose, which is a commonly called mushroom or fungal sugar. Bharadwaj and Maheshwari (1999) chose this enzyme for a comparative investigation in a thermophilic and a mesophilic fungus. The thermophilic fungus *Thermomyces lanuginosus* grows optimally at 50–55°C, whereas the mesophilic fungus *Neurospora crassa* grows optimally between 30 and 35°C. The following conclusions could be made from a study of trehalase from these two fungi:

> Although the upper temperature limit for growth of *N. crassa* and *T. lanuginosus* are quite different (40°C vs. 60°C), trehalase from both fungi had similar temperature optimum for activity (50°C). The temperature optimum of trehalase is not related to the growth temperature of the organisms.
> Trehalase of *T. lanuginosus* and *N. crassa* are structurally very different proteins (monomer vs. tetramer). At 50°C, the $t_{1/2}$ of trehalase was >6 h. Trehalase from both sources was about equally stable.
> The k_{cat}/K_m of *T. lanuginosus* trehalase is one order of magnitude lower than that of *N. crassa* enzyme. From this it can be inferred that trehalase from the mesophilic fungus is a better catalyst than from the thermophilic fungus. Proteinaceous factors in *T. lanuginosus* cell extracts protected trehalase from heat inactivation.

14.4.2 Invertase

Whereas invertase in mesophilic fungi is synthesized constitutively, that in the thermophilic fungi studied is an induced enzyme, produced only in response to added sucrose (Table 14.1). Interestingly, the induced activity begins to diminish even before any substantial quantity of sucrose is utilized or an appreciable biomass is produced (Chaudhuri *et al.*, 1999). Paradoxically, despite this unusual pattern of invertase activity development, neither the growth rate nor the final mycelial yield is affected.

Table 14. 1   Distinctive Properties of Invertase in a Thermophilic and a Mesophilic Fungus

Property	*Thermomyces lanuginosus* (Thermophilic Fungus)	*Neurospora crassa* (Mesophilic Fungus)
Synthesis	Induced by substrates of enzyme (β-fructofuranosides)	Constitutive, synthesized irrespective of carbon source in medium
Distribution	Hyphal tip	Uniformly present along hypha
Relationship to growth	Activity inversely related to biomass	Activity parallels increase in biomass
Glucose repression	Not repressed	Repressed
Stability in cell-free extracts at 0°C	Unstable	Stable for more than one month
Effect of thiol compounds	Activated and stabilized	No effect

Furthermore, invertase in the thermophilic fungi studied (*T. lanuginosus* and *Penicillium duponti*) is a thiol protein. This paradoxical behavior of invertase in thermophilic fungi was explained by postulating that invertase in *T. lanuginosus* is localized in the hyphal tip, which has a reducing environment, and invertase activity depends on the maintenance of a catalytically important sulfhydryl group(s) in the reduced state (Chaudhuri *et al.*, 1999). The number of hyphal tips per unit mass of mycelium is maximal in the early stages of growth, and correspondingly invertase activity is maximal. At later times the number of hyphal tips per unit mass decreases rather abruptly because growth (increase in biomass dry wt) occurred mainly from cell elongation and cell wall thickening. Invertase activity therefore shows an apparent decline during growth. By contrast, trehalase in the same fungus is distributed along the hypha, bound to the cell wall, and its amount steadily increases in proportion with the increase in cell elongation and wall synthesis (Figure 14.8) (Bharadwaj and Maheshwari, 1999).

Invertase in cell-free extracts of *T. lanuginosus* and *P. duponti* was highly unstable The activity of unstable intracellular proteins may be optimized in unexpected ways; for invertase, the strategy

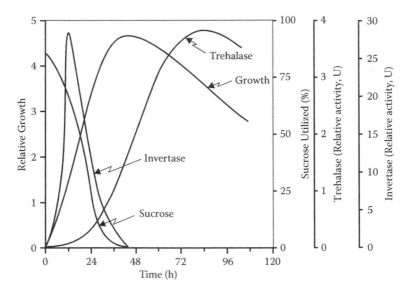

Figure 14.8   Contrasting pattern of development of invertase and trehalase activities in *Thermomyces lanuginosus*. Fungus was grown in a liquid medium containing sucrose as the carbon source. (Redrawn from Chaudhuri, A., Bharadwaj, G. and Maheshwari, R., *FEMS Microbiol. Lett.* 177, 39–45, 1999.)

is its co-localization with sucrose transporter in the growing hyphal tip and its substrate-induced synthesis, such that cellular energy is needed for resynthesis only when necessitated by the availability of sucrose in the environment. The invertase example cautions against generalization that thermophily in fungi is due to thermostability of proteins. Certain enzymes that for physico-chemical reasons are unstable may be continuously synthesized. Because of the energy expended in the resynthesis of certain labile enzymes, the thermophilic fungi exhibit an apparent incongruity in metabolic rate with the van't Hoff rule.

14.5 THERMOTOLERANCE GENE

Because of the ability of *Aspergillus fumigatus* to grow at a temperature ranging from 12 to 52°C and its cellulolytic property, it is a dangerous contaminant in a laboratory, where it grows on cotton plugs of culture tubes, especially in the humid tropical countries. In nature it is a common fungus of composts and of moldy grains. The fungus colony has a gray-green color on agar plates and forms loose conidia that are highly dispersive, capable of entering the lungs as inhaled spores, causing allergies or growing in the lung cavities, causing aspergillosis. Its ability to grow readily at 37°C has made this fungus a significant problem in operating rooms of hospitals, where it can establish infections of the internal organs via surgical wounds, especially during transplant surgery when the patient's immune system is suppressed. Its genome was sequenced in 2005 (Nierman *et al.*, 2005; Ronning *et al.*, 2005). Nearly a third of the predicted genes are of unknown function. *A. fumigatus* has been used to identify gene(s) controlling its thermotolerant character (Chang *et al.*, 2004). By chemical mutagenesis of conidia, a temperature-sensitive mutant was isolated that grew as well as the wild type at 42°C but failed to grow at 48°C. A wild-type genomic library in a cosmid vector containing the hygromycin-resistant (*hph*) gene as the selection marker was used to transform conidia of temperature-sensitive mutant and transformants selected that grow as well as the wild type at 48°C. A 35.5-kb thermotolerance (*THTA*) gene complemented the *ts* phenotype. That *THTA* is required for thermotolerance was shown by gene disruption: The strains carrying the disrupted gene failed to grow at 48°C following colony transfer. Mice injected with wild-type or disrupted strains showed similar survival, demonstrating that *THTA* has an important role in virulence. It is likely that *THTA* is gene encoding for a heat-shock protein.

14.6 HEAT-SHOCK PROTEINS

Heat-shock proteins (HSPs) are generally defined as those whose synthesis is sharply and dramatically induced at high temperatures. They bind to unfolded proteins and prevent their aggregation. The HSP family has been classified into five groups on the basis of their molecular weight: HSP100s, HSP90s, HSP70s, HSP60s, and small HSPs (sHSPs; 12–40 kDa). In the yeast *Saccharomyces cerevisiae* two such proteins, Hsp90 and Hsp 70, are essential for growth at higher temperature. The proteins are induced by a wide variety of other stresses, seem to have very general protective functions, and play a role in normal growth and development. In *Neurospora crassa* Kapoor and Lewis (1987) used a sucrose-sorbose agar medium to restrict mycelial spread. The 60-h-old colonies were subjected for 1 h at 48°C, following which they were replica-plated to determine the survival. Up to 95% of the colonies survived the normally lethal temperature. A plate assay for peroxidase (Figure 14.10) demonstrated that treatments that induced peroxidase in colonies of *N. crassa* also conferred marked heat tolerance. This suggested that peroxidase could be one of the heat-shock proteins but direct proof that it confers thermoresistance is yet lacking.

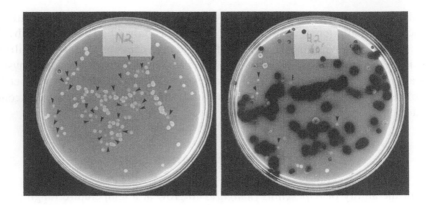

Figure 14.9   Identification of peroxidase as one of the heat-shock proteins in *Neurospora crassa* using a plate assay for peroxidase. (From M. Kapoor, with permission.)

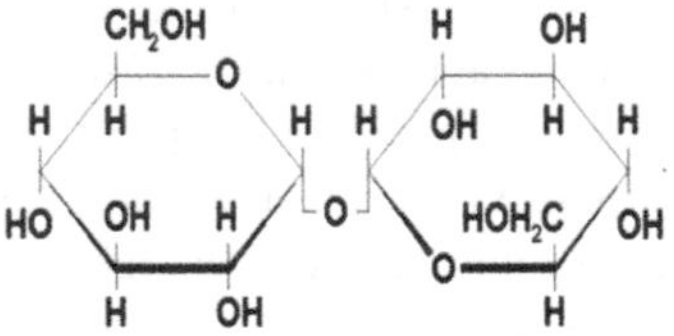

Figure 14.10   Chemical structure of trehalose, a nonreducing disaccharide common in fungi.

14.7 INTRACELLULAR SOLUTES

Some small-molecular-weight compounds known as compatible solutes are known to be synthesized and intracellularly accumulated under stressful conditions—such as low water activity, high temperature, or in the presence of ethanol (Chin *et al.*, 2010)—and allow fungal species to survive and grow at extreme temperatures. Of these, the effect on trehalose has been well explored. Trehalose (Figure 14.10) is composed of two glucose molecules (α-D-glucopyranosyl-α-D-glucopyranoside). This sugar protects macromolecules in the cell through the production of a supersaturated liquid that has the properties of a solid. It is thought to maintain hydrogen bonds within and between macromolecules, thereby maintaining their structures and affording protection against thermal stress.

14.8 CONCLUDING REMARKS

Because thermophilic fungi occur in habitats that are heterogeneous in temperature, and vary in the types and concentrations of nutrients, competing species, and other variables, they have probably also adapted to factors other than high temperature, as shown by the distinctive behavior of invertase in mesophilic *N. crassa* and thermophilic *T. lanuginosus*. Thermophilic fungi can be grown in minimal media with metabolic rates and growth yields comparable to those of mesophilic fungi. Studies of their growth kinetics, respiration, mixed-substrate utilization, nutrient uptake, and protein breakdown rate have provided some basic information not only on thermophilic fungi but also on fungi in general. Thermophilic fungi have a powerful ability to degrade the polysaccharide constituents of biomass. The properties of their enzymes show differences not only among species but also among strains of the same species. Their extracellular enzymes display temperature optima for activity that are close to or above the optimum temperature for the growth of organism and, in general, are more stable than those of mesophilic fungi. Genes of thermophilic fungi encoding lipase, protease, xylanase, and cellulose have been cloned and overexpressed in heterologous fungi, and pure crystalline proteins have been obtained for elucidation of the mechanism of their intrinsic thermostability and catalysis. By contrast, the thermal stability of the few intracellular enzymes that have been purified is comparable to or, in some cases, lower than that of enzymes from the mesophilic fungi. It may well be that gain of thermostability in certain intracellular proteins is not possible without a concomitant loss of catalytic activity, as shown by the example of invertase. There is no single adaptive strategy in fungi inhabiting hot environments. It is a combination of mechanisms that make a thermophilic fungus adapt to a hot environment: Intrinsic thermostability of macromolecules, interaction of proteins with ions and other cellular proteins functioning as chaperonin molecules, self-aggregation, possibly covalent or noncovalent interactions with the cell wall, and rapid turnover of some proteins.

Currently, enzymes from hyperthermophiles that grow at temperatures beyond 80°C are being sought for biotechnology. However, since flexibility of protein conformation is essential for catalysis, the enzymes from the hyperthermophiles will have optimal conformational flexibility at the temperatures at which they are adapted to grow but may become too rigid and have low catalytic rates at temperatures that range from 50 to 65°C. Therefore, in most situations, enzymes from thermophilic fungi may be better suited in biotechnology than enzymes from hyperthermophilic archaea.

REFERENCES

Allen P. J. and Emerson, R. (1949). Guayule rubber, microbiological improvement by shrub retting. *Ind. Eng. Chem.* 41:346–365.

Bharadwaj, G. and Maheshwari, R. (1999). A comparison of thermal and kinetic parameters of trehalases from a thermophilic and a mesophilic fungus. *FEMS Microbiol. Lett.* 181:187–193.

Bhat, K. M. and Maheshwari, R. (1987). *Sporotrichum thermophile* growth, cellulose degradation and cellulose activity. *Appl. Environ. Microbiol.* 53:2175–2182.

Brienzo, M., Arantes, V., and Milagres, A. M. F. (2008). Enzymology of the thermophilic ascomycetous fungus *Thermoascus aurantiacus*. *Fungal Biol. Rev.* 22:120–130.

Brock, T. D. (1995). The road to Yellowstone—and beyond. *Annu. Rev. Microbiol.* 49:1–28.

Chang, Y. and Hudson, H. J. (1967). The fungi of wheat straw compost: I. Ecological studies. *Trans. British Mycol. Soc.* 50:649–666.

Chang, Y. C. C., Tsai, H.-F., Karos, M., and Kwon-Chung, K. J. (2004). *THTA*, a thermotolerance gene of *Aspergillus fumigatus*. *Fung. Genet. Biol.* 41:888–896.

Chaudhuri, A. and Maheshwari, R. (1996). A novel invertase from a thermophilic fungus *Thermomyces lanuginosus*: Its requirement of thiol and protein for activation. *Arch. Biochem. Biophys.* 327:98–106.

Chaudhuri, A., Bharadwaj, G., and Maheshwari, R. (1999). An unusual pattern of invertase activity development in the thermophilic fungus *Thermomyces lanuginosus*. *FEMS Microbiol. Lett.* 177:39–45.

Chin, J. P., Megawa, J., Magilla, C. L., Nowotarski, K., Williams, J. P., Bhaganna, P., Linton, M., Patterson, M. F., Underwood, G. J. C., Mswaka, A. Y., and Hallsworth, J. E. (2010). Solutes determine the temperature windows for microbial survival and growth. *Proc. Natl. Acad. Sci. USA* 107:7835–7840.

Cooney, D. G. and Emerson, R. (1964). *Thermophilic Fungi: An Account of Their Biology, Activities and Classification*. W. H. Freeman, San Francisco.

Eggins, H. O. W., von Szilvinyi, A., and Allsopp, D. (1972). The isolation of actively growing thermophilic fungi from insulated soils. *Internat. Biodeterioration Bull.* 8:53–58.

Gupta, S. D. and Maheshwari, R. (1985). Is organic acid required for nutrition of thermophilic fungi? *Arch. Microbiol.* 141:164–169.

Jack, M. A. and Tansey, M. R. (1977). Growth, sporulation, and germination of spores of thermophilic fungi incubated in sun-heated soil. *Mycologia* 69:109–117.

Kapoor, M. and Lewis, V. (1987). Heat shock induces peroxidase activity in *Neurospora crassa* and confers tolerance towards oxidative stress. *Biochem. Biophys. Res. Com.* 147:904–910.

Maheshwari, R. (1997). The ecology of thermophilic fungi. In K. K. Janardhanan, C. Rajendran, K. Natarajan, and D. L. Hawksworth, eds., *Tropical Mycology*. Oxford & IBH, New Delhi, pp. 278–289.

Maheshwari, R. (2003). Enzymes of thermophilic fungi. In *McGraw-Hill Yearbook of Science and Technology*. McGraw-Hill, New York, pp. 114–116.

Maheshwari, R. and Balasubramanyam, P. V. (1988). Simultaneous utilization of glucose and sucrose by thermophilic fungi. *J. Bacteriol.* 170:3274–3280.

Maheshwari, R., Bharadwaj, G., and Bhat, M. K. (2000). Thermophilic fungi: Their physiology and enzymes. *Microbiol. Mol. Biol. Revs.* 64:461–488.

Miehe, H. (1907). *Die Selbsterhitzung des Heus. Eine biologische Studie*. Gustav Fischer, Jena.

Miller, H. M., Sullivan, P. A., and Shepherd, M. G. (1974). Intracellular protein breakdown in thermophilic and mesophilic fungi. *Biochem. J.* 144:209–214.

Mishra, R. S. and Maheshwari, R. (1996). Amylases of the thermophilic fungus *Thermomyces lanuginosus*: Their purification, properties, action on starch and responses to heat. *J. Biosci.* 21:653–672.

Mohsenzadeh, S., Saupe-Thies, W., Steier, G., Schroeder, T., Fracella, F., Ruoff, P., and Rensing, L. (1998). Temperature adaptation of housekeeping and heat shock gene expression in *Neurospora crassa. Fung. Genet. Biol.* 25:31–43.

Montero-Barrientos, M., Hermosa, R., Nicolás, C., *et al.* (2008). Overexpression of a *Trichoderma* HSP70 gene increases fungal resistance to heat and other abiotic stresses. *Fung. Genet. Biol.* 45:1506–1513.

Nierman, W. C., Pain, A., Anderson, M. J., Wortman, J. R., *et al.* (2005) Genomic sequence of the pathogenic and allergenic filamentous fungus *Aspergillus fumigatus. Nature* 438:1151–1156.

Prasad, A. R. S., Kurup, C. K. R., and Maheshwari, R. (1979). Effect of temperature on respiration of a mesophilic and thermophilic fungus. *Plant Physiol.* 64:347–348.

Rajasekaran, A. K. and Maheshwari, R. (1990). Growth kinetics and intracellular protein breakdown in mesophilic and thermophilic fungi. *Indian J. Exp. Biol.* 29:134–137.

Rajasekaran, A. K. and Maheshwari, R. (1993). Thermophilic fungi: An assessment of their potential for growth in soil. *J. Biosci.* 18:345–354.

Tansey, M. R. and Jack, M. A. (1976). Thermophilic fungi in sun-heated soils. *Mycologia* 68:1061–1075.

Tendler, M. D., Korman, S., and Nishimoto, M. (1967). Effects of temperature and nutrition on macromolecule production by thermophilic Eumycophyta. *Bull. Torrey Bot. Club* 94:175–183.

Populations

CHAPTER 15

Species
Their Diversity and Populations

Knowledge of a flute or a kettledrum is not sufficient to understand all the other instruments in a symphony orchestra or to predict their characteristic. Nor is knowledge of a single species, however complete, adequate for understanding diverse species. Diversity of research in the laboratory must at least dimly reflect the diversity of species in nature if the scope and beauty of evolutionary improvisations are to be appreciated and the genetic manipulations that brought them about are to be understood.

—David D. Perkins (1991)

As every chapter in this book demonstrates, species of fungi have been selected for investigation on the basis of distinctive features: The mode of their growth, their morphology, the places in which they grow, in the method of their reproduction, or in their response to environment. No single species has all the desirable features for research, and species or strains were selected for particular types of investigations. Indeed the research problem itself got defined by some special feature of the fungus or strain of a fungus. Table 15.1 gives the terms used to denote the extant diversity. The diversity comprises both the variations between the species (interspecies variations) as well as within the species (intraspecies variation).

Research is understandably on those forms that can be isolated from nature, induced to sporulate in pure culture, and recognized unambiguously. Much less is known about the distribution of species in space and time. Sooner or later, a student of mycology is confronted with larger questions; for example:

1. What is meant by species?
2. How can the number of species of fungi in nature be estimated?
3. How can variation be detected in virtually indistinguishable morphologically, keeping in mind their extreme phenotypic plasticity?
4. How can the geographic origin of a species be determined?
5. Is there any correlation between type of variation and ecological (air-dispersed versus soil-dispersed) or geographical features of the isolates?
6. How can the relative contributions of asexual and sexual modes of reproduction be determined in those fungi that possess both modes of reproduction?
7. How variable are endophytic or parasitic species?
8. How variable are fungi that occur in special environments, such as high-temperature compost or in deep-sea sediments?
9. Of what practical value is knowledge of fungal diversity?
10. How can the biological species concept be tested with a homothallic fungus?

Table 15.1   Some Terms That Depict the Special Habitat and Diversity of Fungi

Term	Meaning	Common Example(s)
Anthropophilic fungi	Infectious only to man	*Trichophyton rubrum*
Aquatic fungi (water molds)	Fungi resident in aquatic habitats	*Saprolegnia*
Brown-rot fungi	Fungi that rot wood, giving it a dark brown color	*Fistulina, Daedalea, Coniophora*
Coprophilous fungi	Fungi growing on dung of herbivore animals	*Pilobolus, Podospora, Coprinus*
Corticolous fungi	Fungi growing on tree bark	*Xylaria*
Dermatophyte	Fungi that live as parasites on skin, hair, or nails of man and other animals	*Trichophyton interdigitale, Microsporum canis, Arthobotrys* sp.
Edible fungi	Fungi used as food	*Agaricus campestris, Pleurotus versicolor*
Endophytic fungi	Symptomless parasitic fungi in mutualistic association with living plants	*Balansia* sp., *Curvularia* sp.
Entomogenous fungi	Insect-parasitizing fungi	*Entomophthora, Cordyceps, Septobasidium, Beauveria*
Hypogeous fungi	Fungi growing below ground	*Tuber* sp.
Keratinophilic fungi	Fungi growing on feathers, horns	*Onygena equine, Nannizia*
Lichen forming fungi	Fungal symbiont of lichen thallus	*Cladonia cristellata, Xanthoria* sp., *Phaeographina fulgurata*
Marine (saprobic) fungi	Fungi growing and sporulating in marine or estuarine habitats	*Dendryphiella salina, Mycosphaerella*
Mesophilic fungi	Fungi thriving between 10 and 40°C	Vast majority of fungi, e.g., *Aspergillus niger*
Mildew	Fungi producing whitish growth on living plants	Downy mildew (*Peronospora viticola*), Powdery mildew (*Erysiphe graminis*)
Mycorrhizal fungi	Fungi in symbiotic association with living roots	Mostly Basidiomycetous fungi belonging to the families *Agaricaceae, Boletaceae*
Mycoparasites	Fungi parasitic on other living fungi	*Trichoderma* spp., *Piptocephalius* sp., *Gliocladium roseum*
Necrotrophic fungi	Parasitic fungi that kill host cells in advance of its hyphae and derive their organic nutrients from the dead cells	*Pythium, Monilinia fruticola, Penicillium expansum*
Nematophagous (Predacious) fungi	Fungi parasitic on nematodes	*Arthobotrys* sp., *Dactylaria* sp.
Osmotolerant fungi	Fungi capable of growth in solutions of high osmotic pressure	*Aspergillus restrictus, A. flavus, A. amstelodami*
Phylloplane fungi	Fungi growing on aerial surface of living leaves	*Cladosporium herbarum, Alternaria alternata*
Psychrophilic fungi	Fungi growing at <10°C, maximum temperature of growth at 15–20°C	*Mucor* sp., *Fusarium nivale, Typhula idahoensis*
Pyroxyrophilous fungi	Fungi growing on burnt trees, wood, or burnt areas of the ground	*Anthracobia* sp., *Pyronema* sp., *Daldinia* sp., *Neurospora discreta*
Rumen fungus	Anaerobic chytridaceous fungus	*Neocallimastix frontalis*
Rust fungi	Obligate biotrophs causing reddish brown pustules on plants	*Puccinia graminis, Uromyces* spp.
Sewage fungi	Fungi growing in polluted waters	*Leptomitus lacteus, Fusarium aqueductuum*
Slime molds	Fungi with a multinucleate mass of cytoplasm not bound by cell wall moving in amoeboid fashion.	*Physarum* sp., *Didymium* sp.
Soft-rot fungi	Fungi that cause a watery rot in plant	*Penicillium expansum, Monilinia fruticola*
Sooty molds	Fungi that cover the leaves as black sooty mass	*Meliola* sp.

Table 15.1 (*Continued*)   Some Terms That Depict the Special Habitat and Diversity of Fungi

Term	Meaning	Common Example(s)
Sugar fungi	Fungi which utilize only simple organic compounds, lacking the ability to decompose complex organic materials.	*Aspergillus* sp., *Penicillium* sp., *Mucor* sp.
Thermophilic fungi	Fungi that can grow at 45°C or above but not at 20°C	*Thermomyces lanuginosus*, *Mucor miehei*
Water molds	Fungi found in waters	*Saprolegnia, Achlya, Dictyuchus*
Wilt fungi	Fungi causing wilt of plants, trees	*Ceratocystis fagacearum*
Xerotolerant fungi	Fungi growing on jams, salty foods at <0.85 aw	*Aspergillus fumigatus, Cladosporium* sp.

15.1 NUMBER OF FUNGAL SPECIES

Earlier, a species was defined as a group of individuals having common morphological characters. Based on the degree of discrimination adopted by the taxonomist—a scientist who identifies and classifies according to a nomenclature and classification system—approximately 70,000 species of fungi have been described based on morphological features such as the structure of conidiophores, the color and the method of formation of spores, the types of ascocarps, the features of basidium, etc. However, this number is considered to be grossly lower than the total number of fungal species. Hawksworth (1991) estimated the number of species of fungi based on the ratio between the species of vascular plants and the species of fungi in well-studied regions of the world. For the United States, this ratio is 1:1; for Finland, 1:4; for Switzerland, 1:4; and for India, it is 1:0.5. The best-studied region is the British Isles where this ratio is 1:6. These ratios inform about the regions that require mycological studies. For example, the ratio for India is undoubtedly because of the underexploration of the subcontinent's alpine, aquatic, arid, and tropical environments. Extrapolating the 1 vascular plant:6 fungi ratio to the global total of 250,000 species of vascular plants, the total number of fungal species comes close to 1.5 million. Only 5% of this estimated number of fungal species is actually documented. Where are the undiscovered fungi to be found? Unusual fungi lurk in unusual ecological niches and habitats (Subramanian, 1992). Marine fungal diversity has received very little attention, and it is believed that true marine fungi have not been cultured so far (Raghukumar, 2006).

Devising methods for their enumeration is difficult. Traditional isolation and enumeration techniques uses the soil dilution plates, which in essence involves adding soil, or a diluted suspension, to Petri dishes and covering with a suitable agar medium. It has been estimated that there are up to a million fungal spores in a gram of dry soil. Some species may have unusual nutritional requirements, hence are uncultivable on media commonly employed and therefore unidentifiable. There is no simple solution for estimating species diversity.

15.2 VALUE OF STUDYING POPULATION STRUCTURE AND DIVERSITY

Because of biological diversity, no single species can serve as a universal model. Each species has its peculiar strengths and limitations.

—**D. D. Perkins (1991)**

It is through biochemical dissection of the wealth of genetic diversity that we will appreciate the fundamental design principles of living organisms and the mechanisms in their adaptations to the niche in which they live. The currently favored fungi in research are given in Table 15.2.

Table 15.2 Currently Favored Fungi in Research

Fungus	Research
Ascobolus immersus	Meiotic recombination
Ashbya gossypii	
Aspergillus nidulans	Nuclear dynamics and polarized hyphal growth Conidial differentiation, nuclear migration, cytokinesis, secondary metabolism
Aspergillus flavus	Mycotoxins
Aspergillus fumigatus	Genetic variation between clinical and environmental isolates and their role in disease epidemiology
Aspergillus niger	Enzymology and metabolism, including secondary metabolism
Candida albicans	Pathogenicity, dimorphism
Colletotrichum graminicola	Topographical signals in pathogenesis
Coprinus cinereus	Morphogenesis of multihyphal fruiting body
Fusarium oxysporum	Transposons
Magnaporthe grisea	Mechanisms in pathogenicity
Neurospora crassa, N. intermedia, N. tetrasperma	Hyphal growth, chromosomal rearrangements, function of het genes, mating type genes, meiotic drive, epigenetic phenomena, light signaling and biological rhythm, mitochondrial biogenesis, vesicle trafficking, developmental cytogenetics, biochemical regulation, speciation
Phanerochaete chrysosporium	Biochemistry of lignin degradation
Phycomyces blakesleeanus	Photoresponse
Podospora anserina	Senescence, mitochondrial genetics, prions
Saccharomyces cerevisiae	Cell cycle regulation, cell polarity, protein secretion, biotechnology
Saprolegnia ferax	Hyphal morphology and growth
Schizophyllum commune	Hyphal growth and wall structure
Thermomyces lanuginosus	Eukaryotic thermophily
Trichoderma atroviride	Blue light response (conidiation)
Trichoderma reesei	Heterologous protein production, biocontrol

15.2.1 Basic Research

When a culture began to be studied in detail with respect to its general biology—i.e., the type of reproductive structures, the mode of germination of spores or their development, or the major chemical constituents—it began to be realized that though outwardly showing great similarities in hyphal mode of construction and in absorptive mode of nutrition, one group of organisms has a cellulosic wall composition (in contrast to noncellulose wall) and has a chiefly aquatic habitat, rather than the large group of terrestrial filamentous fungi in which the life cycle is for the most part haploid. This group—which includes the potato late blight fungus *Phytophthora infestans*, a "fungus" that actually initiated the study of such organisms—was removed from other hyphal organisms and placed in a new kingdom called Straminipila. This diploid-haploid group of organisms is thought to have been derived from the green alga, though they are achlorophyllous, but which they resemble in forming antheridia and oogonia for sexual reproduction and were set apart from other haploid-diploid groups even though they have an absorptive mode of nutrition (see the Appendix).

On the other hand, some species—for example, *Neurospora crassa* and *N. tetrasperma*—are similar in their general habit but differ in just a quirk, e.g., spindle orientation during delimitation of ascospores leading to a four-spored ascus instead of the more common eight-spored ascus (Chapter 10). Both species are useful for investigation of the mechanisms that control the orientation of the spindle during cell division.

Once intraspecies variation was recognized as a general phenomenon, several practical methods began to be developed for its detection. Detection of variation and identification of strains assumes special significance in medically important fungi causing infections. For this reason, sequence-based identification of fungi is gaining momentum. The results show that the sequence variation in clinical samples is much higher than in samples from type-culture collections.

15.2.2 Applied Research

Most marketable biotechnology products still result from the traditional approach of isolating and screening of microbes from natural habitats and the exploitation of natural variability in populations. All biotechnology begins with a preferred fungal genome drawn from nature—a natural species and its numerous genetic strains. Some examples of inter- and intraspecies differences in enzyme productivity, and properties of enzymes of commercial value can be found in Maheshwari *et al.* (2003). Designing culture media for the selective enrichment and isolation of desired types and manipulating environmental conditions (temperature, aeration, pH, osmotic pressure) from an infinite number of permutations and combinations of environmental variables is a challenge.

15.3 SPECIES RECOGNITION

The 70,000 described fungi were identified based on morphological characters. However, fungi are notorious for their phenotypic plasticity—a characteristic that in the past had led Buxton (1960) remark that fungi are "a mutable and treacherous tribe." Individuals may show striking nongenetic variation depending on changes of the growth environment. Forms that may in fact be closely related may be given the status of different species. On the other hand, a group of individuals may share certain morphological characters but in fact be genetically isolated. Therefore, for species recognition, morphological criteria (morphological species recognition, MSR) alone are not sufficient. A species has been defined variously (see Taylor *et al.*, 2000) as "groups of actually or potentially interbreeding populations which are reproductively isolated from other such groups," "the smallest aggregation of populations with a common lineage that share unique, diagnosable phenotypic characters," or "a single lineage of ancestor–descendant populations which maintains its identity from other such lineages and which has its own evolutionary tendencies and historical fate." An excellent example of species discriminated by mating tests (biological species recognition, BSR) is provided by the fungus *Neurospora*. Shear and Dodge (1927) showed that by mating tests, the morphological species *Monilia sitophila* could be split into three species: *N. crassa*, *N. sitophila*, and *N. tetrasperma*. However, the identification of species mating tests is not easy to apply; in approximately 20% of all fungi grouped as Fungi Anamorphici, the sexual stage either does not occur or the conditions under which it occurs are not known. The use of the mating test (reproductive success in crosses) in species identification is therefore of limited use.

With recombinant DNA technology, long stretches of DNA can be cloned, sequenced, and compared for determining whether two nucleic acid molecules are similar. Nucleic acid sequences are scanned and fitted by computer programs into a branching tree pattern based on maximum parsimony. A computer program is used to measure the summed average index of resemblance between fungi and arrange them into a branching tree diagram, i.e., a tree that unites the specimens having the most features in common to demonstrate the relative relationships in a group of individuals, with adjacent branches depicting the greatest or closest genetic similarity (Burnett, 2003). A phylogenetic system that classifies organisms according to their evolutionary sequence—that is, which enables one to determine at a glance the ancestors and derivatives—is therefore being used in recognizing species. It is particularly useful in those organisms where the mating test cannot be applied. In this phylogenetic diagram, the branch points (or nodes) reflect divergence from a previously common sequence and the length of a branch (or distance) is a measure of the mean number of estimated character changes (substitutions) required to convert one sequence to another. A group of organisms is identified that is an out-group—an independent evolutionary lineage, qualifying as reproductively isolated species (phylogenetic species recognition, PSR). In *Neurospora*, there was correspondence between groups of individuals identified as species by BSR and PSR criteria. However, PSR provided greater resolution (Dettman *et al.*, 2003). Data from six loci (*al-1, frq, gpd, mat a, mat A*, and ITS/5.8S rRNA) were consistent in suggesting that among the five outbreeding species hitherto recognized on the bases of MSR or BSR, *N. discreta* diverged first. More species have been found by the criterion of PSR than by MSR or BSR.

15.4 DISCOVERY OF INTRASPECIES VARIABILITY

15.4.1 Physiological Races

Before the molecular techniques, the concept of variation within a species identified on the basis of spore morphology was provided by E. C. Stakman and his associates. They noted that a variety of crop plants bred for resistance to a particular species of a rust fungus failed to remain resistant to that particular fungus. For example, the species of the black (or stem) rust fungus, an obligate parasite on wheats and grasses, namely *Puccinia graminis* f. sp. *tritici*, could be subdivided into over 200 "physiological races." These could be resolved based on the size and shape of lesions produced on leaves of different varieties of wheat called the "tester" varieties (Figure 15.1). The recognition of physiological races explained why no variety of wheat plant bred for resistance to a pathogen is permanent. This is because a new physiological race can arise through sexual recombination. *Puccinia graminis* f. sp. *tritici* is heteroecious—that is, it has a part of its life cycle on wheat and the other part on an alternate host, barberry. When this was recognized, a barberry destruction program was undertaken in the United States to control rust epidemics with some limited success; new races of wheat rust continue to arise through mutation in the absence of sexual reproduction.

The application of pathogenicity tests showed that intraspecific groups of individuals restricted to a characteristic host (*forma speciales*) is a common feature of several plant pathogenic fungi. For example, *forma speciales* have been discovered in *Phytophthora infestans*, in the powdery mildew (*Erysiphe* spp.), in the downy mildew (*Peronospora* spp.), and in *Fusarium oxysporum*.

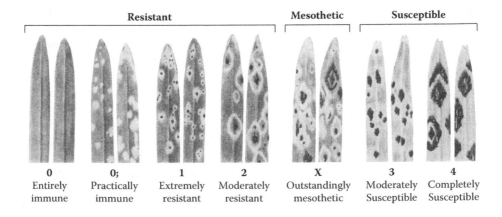

Figure 15.1 Infection types produced by races of wheat stem rust, *Puccinia graminis* var. *tritici*, on differential varieties of wheat selected for identifying races. 0, Entirely immune. 0;, Practically immune. 1, Extremely resistant. 2, Moderately resistant. X, Outstandingly mesothetic. 3, Moderately susceptible. 4, Completely susceptible. Temperature control is necessary for identification of races. (From Stakman, E. C. 1950. *Principles of Plant Pathology*. Ronald Press, New York.)

15.4.2 Vegetative Compatibility

Since the majority of species are nonpathogenic, the pathogenicity test for detecting intraspecies variation as used for the rust fungi is of limited application. In the pre-molecular era, the formation of heterokaryon, which involves hyphal fusion between strains followed by mixing of their proto-plasm, was a commonly used method to determine their genetic relatedness. That is, when two test strains are paired opposite each other on agar medium, there are frequent anastomoses between the hyphae of two closely related fungal strains. If not, the mycelia of the test strains confronted on nutrient medium show an antagonistic reaction consisting of vacuolated, dying, and often pig-mented hyphal cells between them, called the barrage reaction. Fusion of hyphae is controlled by genes called *het*erokaryon incompatibility (*het*) loci or vegetative *in*compatibility (*vic*) loci. For example, 11 and 8 *het* loci in *N. crassa* and *Aspergillus nidulans*, respectively, and 10 *vic* loci in *Fusarium oxysporum* have been identified. The hyphae of two fungal strains fuse to form a homo-geneous mycelium (heterokaryon) only if the alleles at each of the corresponding *het* or *vic* loci are identical. The method is more reliable if the test strains are marked by genetic markers whose defect can be remedied by complementation. In one of the procedures, the test strains are grown in medium containing chlorate. The spontaneous nonutilizing nitrates (nitrate reductase mutants) as chlorate-resistant mutants are selected. These strains are paired against each other on a minimal medium plate with nitrate as the nitrogen source. Strains of the same vegetative compatibility group (VCG) form protrophic heterokaryons, whereas the strains of different VCG are incapable of form-ing prototrophic heterokaryons. Based on this test, Jacobson and Gordon (1990) resolved a morpho-logically identical population of the muskmelon wilt fungus *Fusarium oxysporum* f. sp. *melonis* into several incompatible groups and demonstrated intraspecies variability in this facultative parasite. This method has also been used to detect variation in the population of a honey fungus, *Armillaria mellea* (Basidiomycotina), in Australia. The vegetative compatibility test reflects closeness but not the genotypic identity of the individuals.

15.5 GENERATION OF VARIATION

15.5.1 Mutation and Heterokaryosis

Ultimately all variations arise from mutations. Although a rare event, on the order of 10^{-6} per nucleus, mutation assumes greater importance in fungi than in other organisms because the coeno-cytic fungal mycelium contains thousands or perhaps millions of nuclei. The effect of mutation on phenotype may not be detectable because of masking of the mutant nuclei by the wild-type nuclei in heterokaryon. It is only when the mutant nuclei segregate as uninucleate spores and germinate to produce a growth variant containing descendants of the mutant nuclei that the mutation is revealed (Chapter 2). Such a variant may differ morphologically from the original strain (Adhvaryu and Maheshwari, 2002). In Chapter 3, we referred to the finding that a single multinucleate spore of mycorrhizal fungi contains genetically different nuclei. More than one internal transcribed spacer (ITS) sequence has been obtained from a single multinucleate spore. Since sexual reproduction is unknown in these fungi, heterokaryosis is believed to play an important role in the variability of glomales (Sanders, 1999).

15.5.2 Transposable Elements

Transposable elements are insertion sequences in DNA that have an intrinsic capability of transposing within the host genome. They contribute to genetic variation by both modification of gene sequence and modification of chromosome structure through translocation, deletion, and duplication, and have been identified in fungi (Daboussi, 1996). Transposons are present in multiple copies and consequently can be discovered through DNA sequencing. The plant pathogenic fungi *Fusarium oxysporum* and *Mycosphaerella graminicola* are typical examples wherein various types of transposons have been discovered.

15.6 DETECTION OF GENETIC VARIATION IN POPULATIONS

A population is a pattern of distribution of individuals with distinctive but comparable morphologies and genotypes in space and time (Burnett, 2003). A basic requirement for the study of populations is that the fungus is recognized in nature and enough samples are collected over a wide range of geographical distances for inferences. A fungus that lends itself admirably for population studies is *Neurospora* (Ascomycotina) since it produces distinctive orange conidia that allow its practically unambiguous recognition in nature. Practical techniques have been devised for sampling, transportation, and reliable identification of its species in the laboratory (Perkins and Turner, 1988, 2001). Over 4000 samples have been collected globally and collections preserved, providing the most valuable resource material for population studies (Turner *et al.*, 2001). For fungi whose mycelium is in soil—for example, *Fusarium oxysporum* (Fungi Anamorphici)—the fungus is isolated from soil by dilution plating on a selective medium (Gordon and Okamoto, 1992). For fungi immersed in substrate such as wood, blocks can be taken and plated to isolate the fungus. Using this method, Rayner (1991) and colleagues detected different basidiomycete individuals demarcated by interactive zones of lines in a single branch of an oak tree.

15.6.1 Isozyme Electrophoresis

Variation in strains can be due to variants of proteins that catalyze the same reaction but differ in net charge due to differences in their amino acid composition and can therefore be resolved by electrophoresis. Such variants are called isozymes. The protein extracts of samples is subjected to polyacrylamide gel electrophoresis and the difference in mobilities of a specific enzyme is compared after visualizing them following specific staining reaction. Spieth (1975) surveyed *N. intermedia* for general proteins (acid phosphatase and esterase) and detected a high degree of genetic variability among natural populations. This technique has been widely used to detect variations in all forms of life. The technique ultimately detects variation among DNA sequences that code for isozymes.

The late blight "fungus" *Phytophthora infestans* (Straminipila), collected worldwide, had long been considered as a single asexual clone. However, populations from the highlands of central Mexico were found to harbor more variability than those collected elsewhere. The finding of protein polymorphism (isozyme analysis) in *P. infestans* isolates was consistent with the discovery of heterothallism and sexual reproduction, suggesting that this region is the site of origin of this fungus. In the 1970s–1980s, the analysis of isozymes of glucose-phosphate isomerase in populations of the *P. infestans* from Europe, the United States, and Canada suggested an intercontinental spread of the fungus from Mexico to Europe through transport of diseased potato tubers. As a result of man-mediated migration of strains on diseased potato tubers, sexual reproduction elsewhere has become potentially possible, alerting the plant pathologists to protect the potato crop by continuous resistance breeding.

In Section 15.4.1 we referred to control of wheat rust in the United States by the eradication of barberry, which is the alternative host of this heterothallic fungus on which sexual reproduction occurs (see the Appendix). Protein markers and pathogenicity tests were used to investigate the role of sexual recombination in maintaining variability within populations of *Puccinia graminis* f. sp. *tritici*. By 1975 there was a notable difference in the number of physiological races as the sexual stage has ceased to be functional. In regions where barberry is absent, as in Australia, the pathogen population has more limited isozyme and virulence phenotypes than in regions where opportunities for sexual reproduction exist (Burdon and Roelfs, 1985).

15.6.2 Restriction Fragment Length Polymorphism

Restriction fragment length polymorphisms (RFLP) are based on the hybridization of DNA probes to fragments of DNA that have been produced by cutting with specific restriction endonucleases and size fractionated by electrophoresis. RFLPs (Figure 15.2) are used to detect variation in nucleotide sequence among homologous sections of chromosomes due to restriction-enzyme recognition-site changes along the section of chromosome where the probe hybridizes. Because RFLPs can detect variation in both coding and noncoding regions of DNA, they are much more variable than isozymes and therefore more useful markers (Figure 15.3). RFLP and isozyme methods have been commonly used to analyze populations and to recognize subdivisions within the population (Table 15.3). RFLP technique is a convenient method to study long-range dispersal patterns of plant pathogenic fungi. It suggested the airborne dispersal of the wheat stem rust fungus from India and Africa to Australia across the Indian Ocean. Figure 13.3 is an autoradiogram showing DNA sequence variation as 4 RFLP alleles in 28 *Mycosphaerella graminicola* isolates collected from a single field of wheat.

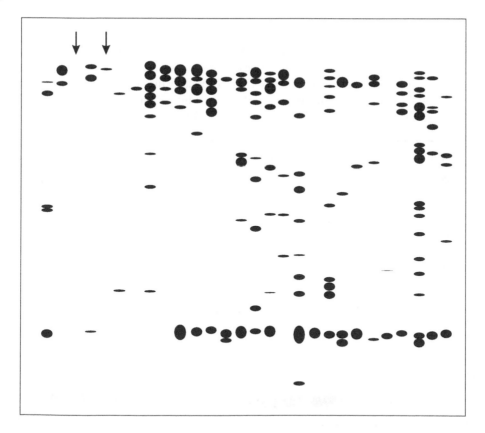

Figure 15.2 RFLPs of 28 *Mycosphaerella graminicola* isolates from a single infected wheat field. The DNA extracted from isolates was separated by gel electrophoresis and hybridized to a cloned radio-labelled DNA fragment (probe) segment and x-ray film was exposed to the gel. Arrows indicate isolates having the same DNA fingerprints are products of asexual reproduction. (Tracing from McDonald, B.A. and McDermott, J.M., *Bioscience* 43, 311–319, 1993.)

Table 15.3   Variations in Fungal Populations

Species	Data Type	Geographic Scale	No. of Populations	No. of Isolates
Agaricus bisporus	RFLP	Intercontinental	2	342
Mycosphaerella graminicola	RFLP	Local	3	512
Neurospora intermedia	Allozyme	Intercontinental	4	145
Schizophyllum commune	Allozyme	Intercontinental	7	136

This RFLP locus is defined by the restriction enzyme XhoI and a DNA probe pSTS196. The first and the last lanes are size markers. From McDonald and McDermott (1993), with permission of the publisher.

15.6.3 Random Amplified Polymorphic DNA

In a modified technique called random amplified fragment length polymorphism, synthetic 10–15-base-long primers are used to amplify random sequences of genomic DNA from the test strains by the polymerase chain reaction. Variation is detected as the presence or absence of amplified DNA sequences. Random amplified polymorphic DNA (RAPD) polymorphisms were detected in 10 random isolates of *Erysiphe graminis* f. sp. *hordei*, which causes powdery mildew of barley, sampled from a single field. The DNA fingerprints were different, suggesting that these isolates were clonal lineages. In a study of the ectomycorrhizal fungus *Suillus granulatus*, Jacobson *et al.* (1993) demonstrated the high resolution of RAPD marker analysis. Some of the isolates of the fungus, which were earlier reported as similar based on vegetative compatibility reaction and considered to belong to same genotype, were found to be genetically dissimilar by RAPD analysis. This method resolved 17 collected strains of *Aspergillus niger* into 15 subgroups (Megnegneau *et al.*, 1993). RAPD method has a finer resolution than either RFLP or isozymes.

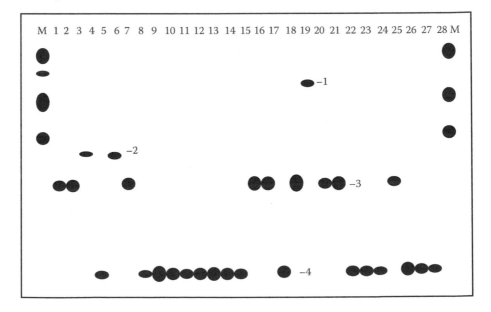

Figure 15.3   Autoradiodiagram of hybridization of a DNA probe with size fractionated DNA fragments of same 28 *Mycosphaerella graminicola* isolates as in Figure 15.2. The first and last lanes are markers. The numbers designate examples of four alleles. (Tracing from McDonald, B.A. and McDermott, J.M., *Bioscience* 43, 311–319, 1993, with permission.)

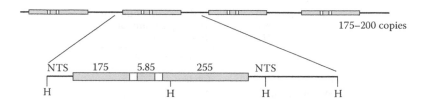

Figure 15.4 Basic organization of one unit of ribosomal DNA.

The RAPD technique was used to investigate population structure of the opportunistic human pathogenic and thermotolerant fungus *Aspergillus fumigatus*. The fungus lives in soil and reproduces only asexually by means of airborne conidia, causing asthma and infections in immuno-compromised patients and resulting in at least 50% mortality. Using 16-base-long primer isolates from Mexico, Argentina, France, and Peru, DNA fragments were analyzed for genetic variability in environmental and clinical isolates (Duarte-Escalante *et al.*, 2009). The clinical isolates had higher genetic variation than the environmental ones. Its genome sequence identified sequences homologous to MAT locus (Chapter12), suggesting cryptic sexual reproduction (Gorman *et al.*, 2009). A worldwide screening of 290 isolates showed nearly 1:1 ratio of MAT1-1:MAT1-2 locus. This led to setting up crosses between opposite mating types. After six months of incubation at 30°C of paired cultures on an oatmeal agar plate, cleistothecia with ascospores were formed, and as per the rules of nomenclature, the teleomorph stage was named *Neosartorya fumigata*.

15.6.4 Ribosomal DNA

The nuclear ribosomal coding cistron (rDNA) has been widely utilized for detecting variation among isolates of a fungus. The rDNA is composed of tandemly repeated units, each unit being composed of 5S, 25S, and 18S rDNA (Figure 15.4). Two noncoding regions exist in each repeat: The internal transcribed spacer (ITS1 and ITS 2) and an intergenic nontranscribed spacer (NTS). The NTS can be preferentially amplified by PCR using primers, cut with restriction enzymes, and the RFLPs separated by electrophoresis and compared.

The fruiting bodies of the split-gill mushroom fungus *Schizophyllum commune* are readily recognized on natural woody substrates, and this tissue was collected for isolation and analyses of genomic DNA (James *et al.*, 2001). Using the primer sequences annealing to the 5S and 18S rDNA genes, the intragenic spacer regions were amplified by PCR technique and analyzed for molecular variation. Of the 195 strains collected from different geographical regions, 145 haplotypes (distinct DNA fingerprints) showed unique IGS1. The sequence data showed that populations from the Eastern Hemisphere, North America, South America, and the Caribbean are genetically discrete.

The ITS region has been proposed as the prime fungal barcode species identification (http://www.allfungi.com/its-barcode.php).

15.6.5 Mitochondrial DNA and Mitochondrial Plasmids

Not all variations are due to changes in nuclear DNA. Within a fungal species, mitochondrial DNA shows a high level of variation at the population level. This variation can be detected as RFLPs. In addition, natural populations of fungi have a variety of linear and circular plasmids in mitochondria (Table 15.4). Some of these plasmids integrate into mitochondrial DNA and cause death of strains that harbor these plasmids (Griffiths, 1995; Souza *et al.*, 2005). By simple subculture tests, natural populations of *Neurospora* have been broadly classified into senescence-prone and immortal strains (Chapter 16).

Table 15.4 Mitochondrial Plasmids Found in Naturally Occurring Strains of _Neurospora intermedia_

Plasmid	Structure	Size (bp)
pVARKUD	Circular	3675
pVARKUD SATELLITE	Circular	881
pFIJI	Circular	5268
pLABELLE	Circular	4070
pHARBIN	Circular	4.9 kb
p KALILO	Linear	8642
p MADDUR	Circular	3614
pMAURICEVILLE	Circular	3581

15.6.6 Karyotype Polymorphism

Although the small size of the fungal chromosomes, between 0.2 and 10 Mb of DNA, has been a deterrent in cytological studies, their small sizes can be resolved by pulsed field gel electrophoresis (Chapter 4) and therefore provide another potentially useful marker for analysis of variation. Chromosome length polymorphism—in the form of variation in chromosome size and/or number—was detected among field isolates of a pea root pathogen, _Nectria haematococca_, from diverse geographic origins. Variability was marked among minichromosomes smaller than 2 Mb. Five isolates of the wheat-blotch pathogen _Mycosphaerella graminicola_ had noticeably different karyotypes (McDonald and Martinez, see Kistler and Miao, 1992).

15.6.7 Spore Killer Elements

Spore killer elements in Ascomycotina are genes that cause the death of ascospores that do not contain the killer (Sk^k) elements. In _Neurospora_, distribution of spore killer allele in a population can be simply made by crossing tester strains that are sensitive to the killer (Turner, 2001). The strains to be tested are used as the male parent, and each group of octads (eight ascospores per ascus) in the sides of the cross tubes are examined under a dissecting microscope to determine if they are comprised of all black or four black and four white (aborted) ascospores. Spore killer elements are known in three other fungi: _Podospora_, _Gibberella_, and _Cochliobolus_ (Raju, 1994). In natural populations of _N. intermedia_, nonkiller strains are frequent; killer strains were found only in samples collected from Borneo, Java, and Papua New Guinea.

15.6.8 Multilocus Strain Typing

While multilocus strain typing (MLST) does not provide complete genome sequences, it generates evidence for similarities and differences between isolates from sequences determined, typically, from independent chromosomal genes. The best example of this method is in the human pathogen _Candida albicans_ where the ability to discriminate between strain types is important for tracing the sources and the routes of transmission of infection outbreaks. In practice, to type an isolate by MLST, DNA is extracted; housekeeping genes are PCR amplified. MLST directly measures the DNA sequence variations in a set of housekeeping genes (commonly seven) and characterizes strains by their unique allelic profiles. The principle of MLST is simple: PCR amplification followed by DNA sequencing and the display of the relatedness of isolates as a dendrogram constructed using the matrix of pair-wise differences between their allelic profiles.

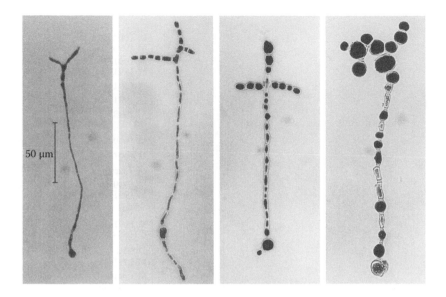

Figure 15.5 Stages in development of a microcycle structure in *N. crassa*, strain Vickramam, grown in suspension culture for 20–24 h. Stained with cotton blue.

15.6.9 Microcycle Conidiation

Following the accidental finding that a strain of *N. crassa* showed an unusual development called microcycle conidiation in submerged shake cultures, several strains of *Neurospora* were analyzed. In this development, conidial germlings enter into asexual reproductive phase bypassing the intervening mycelial phase (Figure 15.5). The phenomenon was found in 12 of the 114 *Neurospora* strains belonging to *N. crassa*, *N. sitophila*, and *N. discreta* from India, Papua New Guinea, Thailand, and the Ivory Coast (Maheshwari, 1991). Whether a strain is microcyclic is based on the phenotype in submerged suspension culture, not on agar-grown culture.

15.7 SPECIATION

Speciation is the process of origin of a new species. For example, *N. intermedia* occur on burned substrates in tropical and subtropical areas in both hemispheres. A type found almost exclusively on nonburned substrates (for example, on ontjom, a food item made by inoculating pressed soya or peanut cakes with *Neurospora*, or on corncobs) in the Eastern Hemisphere is yellow rather than pinkish-orange. The yellow "ecotype" is distinct also in its conidia (size and nuclear number), habitat, and ecology. Although the two types can be coerced to mate in laboratory conditions, there is as yet no evidence that the two types are members of an interbreeding population because of their geographical isolation. The phylogenetic trees constructed based on variation in the nontranscribed spacer suggested that the yellow isolates, mostly obtained from roasted maize kernels, are a separate lineage, distinct from a larger *N. crassa/N. intermedia* clade (Figure 15.6). No definite phylogeny was apparent other than that *N. discreta* was divergent from all other species. Rather, the yellow type is on the threshold of evolving into a distinct species. Though gene homology is identified more precisely than homology based on morphological characters, nevertheless caution is necessary in drawing inferences from molecular studies or phylogenetic trees constructed based on a single locus.

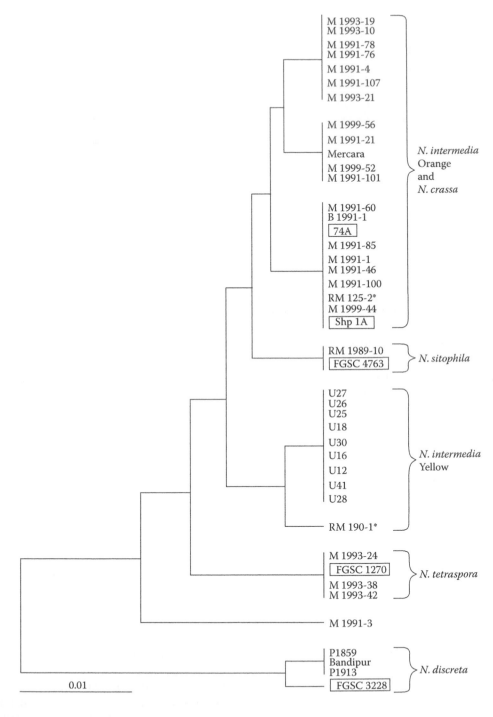

Figure 15.6 A phylogenetic tree of five heterothallic species of *Neurospora* collected from India. The phylogenetic tree is based on variations in the nontranscribed spacer of DNA. Standard species testers based on BSR are boxed. Asterisks are local species testers. The yellow ecotype of *N. intermedia* formed a monophyletic group. The scale indicates genetic distance. (After Adhvaryu, K.K. and Maheshwari, R., *Curr. Sci.* 82, 1015–1020, 2002).

15.8 CONCLUDING REMARKS

Although it is not easy to observe phenotypic differences, at the molecular level the degree of variability in fungi is similar to that displayed by man, mice, and flies. The trend in molecular analysis of populations is the application and development of DNA sequencing and large-scale analysis of single nucleotide polymorphisms using oligonucleotide microarray. The study of intraspecies diversity has provided insights into the role of production of asexual and sexual spores, the mode of spread and survival of fungal pathogens, mechanisms in generation and maintenance of variation, and the adaptive mechanisms in fungi. It has led to recognition of mitochondrial plasmids, of transposons, and of the discovery of senescing strains. Molecular markers allow the determination of phylogenetic relationships among individuals within and between subpopulations, charting the course of the evolution of species, and the understanding of biochemical design and adaptation in particular habitats. With the advent of recombinant DNA methodology, a new era in biotechnology has begun, but new products of industrial or pharmaceutical applications still result from the traditional approach of isolation and exploitation of variants in populations.

REFERENCES

Adhvaryu, K. K. and Maheshwari, R. (2002). Heterogeneity in NTS of rDNA in localized populations of *Neurospora*. *Curr. Sci.* 82:1015–1020.

Burdon, J. J. and Roelfs, A. P. (1985). The effect of sexual and asexual reproduction of the isozyme structure of populations of *Puccina graminis*. *Phytopathology* 75:1068–1075.

Burnett, J. (2003). *Fungal Populations and Species*. Oxford University Press, Oxford.

Buxton, E. W. (1960). Heterokaryosis, saltation, and adaptation, pp. 359–405. In: Plant Pathology, Vol. II, J. G. Horsfall and A. E. Dimond (eds.). Academic Press, New York.

Daboussi, M. J. (1996). Fungal transposable elements. *J. Genet.* 75:325–339.

D'Souza, A. D., Sultana, S., and Maheshwari, R. (2005). Characterization and prevalence of a circular mitochondrial plasmid in senescence-prone isolates of *Neurospora intermedia*. *Curr. Genet.* 47:182–193.

Dettman, J. R., Jacobson, D. J., Turner, E., Pringle, A., and Taylor, J. W. (2003). Reproductive isolation and phylogenetic divergence in *Neurospora*: Comparing methods of species recognition in a model eukaryote. *Evolution* 57:2721–2741.

Dettman J. R., Jacobson, D. J., and Taylor, J. W. (2006). Multilocus sequence data reveal extensive phylogenetic species diversity within the *Neurospora discreta* complex. *Mycologia* 98:436–446.

Dodge, B. O. (1927). Nuclear phenomena associated with heterothallism and homothallism in the ascomycete *Neurospora*. *J. Agric. Res.* 35:289–305.

D'Souza, A. D., Sultana, S., and Maheshwari, R. (2005). Characterization and prevalence of a circular mitochondrial plasmid in senescence-prone isolates of *Neurospora intermedia*. *Curr. Genet.* 47:182–193.

Duarte-Escalante, E., Zúniga, G., Ramirez, O. N., *et al.* (2009). Population structure and diversity of the pathogenic fungus *Aspergillus fumigatus* isolated from different sources and geographic origins. *Mern Inst. Osdwaldo Cruz* 104:427–433.

Gordon, T. R. and Okamoto, D. (1992). Population structure and the relationship between pathogenic and non-pathogenic strains of *Fusarium oxysporum*. *Phytopathology* 82:75–77.

Griffiths, A. J. F. (1995). Natural plasmids of filamentous fungi. *Microbiol. Rev.* 59:673–685.

Hawksworth, D. L. (1991). The fungal dimension of biodiversity: magnitude, significance, and conservation. *Mycol. Res.* 95:641–655.

Hawksworth, D. L. (2001). The magnitude of fungal diversity: The 1.5 million species estimate revisited. *Mycol. Res.* 112:1422–1432.

Jacobson, D. J. and Gordon, T. R. (1990). Further investigations of vegetative incompatibility within *Fusarium oxysporum* f. sp. *melonis*. *Can. J. Bot.* 68:1245–1248.

Jacobson, K. M., Miller Jr., O. K., and Turner, B. J. (1993). Randomly amplified polymorphic DNA markers are superior to somatic incompatibility tests for discriminating genotypes in natural populations of the ectomycorrhizal fungus *Suillus granulatus*. *Proc. Natl. Acad. Sci. USA* 90:9159–9163.

James, T. Y., Moncalvo, J.-M, Li, S., and Vilgalys, R. (2001). Polymorphism at the ribosomal DNA spacers and its relation to breeding structure of the widespread mushroom *Schizophyllum commune*. *Genetics* 157:149–161

Kistler, H. C. and Miao, V. P. W. (1992). New modes of genetic change in filamentous fungi. *Annu. Rev. Phytopathol.* 30:131–153.

Maheshwari, R. (1991). Microcycle conidiation and its genetic basis in *Neurospora crassa*. *J. Gen. Microbiol.* 137:2103–2106.

Maheshwari, R. (2003). Enzymes of thermophilic fungi. In *McGraw-Hill Yearbook of Science and Technology*, pp. 114–116, McGraw-Hill, New York.

McDonald, B. A. and McDermott, J. M. (1993). Population genetics of plant pathogenic fungi. *Bioscience* 43:311–319.

Megnegneau, B., Debets, F. and Hoekstra, R. E. (1993). Genetic variability and relatedness in the complex group of black *Aspergilli* based on random amplification of polymorphic DNA. *Curr. Genet.* 23:323–329,

Nierman, W. C., Pain, A., Anderson, M. J., *et al.* (2005). Genomic sequence of the pathogenic and allergenic filamentous fungus *Aspergillus fumigatus*. *Nature* 438:1151–1156.

O'Gorman, C. M., Fuller, H. T., and Dyer, P. S. (2009). Discovery of a sexual cycle in the opportunistic fungal pathogen *Aspergillus fumigatus*. *Nature* 457:471–474.

Perkins, D. D. and Raju, N. B. (1986). *Neurospora discreta*, a new heterothallic species defined by its crossing behavior. *Exp. Mycol.* 10:323–328.

Perkins, D. D. and Turner, B. C. (1988). *Neurospora* from natural populations: Toward the population biology of a haploid eukaryote. *Exp. Mycol.* 12:91–131.

Powell, A. J., Jacobson, D. J., Saltere, L., and Natvig, D. O. (2003). Variation among natural isolates of *Neurospora* on small spatial scales. *Mycologia* 95:809–819.

Perkins, D. D. (1991). In praise of diversity. In J. W. Bennett and L. S. Lasure, eds., *More Gene Manipulations*. Academic Press, New York.

Raghukumar, S. (2006). Marine microbial eukaryotic diversity, with particular reference to fungi: Lessons from prokaryotes. *Indian J. Marine Sci.* 35:388–398.

Raju, N. B. (1994). Ascomycete spore killers: Chromosomal elements that distort genetic ratios among the products of meiosis. *Mycologia* 86:461–473.

Rayner, A. D. M. (1991). The phytopathological significance of mycelial individualism. *Annu. Rev. Phytopathol.* 29:305–323

Sanders, I. R. (1999). No sex please, we're fungi. *Nature* 399:737–739.

Shear, C. L. and Dodge, B. O. (1927). Life histories and heterothallism of the red bread –mold fungi of the *Monilia sitophila* group. *J. Agric. Res.* 34:1019–1042.

Spieth, P. T. (1975). Population genetics of allozyme variation in *Neurospora intermedia*. *Genetics* 80:785–805

Stakman, E. C. (1950). *Principles of Plant Pathology*. Ronald Press, New York.

Subramanian, C. V. (1992). Tropical mycology and biodiversity. *Curr Sci.* 63:167–172.

Summerwell, B. A., Laurence, M. H., Liew, E. C. Y., and Leslie, J. F. (2010). Biogeography and phylogeography of *Fusarium*: A review. *Fungal Diversity* 44:3–13.

Tavanti, A., Gow, N. A. R., Senesi, S., Maiden, M. C. J., and Odds, F. C. (2003). Optimization and validation of multilocus sequence typing for *Candida albicans*. *J. Clin. Microbiol.* 41:3765–3776.

Taylor, J. W., Jacobson, D. J., Kroken, S., Kasuga, T., Geiser, D. M., Hibbett, D. S. and C. Fisher, M. S. (2000). Phylogenetic species recognition and species concepts in Fungi. *Fung. Genet. Biol.* 31:21–32.

Turner, B. C. (1987). Two ecotypes of *Neurospora intermedia*. *Mycologia* 79:425–432.

Turner, B. C. (2001). Geographic distribution of *Neurospora* spore killer strains and strains resistant to killing. *Fung. Genet. Biol.* 32:93–104.

Turner, B. C., Perkins, D. D., and Fairfield, A. (2001). *Neurospora* from natural populations: A global study. *Fung. Genet. Biol.* 32:67–92.

Xu, J. (2006). Fundamentals of fungal molecular population genetic analyses. *Curr. Issues Mol. Biol.* 8:75–90.

CHAPTER **16**

Senescence

The fungi ... are progressive, ever changing and evolving rapidly in their own way, so that they are capable of becoming adapted to every condition of life. We may rest assured that as green plants and animals disappear one by one from the face of the globe, some of the fungi will always be present to dispose of the last remains.

—**B. O. Dodge (1939)**

The ability to continuously regenerate hyphal tips gives fungi the potential for indefinite growth. This is manifested by the circular development of aerial fruiting bodies (basidiocarps) known as the "fairy ring" (Figure16.1) from underground mycelium. According to a myth, the circular pattern of basidiocarps represents the path of fairies dancing in the night. Actually, the fairy ring develops from the perennial, subterranean mycelium that extends outward and produces ephemeral fruit bodies, year after year at the periphery of the mycelium. From the rate of expansion of the diameter of the fairy rings, the age of mycelium in some cases has been estimated to be several hundred years. In Chapter 1, we learned about a colony of *Armillaria bulbosa* that is more than a millennium old. The most important reason for the immortal nature of fungi is the cytoplasmic continuity of the hyphal compartments. Aged or dysfunctional nuclei and mitochondria can be recycled and replaced by the migration of functional copies from other cellular compartments. Additionally, the multinuclear condition of hypha enables any potentially deleterious nuclear gene mutation to be complemented by its functional allele in other nuclei. However, in the rare cases of senescing cultures, cytoplasmic continuity of hyphal compartments enables the altered mitochondria to populate the mycelium, resulting in respiratory deficiency and death of the culture.

16.1 DISCOVERY

Podospora and *Neurospora* (Ascomycotina) are commonly used in research. Their stocks are maintained by regular subculturing. Not surprisingly, the phenomenon of senescence is best documented in these two fungi. In *Podospora anserina* the mycelium grows to a limited extent following the germination of the ascospore. Therefore, in this species senescence is a part of its normal development. On the other hand, cultures of *Neurospora* used by Beadle and Tatum nearly 70 years ago are still alive after numerous subcultures. However, nearly 30% of strains of *N. intermedia* collected from burned vegetation (Chapter 10) in Hawaii or from Maddur in peninsular India died in five or more subcultures. Indeed, obtaining a Methuselah strain from this place proved difficult. Initially, a senescent strain is morphologically indistinguishable from a long-living strain. However, as subculturing is continued, the quantity of aerial mycelium is reduced, conidia are not formed, the levels of cytochrome decrease and the respiratory activity diminishes (Figure 16.2). Eventually, growth

Figure 16·1   Fairy ring. (From Angela B. Shiflet, *Introduction to Computational Science*, Princeton University Press, Princeton, N.J., 2006, with permission.)

ceases irreversibly, often abruptly, and the culture is regarded to have died. Thus, if a new isolate is likely to be used for investigation of senescence phenomenon, parallel subcultures from the initial isolates should be metabolically immobilized by storage in anhydrous silica gel.

16.2 TERMINOLOGY OF SENESCENCE

Investigations on senescence in fungi have a strong genetic approach. Therefore, we review the symbols used in reporting results on senescence phenomenon in fungi. The wild-type gene is in three-letter lowercase italics with the superscript "+"; it is infrequently in two letters, for example, the wild-type allele *natural death* is in lowercase italics (*nd*). Its putative protein product is in nonitalic capital letters, e.g., ND. A heterokaryon between a senescing and nonsenescing strain is denoted by enclosing symbols of the component nuclei, separated by a plus sign, in parentheses, as for example (*nd* + *nd*$^+$) or, in case of the mutant *senescent* as (*sen* + *sen*$^+$). The mitochondrial mutants are in brackets, for example (*poky*). Several cases of senescence are due to mitochondrial plasmids. The plasmid name is in capital letters prefixed with a lower case "p," e.g., pKALILO, abbreviated pKAL (names and symbols of plasmids not italicized). The free autonomously replicating pKAL is denoted as AR-kalDNA, and the pKALILO inserted into mitochondria as mtIS-kalDNA. The mutant mitochondrial genomes are italicized. For example, the *stopper* mutant, which has an irregular stop-start growth phenotype due to defective mitochondrial DNA (mtDNA) molecules, is abbreviated as [*stp*].

16.3 NUCLEUS- OR MITOCHONDRIA-BASED SENESCENCE

Since both nuclei and mitochondria contain DNA, the cellular location of the factor determining senescence can be analyzed from its mode of inheritance.

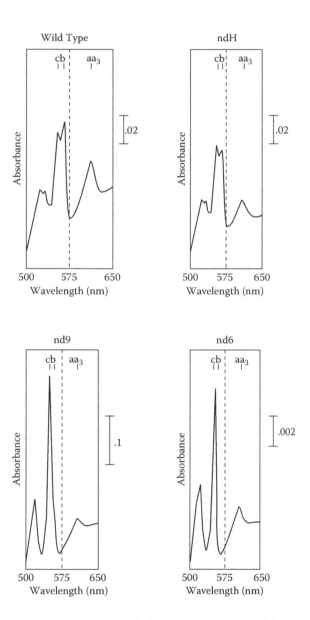

Figure 16.2 Cytochrome content in wild-type and senescent *natural death (nd)* strains of *Neurospora crassa*. The cytochromes were identified by recording the difference in absorbance between dithionite-reduced and ferricyanide-oxidized mitochondria at wavelengths of light from 500 to 650 nm. The absorbance maxima at 550, 560, and 608 nm correspond to cytochrome *c, b,* and cytochrome oxidase *aa₃* and *c*. The senescent strain maintained in a heterokaryon (*ndH*) has spectra similar to the wild type. (Redrawn from Seidel-Rogol, B.L., King, J., and Bertrand, H., *Mol. Cell Biol.* 9, 4259–4264, 1989.)

16.3.1 Genetic Cross

When two parents are crossed, generally the cytoplasm (mitochondria) from the maternal (female) parent is transmitted to the progeny; rarely is there transfer of cytoplasm from the paternal (male) parent. In *P. anserina* (Ascomycotina) a cross between "exceptional" nonsenescent female and senescent male yielded progeny ascospores that grew into nonsenescent cultures, whereas the reciprocal cross senescent female × nonsenescent male (exceptional) yielded senescent progeny. It was inferred that the senescence-determining factor is inside the mitochondria. In designating a genetic cross, the perithecial (female) parent is written on the left side of the cross (×) symbol.

16.3.2 Heterokaryon Test

In this test, germinating conidia or hyphal cells of a senescent strain are mixed with that of a genetically related but nonsenescent strain. The nuclei and mitochondria in the fused cell multiply and become mixed and the senescent nuclei are sheltered indefinitely. The nuclei from a senescent strain can be preserved in heterokaryons with nonsenescent type nuclei (Figure 16.3). To "force" heterokaryotic growth, auxotrophic markers—for example, genetic markers pantothenic acid (*pan*) and leucine (*leu*)—are incorporated into nuclei of the test strains by prior crossing. The two nuclear types can be extracted from the conidiating heterokaryotic mycelium by plating conidia on selective media. Colonies derived from conidia are homokaryotic. If *pan* colonies senesce on medium supplemented with pantothenic acid medium, the senescent determinant is in the cytoplasm (mitochondria).

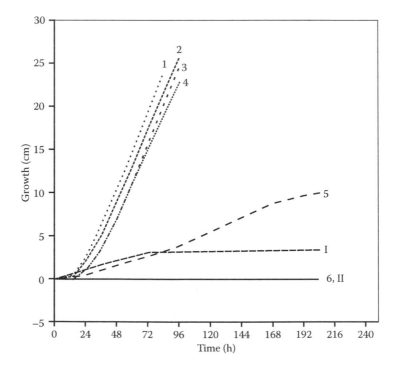

Figure 16.3   Growth curve of a senescing mutant. Ascospore-derived culture was subcultured (subculture number in parentheses) in race tubes at 22°C (1 to 6) and 34°C (I and II). (From Navaraj, A., Pandit, A., and Maheshwari, R., *Fung. Genet. Biol.* 29:165–173, 2000.)

16.4 SENESCENCE AND OTHER DEATH PHENOMENA

The hallmark of fungal senescence is the progressive loss of growth potential of mycelium, culminating in total cessation of growth, when the culture is considered as dead. During senescence the respiration of mycelial suspensions or conidia measured as oxygen uptake switches from cyanide-sensitive, cytochrome-mediated to a KCN-insensitive, salicylhydroxamic acid (SHAM)-sensitive, alternate pathway mediated by alternate oxidase and is associated with deficiencies of cytochrome *a* and *b* (Navaraj *et al.*, 2000). Because of the interconnected hyphal cells and cytoplasmic movement through the septal pores, genetically altered mitochondria that arise spontaneously accumulate throughout the mycelium—a condition called suppressiveness (Griffiths, 1998). The suppressiveness of mutant mitochondrial DNA is explained on the basis of their smaller size due to deletions such that the mutant DNA molecules become dominant with progressive culturing of culture, ultimately in severe deficiency of ATP.

Different ultrastructural abnormalities are reported in senescing hyphae: Glycogen-like material and an almost electron transparent cytoplasm (Niagro and Mishra, 1990), vacuolization, breakdown of nuclear and mitochondrial membranes, loss of cristae, and accumulation of dense material in the mitochondrial matrix (Bok *et al.*, 2003). Since there is neither typical DNA fragmentation nor the release of cytochrome *c*, senescence is distinct from apoptosis. Senescence is not equivalent to autolysis because there is no hyphal fragmentation or lysis of the cell wall. Senescence is perhaps not analogous to necrosis either, because there is no loss of plasma membrane integrity or the release of cellular constituents. In the strains bearing pKALILO, the senescing mycelium showed large vacuoles. An expected subcellular alteration in senescing mycelium would be extensive depolymerization of the actin-cytoskeleton.

The lifespan of senescing strains collected from the same place may vary. However, the replicate subcultures from the same parent stock die in about the same number of subcultures. Interestingly, the death of duplicate senescing cultures of *Neurospora* after these had been stored at −15°C for 12 months occurred in fewer subcultures, indicating that degenerative changes may continue even in the frozen state (Maheshwari and Navaraj, 2008). Though few fungal cultures have been tested by prolonged serial subculturing, senescence in fungi is an exception rather than a rule.

16.5 CYTOPLASMIC MUTANTS

How did the idea of fungi senescence come about? In 1965, E. L. Tatum's group described spontaneous mutants of *N. crassa* that showed stop-start growth on agar media. Microinjection of mitochondria preparations from the normal strain into the slow-growing strain restored normal growth but microinjection of DNA from nuclei had no effect, indicating that the abnormal phenotype is determined by mitochondria. The mitochondrial DNA (mtDNA) of these mutants—for example, (*stopper*)—has deletions and exhibits severe deficiencies of cytochrome *b* and *aa3* (Almasan and Mishra, 1991). The defective mitochondria are maintained in a heteroplasmic state with normal mtDNA, as in several human diseases (Wallace, 1999).

16.6 PRESERVATION OF SENESCING STRAINS

In *N. crassa*, an ultraviolet-induced mutant was obtained that died in two to four subcultures. The mutant was named *natural death* or *nd* for short. Despite its limited life, the mutant *nd* could be, and has been, maintained in combination with nuclei having the wild type (*nd⁺*) allele in a heterokaryon. Genetic markers—for example, *albino*—can be introduced in mutants by crossing and

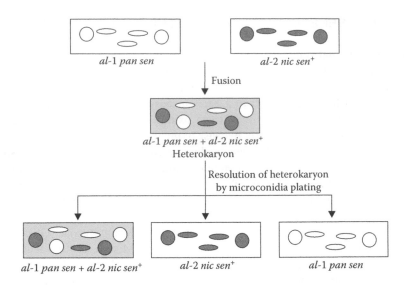

Figure 16.4 Heterokaryon technique of preservation and use of a recessive lethal nuclear gene mutant of *Neurospora crassa*. The mycelia are represented by rectangles, nuclei by open circles, and mitochondria by ovals. Shaded circles or ovals represent sen⁺; open circles or ovals represent sen. The (sen + sen⁺) heterokaryon grew on minimal medium and produced orange conidia. It was resolved into components by plating macroconidia. White cultures that grew on a nicotinamide-supplemented medium but not on minimal or pantothenate medium were tested for senescence. The heterokaryon and the resolved sen⁺ cultures were analyzed for mtDNA deletions.

following plating of conidia, the white *nd* colonies carrying the *al* marker allele selected. When homogeneous *nd* cultures are desired, a mutant *nd* is recovered by plating conidia formed by the heterokaryon. After extraction, the *nd* homokaryon dies again in two to four subcultures, but this time having made sexual crosses and mycelium to be grown for biochemical analysis.

The first nuclear gene mutant of *Neurospora* derived from nature was named *senescent* (Navaraj *et al.*, 2000). Actually *sen* (three-letter abbreviation generally followed) was derived from a phenotypically normal strain that was collected from a sugarcane field in southern India. The recessive, lethal nuclear gene *sen* was introgressed into wild-type *N. crassa* through several backcrosses to make the strain fusion (heterokaryon) compatible. The heterokaryon grows and forms conidia. When required, the homokaryotic *sen* nuclei are extracted in the form of conidia by plating macroconidia (spores). The selection is aided if the nuclei are genetically marked with a color marker gene. Refer to Figure 16.4, which illustrates a technique in which color markers *al-1* or *al-2* were introduced into senescing strain before fusion (heterokaryon formation). The orange-colored heterokaryon due to complementation of *al-1* and *al-2* could be preserved indefinitely, and when required for experimentation the albino senescing mutant is extracted from the heterokaryon by plating macroconidia (D'Souza *et al.*, 2005b). The heterokaryon containing either senescing *nd* or *sen* has a cytochrome spectrum similar to that of wild type, whereas *nd* and *sen* homokaryon have reduced levels of cytochrome *c* or cytochrome oxidase (Navaraj *et al.*, 2000).

16.7 INSTABILITY OF MITOCHONDRIAL DNA

From Section 16.6, we note that in association with its normal allele in other nuclei of the heterokaryon, the culture containing the recessive nuclear gene is indefinitely stable. However, when *nd* or *sen* nuclei are separated from wild-type nuclei, mtDNA in the senescing cultures undergoes rapid deletions.

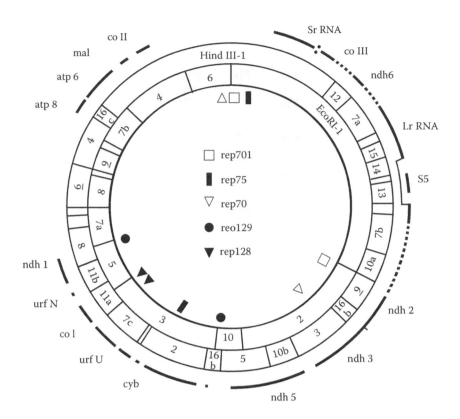

Figure 16.5 Map of wild-type *Neurospora crassa* mitochondrial DNA. The outer circle shows the fragments produced by *Hind*III restriction enzyme. The inner circle shows the fragments produced by *Eco*RI restriction enzyme.

The sequence of 68.5-kb circular mitochondrial DNA molecule of *N. crassa* (Figure 16.5) is known. The enzyme *Eco*RI cuts mtDNA into 11 electrophoretically separable bands (Figure 16.6). A comparison of the restriction digest profiles of mitochondrial DNA from *nd* and *nd⁺* mycelia showed unique fragments in the *nd* homokaryon recovered from the heterokaryon that are not present either in the wild *nd⁺* or in the [*nd* + *nd⁺*] heterokaryon (Seidel-Rogol *et al.*, 1989; Bertrand *et al.*, 1993). The unique *Eco*RI restriction fragments from mitochondrial DNA were cloned and sequenced. Results showed that mtDNA in *nd* had suffered deletions and gross sequence rearrangements. The nucleotide sequence suggested intramolecular recombination. The mitochondrial DNA has stretches of palindrome sequences (5′-CCCTGCAGTACTGCAGGG-3′), which can potentially form hairpin structures (Almasan and Mishra, 1991). These hairpin structures could promote intramolecular recombination between homologous repeats, resulting in deletions. Thus defective growth mutants of fungi have revealed that mtDNA is innately unstable (prone to recombination events and deletions). These findings suggest that the structure of mitochondrial DNA is by default susceptible to self-degeneration, being preserved only because of nuclear-encoded proteins in the cytoplasm that are translocated inside mitochondria for preservation of the mitochondrial genome.

In *P. anserina*, genetic tests localized the senescence determinant in mitochondria. A restriction enzyme analysis revealed that mitochondrial DNA (Figure 16.7) from the senescing cultures contained unique circular DNA molecules comprised of head-to-tail monomer (2.05 µM), dimer, trimer, tetramer, or pentamer sizes that were termed the senDNA (Jamet-Vierny *et al.*, 1980; Wright and Cummings, 1983). SenDNA hybridized to restriction fragments of the mitochondrial DNA but not of nuclear DNA, revealing that it is homologous to mitochondrial DNA. Several sen DNA, named α,

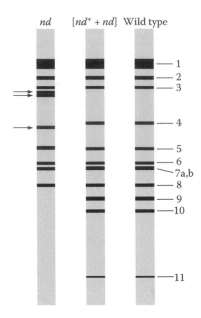

Figure 16.6 Diagram of restriction fragment patterns from mitochondrial DNA of wild-type (*nd⁺*) and *nd* mutants, and a [*nd⁺* + *nd*] heterokaryon. (Adapted from Bertrand, H., Wu, Q., and Seidel-Rogol, B.L., *Mol. Cell. Biol.* 13, 6778–6788, 1993.)

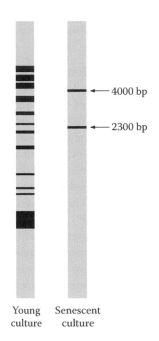

Figure 16.7 Diagram of electrophoretic patterns of mitochondrial DNA from young and senescent cultures of *Podospora anserina*. Arrows point to unique 4000-bp and 2000-bp fragments in senescent strain after *Hae*III digestion. (Based on Jamet-Vierny, C., Begel, O., and Belcour, L., *Cell* 21, 189–194, 1980.)

Table 16.1 Mitochondrial Plasmids in *Neurospora*

Plasmid Name	Type	Size (kbp)	Source
Mauriceville	Circular	3.6	*N. crassa*
Maddur	Circular	3.6–3.7	*N. intermedia*
Varkud	Circular	3.8	*N. intermedia*
Labelle	Circular	4.07	*N. intermedia*
Fiji	Circular	5.2	*N. intermedia*
Kalilo	Linear	9	*N. intermedia*
Maranhar	Linear	7.2	*N. intermedia*

β, γ, etc., are produced by the deletion and amplification of separate regions of mitochondrial DNA, but senDNAα is produced regularly. Its production was traced to site-specific deletion and amplification of intron α, which is the first intron of the mitochondrial *COX1* gene, which encodes subunit I of the respiratory enzyme, cytochrome *c* oxidase. Wright and Cummings (1983) reported that senDNAα probes hybridized to nuclear DNA from the senescing mycelium and hypothesized that a mitochondrial genetic element is transposed to the nucleus and is integrated into nuclear DNA.

Although senescence in *P. anserina* shows extranuclear (maternal) inheritance, in certain nuclear gene mutants the time of senescence is altered, suggesting that nuclear genes control integrity of mitochondrial DNA. For example, in contrast to the wild type that died in less than 21 days, the double mutant *incoloris vivax* (*i viv*) remained alive for at least four years (Tudzynski and Esser, 1979). Whereas the mutant *AS1-4* dies prematurely in five to six days (Belcour *et al.*, 1991), the mutant *grisea* has an extended lifespan (Borghouts *et al.*, 1997). These observations suggested that several nuclear genes control the synthesis of factors that are imported from the soluble phase of cytoplasm (cytosol) into mitochondria where they function to stabilize mitochondrial genome.

16.8 SENESCENCE-INDUCING PLASMIDS

In fungi, plasmids were discovered in a screen of natural populations of *Neurospora* for structural variants of the mitochondrial chromosome. The presence of plasmid is manifested by a brightly staining band on gels after electrophoresis of restriction enzyme digests of mitochondrial DNA preparations stained with ethidium bromide (Collins *et al.*, 1981). The mere presence of plasmid does not identify a strain as senescence-prone because strains can harbor harmless plasmids.

Plasmids (Table 16.1) are implicated in senescence when they are co-inherited maternally with the senescence character and are integrated into mitochondrial DNA in senescing mycelium (Griffiths, 1995).The first plasmid discovered was MAURICEVILLE, named after the place where the host *Neurospora* strain was collected. pMAU is a closed-circular DNA molecule (a concatamer of up to six repeats of a monomer of about 3.6 kb). In Southern hybridization, pMAU hybridized neither to mitochondrial DNA nor to nuclear DNA, refuting its origin from either of these DNA molecules.

The majority of *N. intermedia* strains collected from Kauai (in Hawaii) are senescent and were named "kalilo," which in the Hawaiian language means "dying." Some wild strains of *N. crassa* collected from Aarey near Mumbai, India, died in about 20 subcultures and these strains were named "maranhar," which in the Hindi language means "prone to death" (Court *et al.*, 1991). Kalilo and maranhar strains contain 9- and 7-kb plasmids, respectively. The structural features of the plasmids have been deduced from electron microscopy, gel electrophoresis, sensitivity to 5′ and 3′ exonuclease, and sequencing. Both pKAL and pMAR are linear plasmids having terminal regions with inverted repeats of base sequences. A protein is bound to the 5′ termini inverted repeats as indicated by resistance to 5′ exonuclease digestion (Griffiths, 1995). There are two open reading frames whose amino acid sequences suggest that they encode DNA and RNA polymerases. pKAL and pMAR do

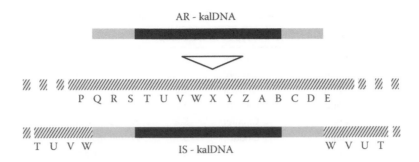

Figure 16.8   Integration of KALILO plasmid into mitochondrial genome. The letters for regions of mtDNA are arbitrarily labeled by letters. AR = autonomously replicating plasmid; IS = insertion sequence of plasmid. mtDNA is linearized for purpose of illustration. (Adapted from Griffiths, A.J.F., *Microbiol. Revs.* 59, 673–685, 1995.)

not have sequence homology. Based on detection by Southern hybridization, KALILO or homologous plasmids were found in *Neurosporas* collected from different regions of world.

Radiolabeled DNA sequences of pKAL and pMAR hybridized only to mitochondrial DNA from senescing cultures, showing that plasmids inserted into mitochondrial genome, consistent with the maternal transmission of the plasmid DNA (Myers *et al.*, 1989; Court *et al.*, 1991). The full-length plasmid, flanked by inverted terminal repeats, inserts at a single site into mitochondrial DNA. The plasmids therefore exist in mitochondria in two forms: An autonomously replicating (AR) form and as an insertion sequence (IS), which for the kalilo are denoted mtAR-kalDNA and mtIS-kalDNA, respectively. Plasmids could induce senescence by inserting into mtDNA and inactivating an indispensable gene (Figure 16.8).

16.8.1 Variant Plasmids

During continuous growth in race tubes or during sequential conidial transfers in agar slants, the Mauriceville strain of *N. crassa* and the Varkud strain of *N. intermedia* exhibited stop-start growth or senesced. The variant strains had deficiencies of cytochrome aa_3 and *b*. Densitometry of ethidium bromide stained gel following electrophoresis of total mtDNA showed plasmid accumulation relative to the mitochondrial DNA, i.e., they had become suppressive. DNA sequencing revealed the insertion of a mitochondrial tRNA sequence and short deletions in the plasmid DNA. As both plasmids encode reverse transcriptase, it was hypothesized that during replication, the variant forms of plasmids arise from the benign forms via an RNA intermediate and reverse transcription of full-length plasmid transcript that integrates into mitochondrial DNA by homologous recombination.

16.8.2 Spread of Plasmids

Related plasmids have been discovered in strains collected globally (Yang and Griffiths, 1993). For example, all isolates of *N. intermedia* collected from Maddur in southern India were senescent and contained pMAD that was 90% homologous to pMAU first found in the Mauriceville strain of *N. crassa* collected from Texas. This raises the interesting question as to how plasmids spread. One possibility is that when multiple strains are growing together as in burned sugarcane (see Figure 10.1), plasmid-harboring and plasmid-lacking strains fuse and the mitochondria become mixed. To test this, a senescing kalilo strain was constructed having the nuclear marker genes *nic-1*, *al-2* (Griffiths *et al.*, 1990). This was mixed with a nonsenescent strain having the marker genes *ad-3B*, *cyh-1*. This heterokaryon was senescent, demonstrating that hyphal fusion could allow the plasmid to spread in nature (horizontal transmission). In some cases, vertical transmission of

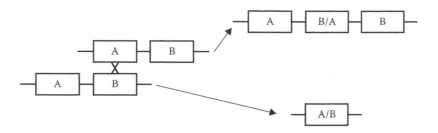

Figure 16.9   Diagram of unequal crossing over between repeat sequences. (Based on Bertrand, H., Wu, Q., and Seidel-Rogol, B.L., *Mol. Cell. Biol.* 13, 6778–6788, 1993.)

plasmid has been demonstrated following limited sexual recombination (Bok and Griffiths, 1999). Further investigations are required to determine how the spread of plasmids in nature is checked.

16.9 NUCLEAR GENE MUTANTS

The biogenesis of mitochondria involves the parallel development of a mitochondrial membrane system and the proteins of oxidative phosphorylation. The large number of proteins are synthesized on the cytoplasmic ribosomes from nuclear mRNA and imported from the cytosol into the mitochondria via multiprotein complexes *t*ranslocase of *o*uter *m*embrane (Tom) and *t*ranslocase of *i*nner *m*embrane (Tim) A number of nuclear gene mutants are known with cytochrome aa_3 deficiency showing very slow or sparse growth (Perkins *et al.*, 2001). Two well-characterized mutants are discussed in Sections 16.9.1 and 16.9.2. The finding of these mutants raises the possibility that more cases might be found that are due to nuclear gene mutations.

16.9.1 Natural Death

In *N. crassa*, following ultraviolet irradiation of conidia, a single nuclear gene mutant was obtained that died abruptly after two to four sequential transfers. It is preserved in a heterokaryon in association with the wild type, as already described. This technique of preserving potentially lethal mutations has made *N. crassa* the most attractive fungus for investigation of the phenomenon of aging and death. Although the *nd* gene has not yet been cloned, some results concerning the effect of this gene may be summarized.

On the basis of their migration in agarose gels, 11 bands in *Eco*RI digest of mtDN can be distinguished (see Figure 16.6). A comparison of the restriction digest profiles of mitochondrial DNA from *nd* and *nd*[+] mycelia showed unique fragments in the *nd* homokaryons recovered from the heterokaryon that are not present either in the wild-type *nd*[+] or in the [*nd* + *nd*[+]] heterokaryon (Seidel-Rogol *et al.*, 1989; Bertrand *et al.*, 1993). The unique *Eco*RI restriction fragments from mitochondrial DNA were cloned and sequenced. Comparison of its sequence with wild-type mtDNA revealed that *nd* had suffered short deletions and its nucleotide sequences were rearranged. The deleted segments were those that have a palindrome sequence, suggesting that this sequence gets recognized and is removed by specific endonucleases and the separated fragments juxtaposed. It was hypothesized that unequal crossing-over between repeat sequences results in deletions followed by recombination of distant nucleotide sequences (Figure 16.9).

16.9.2 Senescent

The nuclear gene mutant *senescent* was derived from a naturally occurring isolate that was phenotypicaly indistinguishable from the wild type. The technique used was to extract nuclei from mycelium in the form of uninucleate microconidia (Navaraj *et al.*, 2000) and to grow them into homokaryotic cultures (Figure 16.10). One out of 159 homokaryotic cultures exhibited the "death" phenotype. In crosses of senescent × wild type, the mutant and wild type segregated in a 1:1 ratio, demonstrating that *senescent* is a single nuclear-gene mutant (Figure 16.11). Restriction fragment analysis of mitochondrial DNAs from *sen*+ showed that senescing mycelia had suffered deletions and gross sequence rearrangements.

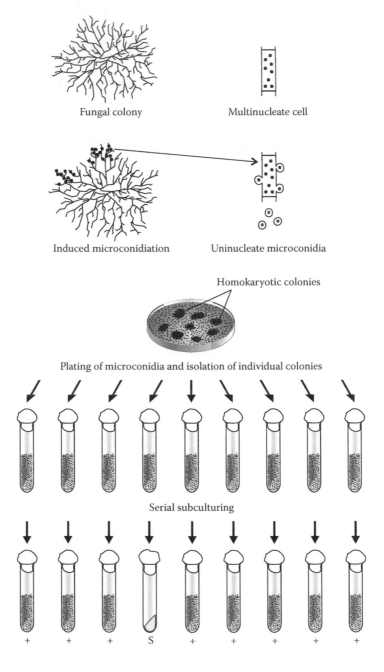

Figure 16.10 Derivation of *senescent* mutant from a heterokaryotic wild isolate of *Neurospora* using microconidia.

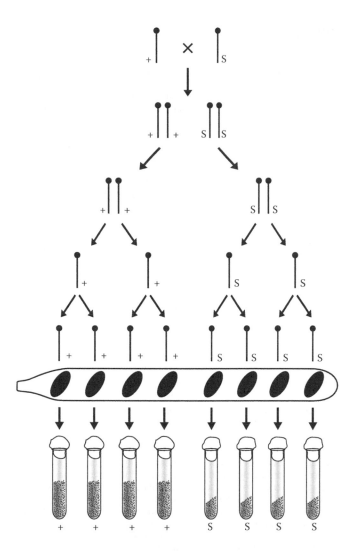

Figure 16.11 Inheritance of a single nuclear gene senescent mutant of *Neurospora*. Mutant strain was crossed to wild-type strain and ascospore progeny from a single octad analyzed for growth.

The time of death of the *sen* and *nd* mutants is highly influenced by temperature. At 34°C, *sen* progeny from cross could be scored immediately or in one subculture. At 26°C, four to six subcultures were required for identifying the *sen* progeny. In *N. crassa* several other mutants referred to as *unknown* exhibit thermosensitive growth. To explain marked thermosensitivity, a hypothetical model has been proposed (Navaraj and Maheshwari, 2005). It is speculated that SEN protein synthesized in the cytosol is an essential component of the multiprotein translocase of outer mitochondrial membrane (Tom) involved in recognizing proteins made on cytoplasmic ribosome and their import into mitochondria (Neupert, 1997). Presumably, depending on the growth temperature, SEN assumes two metastable structures and is inserted into the mitochondrial membrane (Figure 16.12). Once the mutant form of SEN is trapped in the mitochondrial membrane, conformational interconversion is unaffected by temperature.

The discovery of *nd* and *sen* mutants raises the possibility that other cases of senescence would be found that are due to nuclear mutations in genes encoding components of the Tom or the Tim complex.

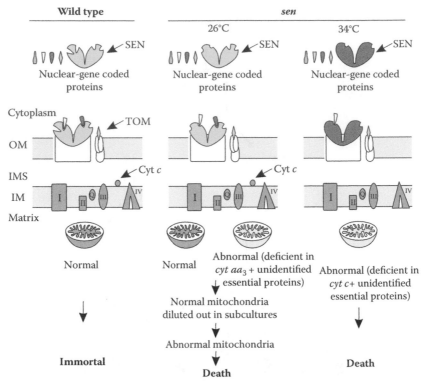

OM - Outer membrane IMS - Inter - Membrane Space IM - Inner Membrane
TOM - Translocase of Outer Membrane Cyt c - Cytochrome C

Figure 16.12 A hypothetical model to explain thermosensitivity of mutant strains. The wild-type *sen+* is assumed to encode a polypeptide component of the mitochondrial outer membrane receptor complex and is involved in transport of several nuclear-encoded proteins, including cytochrome *c* heme lyase (oval) and subunit polypeptides of cytochrome oxidase (triangle). The mutant SEN protein exists in two conformational states depending on the growth temperature. The conformational states of the mutant SEN protein are stabilized by insertion into the membrane. At 26°C, the conformation of SEN approximates the native state but has reduced binding affinity for some precursor proteins destined for mitochondrial compartments. As a consequence some mitochondria formed are deficient in some mitochondrial proteins (including cytochrome oxidase). The deficient (abnormal) mitochondria eventually dominate the normal mitochondria and cultures senesce with passages. At 34°C, a severe alteration in the conformation of SEN affects binding of several cytosolic peptides, and death occurs faster.

16.10 LINK WITH AEROBIC RESPIRATION

Senescence is correlated with cyanide-insensitive alternative respiratory pathway: The electron flow at the level of ubiquinone (coenzyme Q) is bypassed and the enzyme alternative oxidase transfers the electron directly to molecular oxygen ($O_2 + e^- \rightarrow O_2^-$), giving rise to superoxide anion. (A superoxide anion is highly destructive and leads to formation of hydrogen peroxide H_2O_2 and hydroxyl radicals OH^-, which are collectively called *reactive oxygen species*, ROS. ROS oxidize and damage nucleic acids, lipid, and protein molecules, and damage cells and tissues, resulting in aging and ultimately death.) Since alternative oxidase (Figure 16.13) is located upstream of complex III, the production of ATP is restricted to complex I and consequently the production of ATP is lowered. To determine which of the aforementioned biochemical modifications is the primary cause of senescence, Dufour *et al.* (2000) examined the effects of specific inactivation of

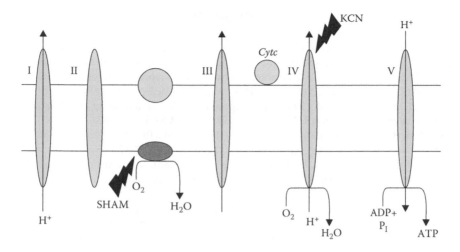

Figure 16.13 Mitochondrial respiratory assemblies that transfer electrons to oxygen. AOX, alternative oxidase, SHAM, salicyl hydroxamate.

cytochrome oxidase on lifespan, ROS formation, and senDNA accumulation using *P. anserina* since it has (i) a short life-span, (ii) a cytochrome-mediated aerobic respiratory pathway, (iii) an alternative respiratory pathway linking the oxidation of ubiquinol directly to the reduction of oxygen to water and playing a crucial role in protecting against the lethal effects of ROS, and (iv) well-developed genetic and molecular techniques. The functional cytochrome gene was replaced by transformation with a plasmid construct that contained the cytochrome oxidase gene lacking 169 bp of the *COX5* gene. Replacement of the functional wild-type *COX5* gene by a defective copy of synthetic gene prevented the accumulation of senDNA, decreased the formation of ROS, and increased the lifespan of the fungus. Reintroduction of a functional *COX5* gene into the deleted strain restored the original phenotype, i.e., the original growth rate, reduced lifespan, loss of fertility, and increased formation of ROS without the accumulation of senDNA. It was hypothesized that aerobic respiration mediated by cytochrome oxidase is the general cause of aging and senescence of eukaryotic organisms.

16.11 CONCLUDING REMARKS

Fungi normally do not senesce, but in some species mitochondrial plasmids induce senescence. Studies of senescence in fungi provide a basis for the endosymbiotic hypothesis of eukaryotic cells. The finding of nucleus-gene senescent strains supports the notion that nuclear gene-encoded factor(s) protect the mitochondrial genome from deletions and recombination events. The nucleotide sequence of mitochondrial DNA revealed palindromic sequences that have the potential of forming secondary structures prone to intramolecular recombination and deletions. It is likely that certain nuclear-encoded proteins protect mtDNA from deletions in different regions of mtDNA. A nuclear gene mutation that affects the proteinaceous protective factors or that affects the component of the multiprotein translocase in import of proteins into mitochondria will be at a severe disadvantage in the cell and affect viability. It would be important to find evidence for proteins capable of recognizing palindromic sequences and/or altered components of mitochondrial protein import machinery. Senescing strains support the idea that heteroplasmons and heterokaryons representing mixed populations of mitochondria and nuclei are crucial to longevity and survival of fungi.

REFERENCES

Akins, R. A., Kelley, R. L. and Lambowitz, A. M. (1989). Characterization of mutant mitochondrial plasmids of *Neurospora* spp. that have incorporated tRNAs by reverse transcription. *Mol. Cell Biol.* 9:678–691.

Almasan, A. and Mishra, N. C. (1991). Recombination by sequence repeats with formation of suppressive or residual mitochondrial DNA in *Neurospora*. *Proc. Natl. Acad. Sci. USA* 88:7684–7688.

Belcour, L., Begel, O., and Picard, M. (1991). A site-specific deletion in mitochondrial DNA of *Podospora* is under the control of nuclear genes. *Proc. Natl. Acad. Sci. USA* 88:3579–3583.

Bertrand, H., Chan, B. S. S., and Griffiths, A. J. F. (1985). Insertion of a foreign nucleotide sequence into mitochondrial DNA causes senescence in *Neurospora intermedia*. *Cell* 47:877–884.

Bertrand, H., Griffiths, A. J. F., Court, D. A., and Cheng, C. K. (1986). An extrachromosomal plasmid is the etiological precursor of kalDNA insertion sequences in the mitochondrial chromosome of senescent *Neurospora*. *Cell* 47:829–837.

Bertrand, H., Wu, Q., and Seidel-Rogol, B. L. (1993). Hyperactive recombination in the mitochondrial DNA of the natural death mutant of *Neurospora crassa*. *Mol. Cell. Biol.* 13:6778–6788.

Bok, J.-W. and Griffiths, A. J. F. (1999). Transfer *of Neurospora* kalilo plasmids among species and strains by introgression. *Curr. Genet.* 36:275–281.

Bok, J. W., Ishida, K. I., and Griffiths, A. J. F. (2003). Ultrastructural changes in *Neurospora* cells undergoing senescence-induced by kalilo plasmids. *Mycologia* 95:500–505.

Borghouts, C., Kimpel, E., and Osiewacz, H. D. (1997). Mitochondrial DNA rearrangements of *Podospora anserina* are under the control of nuclear gene *grisea*. *Proc. Natl. Acad. Sci. USA* 94:10768–10773.

Collins, R. A., Stohl, L. L., Cole, M. D., and Lambowitz, A, M. (1981). Characterization of a novel plasmid DNA found in mitochondria of *N. crassa*. *Cell* 24:443–452.

Court, D. A., Griffiths, A. J. F., Kraus, S. R., Russell, P. J., and Bertrand, H. (1991). A new senescence-inducing mitochondrial linear plasmid in field-isolated *Neurospora crassa* strains from India. *Curr. Genet.* 19:129–137.

Debets, F., Yang, X., and Griffiths, A. J. F. (1995). The dynamics of mitochondrial plasmids in a Hawaiian population of *Neurospora intermedia*. *Curr. Genet.* 29:44–49.

Diacumakos, E. G., Garnjobst, L., and Tatum, E. L. (1965). A cytoplasmic character in *Neurospora crassa*: The role of nuclei and mitochondria. *J. Cell Biol.* 26:427–433.

Dodge, B. O. (1939). Some problems in the genetics of fungi. *Science* 90:379–385.

D'Souza, A. D., Sultana, S., and Maheshwari, R. (2005a). Characterization and prevalence of a circular mitochondrial plasmid in senescence-prone isolates of *Neurospora intermedia*. *Curr. Genet.* 47:182–193.

D'Souza, A. D., Bertrand, H., and Maheshwari, R. (2005b). Intramolecular recombination and deletions in mitochondrial DNA of *senescent*, a nuclear-gene mutant of *Neurospora crassa* normally exhibiting "death" phenotype. *Fung. Genet. Biol.* 42:178–190.

Dufour, E., Boulay, J., Rincheval, V., and Sainsard-Chalet, A. (2000). A causal link between respiration and senescence in *Podospora anserina*. *Proc. Natl. Acad. Sci. USA* 97:4138–4143.

Garnjobst, L., Wilson, J. F., and Tatum, E. L. (1965). Studies on a cytoplasmic character in *Neurospora crassa*. *J. Cell Biol.* 26:423–425.

Griffiths, A. J. F. (1995). Natural plasmids of filamentous fungi. *Microbiol. Revs.* 59:673–685.

Griffiths, A. J. F. (1998). The *kalilo* family of fungal plasmids. *Bot. Bull. Acad. Sin.* 39:147–152.

Griffiths, A. J. F., Kraus, S. R., Barton, R., Court, D. A., Myers, C. J., and Bertrand, H. (1990). Heterokaryotic transmission of senescence plasmid DNA in *Neurospora*. *Curr. Genet.* 17:139–145.

Jamet-Vierny, C., Begel, O., and Belcour, L. (1980). Senescence in *Podospora anserina*: Amplification of a mitochondrial DNA sequence. *Cell* 21:189–194.

Jamet-Vierny C., Boulay, J., and Briand, J.-F. (1997a). Intramolecular cross-overs generate deleted mitochondrial DNA molecules in *Podospora anserina*. *Curr. Genet.* 31:162–170.

Jamet-Vierny, C., Boulay, J., Begel, O., and Silar, P. (1997b). Contributions of various classes of defective mitochondrial DNA molecules to senescence in *Podospora anserina*. *Curr. Genet.* 31:171–178.

Jamet-Vierny, C., Rossignol, M., Haedens, V., and Silar, P. (1999). What triggers senescence in *Podospora anserina*? *Fung. Genet. Biol.* 27:26–35.

Maheshwari, R. and Navaraj, A. (2008). Senescence in fungi: The view from *Neurospora FEMS Microbiol. Lett.* 280:135–143.

Myers, C. J., Griffiths, A. J. F., and Bertrand, H. (1989). Linear kalilo DNA is a *Neurospora* mitochondrial plasmid that integrates into the mitochondrial DNA. *Mol. Gen. Genet.* 220:113–120.

Navaraj, A., Pandit, A., and Maheshwari, R. (2000). *senescent*: A new *Neurospora crassa* nuclear gene mutant derived from nature exhibits mitochondrial abnormalities and a "death" phenotype. *Fung. Genet. Biol.* 29:165–173.

Neupert, W. (1997). Protein import into mitochondria. *Annu. Rev. Biochem.* 66:863–917.

Niagro, F. D. and Mishra, N. C. (1990). Biochemical, genetic and ultrastructural defects in mitochondrial mutant (*ER-3*) of *Neurospora crassa* with senescence phenotype. *Mech. Ageing Dev.* 55:15–37.

Osiewacz, H. D. and Esser, K. (1984). The mitochondrial plasmid of *Podospora anserina*: A mobile intron of a mitochondrial gene. *Curr. Genet.* 8:299–305.

Perkins, D. D., Radford, A., and Sachs, M. (2001). *The Neurospora Compendium.* Academic Press, San Diego.

Seidel-Rogol, B. L., King, J., and Bertrand, H. (1989). Unstable mitochondrial DNA in natural-death nuclear mutants of *Neurospora crassa*. *Mol. Cell Biol.* 9:4259–4264.

Sellem, H., Lecelier, G., and Belcour, L. (1993). Transposition of a group II intron. *Nature* 366:176–178.

Shiflet, A. B. (2006). *Introduction to Computational Science.* Princeton University Press, Princeton, N.J.

Tudzynski, P. and Esser, K. (1979). Chromosomal and extrachromosomal control of senescence in the ascomycete *Podospora anserina*. *Mol. Gen. Genet.* 173:71–84.

Wallace, D. C. (1999). Mitochondrial diseases in man and mouse. *Science* 283:1482–1488.

Wright, R. M. and Cummings, D. J. (1983). Integration of mitochondrial gene sequences within the nuclear genome during senescence in a fungus. *Nature* 302:86–88.

Yang, X. and Griffiths, A. J. F. (1993). Plasmid suppressors active in the sexual cycle of *Neurospora intermedia*. *Genetics* 135:993–1002.

A Glossary of Mycological and Interdisciplinary Terms

A

Acervulus: Mass of short conidiophores under epidermis.

Acropetal: Chain of conidia in which the youngest conidium is at the tip and the oldest is at the base.

Actin: Cytoskeleton track for movement of the vesicles carrying enzymes and precursors for cell membranes and cell wall.

Actinomycetes: Bacteria that grow and replicate like filamentous fungi.

Active site: Region of an enzyme molecule where substrates are bound when they undergo the reaction catalyzed by the enzyme.

Adaptation: Process that improves an organism's chances to survive and reproduce in habitat.

Aeciospore: Binucleate spore of rust fungi that leads to formation of urediospores or teliospores.

Aerial mycelium: Hyphae projecting above the surface of the medium.

Aflatoxin: Toxin produced by species of *Aspergillus* and related species.

Agar: Polysaccharide material derived from algae for preparing semi-solid culture medium.

Albino: Pigmentless phenotype.

Aleuriospore: Thick-walled spore produced at the tip or the sides of hypha.

Allele: Alternate form of gene present in chromosomes.

Allergen: Antigen that induces hypersensitivity reactions through induction of a certain class of antibodies.

Allozyme: Variants of the enzyme formed by different allelic alternatives in the genome.

Amino acid: Building block of protein.

Analogue: Nonhomologous genes/proteins having similar functions.

Anamorph: Phase in life of a fungus in which spores are produced by mitosis.

Anaphase: Phase of cell division in which the chromosomes move to opposite poles of the cell.

Anastomosis: Fusion of two cells resulting in mixing of their contents.

Anisogamy: Fusion of gametes that are morphologically similar but different in size.

Annotation: Finding location of gene and the nature of its product in genome sequence.

Antheridia: Male gametangia.

Antibiotic: Naturally occurring compound produced by one organism that exerts an inhibitory effect on growth of other microorganisms.

Antibody: Type of protein produced by certain blood cells in mammals that specifically recognizes a foreign antigen.

Antigen: Molecule that induces the production of antibodies.

Antimetabolite: Structural analogue of a metabolite.

Antisense: Complementary strand of a coding sequence (gene).

Aplanospore: Nonmotile spores.

Apothecium: Open or "cup-shaped" ascocarp.

Appressorium: Cell formed at the tip of spore germ tube for adhering to host surface for its forced entry inside host.

Arbuscular mycorrhiza: Mycorrhizal fungi in which the hyphae develop spore-like structures in the root cortex.

Arbuscule: Mycorrhizal hypha between the cell wall of host root and plasma membrane; branched like tree.

Archaebacteria: Unicellular organism without a nucleus.

Arthroconidium: Conidium formed by septation in hypha.

Arthrospore: Spore formed by septation of hypha and disarticulation of cells.

Ascocarp: Fruiting body containing asci.

Ascomycetes: Division in the Kingdom Fungi that reproduce by the formation of ascospores enclosed in a sac-like ascus.

Ascospore: Spores produced inside ascus following meiosis.

Ascus (pl., asci): Sac-like structure containing meiotic products (ascospores).

Aseptate: Hyphae with no cross walls.

Asexual: Type of reproduction that does not require the union of female and male gametes.

Autoecious: Rust fungi that complete their life cycle on single host.

Autolysis: Process of self-digestion of hypha.

Autophagy: Intracellular breakdown of cellular components to provide substrates for metabolism during limitation of external sources of energy.

Autotroph: An organism that obtains its carbon nutrition from photosynthesis only.

Auxotroph: Strain lacking the ability to produce certain nutritional substances required for growth.

Avirulent: Parasite lacking the ability to attack a plant.

Axenic: Pure culture without another organism being present.

B

Baker's yeast: Yeast *Saccharomyces cerevisiae* used as a leavening agent or for production of alcohol by fermentation.

Barrage reaction: Lysis of cells at the growing edge when mycelia of two incompatible strains are oppositely grown on medium in a Petri dish.

Basidiomycetes (Basidiomycotina): One division in Kingdom Eumycota comprised of fungi distinguished by haploid basidiospores produced on basidium.

Basidiospore: Spores produced as a result of meiosis in the Basidiomycotina.

Basidium (pl., basidia): Club-shaped cell in Basidiomycotina that contains the diploid zygote nucleus that gives rise to basidiospores after a meiotic division.

Batch culture: Flask culture containing an initial amount of nutrient.

Biodegradation: Microbial decomposition of organic compounds.

Biodiversity: Numbers and relative abundances of different genera and species in a particular area.

Biofuel: Fuel produced from biological raw materials.

Bioinformatics: Analysis of DNA sequence data.

Biomass: Nature's organic leftover.

Biopharming: Use of genetically transformed crop plants to produce pharmaceuticals.

Biopolymers: Long-chain compounds composed of organic subunits.

Biotechnology: Any research or technique developed for the use of living organisms to produce a useful product.

Biotroph: Pathogen that obtains nutrients only from living tissue.

BLAST (Basic Local Alignment and Search Tool): Computer program for finding sequences in a database that identify a query sequence.

Blastic: Form of conidial development in which a conidial initially swells before delimitation by a septum.

Blastoconidia: Asexual spores formed by budding.

BSR (Biological Species Recognition): Species recognition based on compatibility in crosses.

Budding: Production and sepation of a small outgrowth or bud.

C

Catabolism: Energy-yielding degradation of nutrient molecules.

cDNA: DNA sequence that is derived from mRNA.

cDNA library: Collection of DNA sequences containing only protein-coding genes.

Cell: Unit of structural organization containing nucleus and bounded by a membrane.

Cell wall: Multilayered structure made of carbohydrate that provides rigidity to a cell.

Central dogma: Tenet that states that genes are transcribed into messenger RNA molecules, which are then translated into proteins.

Centriole: Organelles found near the nucleus in pairs that aid in cellular division.

Centromere: Region in chromosome that is attached to spindle during mitosis or meiosis.

Chaperonin: Molecule that assists other molecules to fold.

Chitin: Polymer of *N*-acetyl glucosamine.

Chlamydospore: Thick-walled spore formed within the vegetative hyphae.

Chromatid: One of the two strands resulting from the duplication of chromosome during prophase of nuclear division.

Chromatin: Protein wrapping DNA.

Chromosome: Condensed form of chromatin visible under light microscope.

Circadian: Rhythm with a periodicity of ~24 h.

Clade: Group of organisms based on genetic similarities.

Clamp connection: Hook-shaped cell over a septum.

Classification: Grouping of organisms into classes, families, orders, genera, and species.

Cleistothecium: Closed fruiting body containing asci without an opening for release of ascospores.

Clone: Visible clone of cells.

Cloning: Use of DNA manipulation such that a cell harbors one or more DNA fragments.

Clustered genes: Tandemly arrayed genes within segments of DNA.

Codon: Triplet of nucleotides in a DNA that codes for one of the 20 amino acids.

Coenocytic: Multinuclear cell.

Comparative genomics: Comparing the genome organization of different organisms.

Compatible interaction: Interaction between susceptible host and virulent pathogen.

Compatible solute: Organic solutes that accumulate in cell under conditions of water stress.

Complementary DNA: DNA synthesized from a messenger RNA molecule.

***con-10*:** One of the conidiation genes expressed during sporulation pathways in *Neurospora*.

Conidiophore: Hypha that grows into air to form conidia for dispersal into air.

Conidium: Single-celled asexual fungal spore.

Contig: Set of overlapping segments of DNA.

Coprophilous: Growing on dung.

Corticolous: Growing on bark.

Cross: Deliberate mating of two parents.

Crossing over: Exchange of parts of chromosome by breakage and fusion during meiosis.

Cyst: Nonmotile phase of zoospores prior to germination.

Cytology: Microscopic study of cell structure.

Cytoplasm: Cytosol and all the organelles contained in it outside the nucleus and within the plasma membrane.

Cytoskeleton: Network of microtubules.

Cytosol: Supernatant obtained by centrifugation of cell homogenate at high speed (10,000 × g).

D

Dalton: Relative molar mass (M_r) of a compound.

Database: Sequence information from several species.

Decomposition: Return of carbon from organic matter to the atmosphere due to microbial activity.

Defense signaling pathways: Defense pathway activated when a pathogen enters a plant.

Deliquescent: Becoming liquidish.

Dematiaceous: Pigmented spores and mycelium.

Deuteromycetes: Class of fungi that accommodates those fungi whose sexual state is unknown.

Diauxic growth: Growth in a combination of two sugars in which glucose is utilized faster than the other sugar.

Dichotomous: Branching of hypha into two equal forks.

Dikaryon: Hyphal cell in which nuclei are maintained as "n + n" without nuclear fusion.

Dimorphic: Fungus able to grow either in yeast form or in mycelial form.

Diploid: Cell having two chromosome sets, or an individual having two chromosome sets in each of its cells.

DNA: Deoxyribonucleic acid; the polymer composing genes.

Discomycetes: Ascomycotina that form apothecia.

Disease: Condition that affects the ability of plants to photosynthesize and make food.

Disjunctor cell: Empty cell that fragments and/or undergoes lysis to release a conidium.

DNA chip: High-density array of short DNA molecules bound to a solid surface for determining gene expression, marker pattern, or nucleotide sequence of DNA/RNA.

DNA fingerprint: Pattern of DNA bands on a gel.

DNA library: Pool of DNA fragments cloned in a vector.

DNA microarray: Solid support where single- or double-stranded DNA fragments have been spotted at a high density.

DNA probe: Single-stranded DNA molecule used to detect the presence of a complementary sequence among a mixture of other single-stranded DNA molecules.

DNA profile: Pattern of DNA restriction fragments or the PCR products for identification of fungal isolates.

DNA sequencing: Determining the order of nucleotides in a specific DNA molecule.

Dolipore septum: Barrel-shaped septum that flares out near the pore to form an elongate channel.

Dominant: Phenotype expressed in organisms that are either homozygous or heterozygous for the corresponding allele.

Dormancy: Physiological state of a spore in which the block in germination is not removed by the mere addition of nutrients.

Double septum: Two-layered septum that may undergo centripetal separation to release a conidium.

Downy: Having a short and dense mycelial texture.

Dynein: Microtubule motor protein that transports nuclei along microtubules.

E

Ecosystem: Assemblage of organisms and the associated physical environment.

Ecotype: Genetically differentiated subpopulation that is restricted to a specific habitat.

Ectomycorrhiza: Symbiosis between higher plants and fungi in which hyphae form a sheath surround the root tips.

Ectopic: Gene integrated by transformation at other than its normal position.

Effector: Protein molecule secreted by a fungus that alters plant response to infection.

Electrophoresis: Method by which substances, especially proteins and nucleic acids, can be separated from each other on the basis of their electric charge and molecular weights.

Elicitors: Molecules produced by parasitic fungus that trigger defense reactions in the host plant.

Endocytosis: Uptake of extracellular materials by invagination of plasma membrane.

Endogenous: Produced or originating within an organism.

Endomycorrhiza: Mycorrhiza in which hyphae penetrate inside living cells of the root epidermis and cortex.

Endophyte: Nonpathogenic fungus living inside plant tissue.

Endosymbiotic theory: Eukaryotic cell containing an energy-generating endosymbiont in the cytoplasm having its own DNA.

Enrichment: Technique of encouraging the growth of a desired organism.

Entrainment: Synchronization.

Enzyme: Protein molecule that catalyzes a chemical reaction.

Epibiotic: Living on the substratum or host.

Epidemiology: Study of factors influencing the development and spread of infectious disease in populations of plants.

Epigenetic: Heritable phenotypic change not involving change in DNA sequences.

Epiphyte: Living on the surface of plant parts.

Epiphytotic: Plant disease that is equivalent to an epidemic of human disease.

Eubacteria: Commonly known bacteria, including those responsible for diseases.

Eukaryotes: Group of unicellular and multicellular organisms having a membrane-bounded nucleus.

Evolution: Gradual development of organisms from preexisting organisms since the beginning of life.

Exocytosis: Process by which contents of the vesicle are discharged into the extracellular membrane.

Expressed sequence tag (EST): Sequence of a cDNA consisting of only part of the full-length transcript or cDNA sequence.

Expression cloning: Cloning a gene from a genomic or cDNA library by assaying for the expression of the protein of interest by the host organism and isolating the host transformant expressing the desired protein.

Expression vectors: Expression systems containing nucleotide sequences encoding a subunit of a heterologous heterodimer.

Extraradical mycelium: Hyphal network around the root that absorbs and transfers inorganic nutrients to the plant.

F

Fab fragments: Antigen-binding domains of an antibody molecule.

Facultative: Capable of changing lifestyle, e.g., from saprophyte to parasite or the reverse.

Facultative parasite: Organism that is normally saprophytic but is capable of being parasitic.

Facultative saprophyte: Organism that is normally parasitic but is capable of being saprophytic.

Family: A group comprising several genera.

Fermentation: Process of conversion of grape juice by yeast cells to create wine. In biochemistry, the oxidation of certain organic substances in the absence of molecular oxygen. In biotechnology, fermentation refers to the process of growing microorganisms to produce biochemical products. Also, a process for the production of product by the mass culture of a microorganism.

Filamentous fungus: Composed of thread-like hyphae, a mold.

Flagellum: Cellular appendage that enables movement of cell through whip-like motions.

Fluence: Energy density from an optical source impingent on a sample.

Fluorescence: Absorption of high-energy radiation and subsequent emission of light of a longer wavelength, typically visible light.

Folicolous: Growing on leaves.

forma specialis **(abbreviated f. sp.):** Specialized forms of a pathogenic species of fungus that are adapted to particular host plants can be subdivided into physiologic races by the use of host varieties as differential hosts.

Forward genetics: Identification of a mutant and the subsequent cloning of the mutated gene to identify the wild-type sequence responsible for the process being investigated.

Free radical: Highly reactive chemical molecule that has at least one unpaired electron.

Fructification: Spore-bearing organs in both macro- and microfungi.

Fruiting body: Multicellular fungal structure in which sexual spores are produced.

Functional genomics: Determining the function of genes and gene products of an organism from its DNA sequence.

Functional group: Substituent that interacts with biological targets for the specific recognition and interaction with targets.

Fungal endophyte: A fungus that lives asymptomatically inside plant parts for all or a significant part of its life cycle.

Fungi: Kingdom of unicellular or multicellular organisms that obtain their food from external sources and reproduce sexually or asexually.

Fungi Imperfecti: Group of fungi that have no known sexual stage. Also called Deuteromycetes.

Fungicide: Chemical agent that kills or inhibits the growth of fungi.

Fungistasis: Inhibition of fungal growth, sporulation, or spore germination but not death.

G

G proteins: Signaling proteins.

Gametangium: Cell containing haploid gamete nuclei.

Gamete: Cell whose nucleus and cytoplasm fuse with those of another during fertilization.

Gel electrophoresis: Separation of substances on a gel by their rate of movement through an electrical field.

Gene: Functional component unit within a chromosome that controls heritable characteristics of an organism.

Gene bank: Annotated collection of all publicly available DNA sequences.

Gene cloning: Isolating a gene and making many copies of it by inserting the DNA sequence into a vector, then into a cell, and allowing the cell to reproduce and make many copies of the gene.

Gene conversion: Nonreciprocal process in which a gene is replaced by its homologue during meiosis.

Gene expression: Transcription of genes into mRNA followed by its translation into proteins.

Gene flow: Movement of genes from one population to another.

Gene-for-gene: Hypothesis that for every gene that conditions resistance in the plant, there is a corresponding and complementary gene that conditions avirulence in the fungus pathogen.

Gene-for-gene relationship: For every resistance gene in the host there is a corresponding avirulence gene in the pathogen.

Gene library: Collection of DNA fragments representing the total DNA of a cell.

Gene map: Distances of various genes or loci from their centromeres.

Gene mapping: Determining the relative locations of different genes on a chromosome.

Gene pool: All the genes in all the organisms belonging to a population.

Gene regulation: Process of controlled synthesis of gene products in specific cells or tissues.

Gene splicing: Joining pieces of DNA from different sources using recombinant DNA technique.

Generation time: The time required for a population (number of cells or biomass) to double.

Genet: Single genetic individual.

Genetic: Relating to heredity; referring to heritable characteristics.

Genetic code: System of triplet codons composed of nucleotides of DNA or RNA that determine the amino acid sequence of a protein.

Genetic drift: Random changes that become established by chance.

Genetic engineering: Technology of isolating a gene from an organism, modifying it, and inserting it into another organism.

Genetically modified fungus: Fungus containing a transferred gene encoding a protein/enzyme or some cellular metabolite for production in large amounts.

Genome: Total genetic material within a cell of an organism.

Genome size: Total amount of DNA within one copy of genome measured as number of nucleotide base pairs.

Genomics: Determination and use of the genome sequence of organisms to identify genes and functionally important noncoding regions of the genome.

Genotype: Specific combination of alleles present at a single locus in the genome.

Genus (pl., genera): Group of closely related (structurally or phylogenetically) species; the genus or generic name is the first name in a Latin binomial.

Germ cells: Sex cells, sperm or egg

Germ tube: Elongated protrusion from a spore that is the precursor to a hypha.

Germination: Beginning of growth of a spore.

Germling: Hypha of a germinating spore.

Gleba: Fertile portion of a basidiocarp.

Glucan: Polysaccharide containing β-1,3, and β-1,6 linked glucose residues.

Golgi/Golgi apparatus: Multilayered organelle near the nucleus in which oligosaccharide is added to a secretory protein.

Green revolution: Enhanced production of food through methods of plant breeding.

Growth: Elongation and branching of hyphae.

Growth factor: Compound required in minute amount for growth.

H

Habitat: Locality or place where a fungus lives.

Haploid: Possessing a chromosome set composed of only one chromosome of each kind.

Haplophase: Part of life cycle in which the cells contain half the number of chromosomes present in zygote.

Haplotype: Combination of DNA sequences on the chromosome that are transmitted together.

Hartig net: Weft of hyphae on the surface of the root.

Haustorium: Cellular structure formed by a plant parasite that invaginates into the host cell plasma membrane for absorbing nutrients.

Hemiascomycetes: Fungi that produce asci which are not enclosed in an ascocarp.

Hemicellulose: Noncellulosic plant cell-wall polysaccharide (xylan, glucuronoxylan, arabinoxylan, arabinogalactan, glucomannan, xyloglucan, and galactomannan).

Heterochromatin: Condensed inactive DNA.

Heterodimer: Molecule having two or more distinct subunits.

Heteroecious: Rust fungi (Basidiomycotina, Uredinales) that require two unrelated host plants for completion of their life cycle.

Heterokaryon: Mycelium that contains two or more genetically different types of nuclei.

Heterokaryosis: Condition in which the same mycelium contains genetically different nuclei.

Heterologous expression: Expression of gene products originating from different fungi.

Heterologous genes: Genes in a species that are structurally different.

Heterologous protein: Non-native protein produced in the organism by recombinant DNA technique.

Heteroplasmy: Situation where mutant and wild-type mtDNA molecules are found in the same tissue.

Heterothallic: Fungus of opposite mating types.

Heterothallism: Condition in which sexual reproduction occurs in the presence of mycelia of opposite mating types.

Heterotroph: Organism that obtains its carbon from organic carbon compounds.

Heterotrophy: Dependence of an organism on organic foodstuffs for survival.

Heterozygous: Situation where the two alleles at a specific genetic locus are different.

Holobasidium: One-celled structure producing basidiospores.

Holomorph: Name of pleomorphic fungus that includes all stages of the life cycle, sexual and asexual.

Homokaryon: Mycelium of one genetic type.

Homokaryotic: Mycelium containing nuclei of only one genetic type.

Homologous: DNA sequences that are very similar.

Homologous chromosome pairing: Pairing of chromosomes that are identical with respect to both number and order of genes.

Homologous recombination: Exchange of DNA molecules between paired chromosomes during meiosis.

Homologues: Proteins that share a common ancestor.

Homology: Relationship of any two characters that have descended from a common ancestral character.

Homothallic: Fungi in which progeny ascospore cultures are self-fertile.

Homozygous: Situation where two alleles at a specific genetic locus in two homologous chromosomes are identical.

Horizontal (lateral) gene transfer: Movement of genetic material between species (or genera).

Host plant: Living plant from which the parasitic fungus obtains its nourishment.

Hülle cells: Thick-walled cells in *Aspergillus* that occur in association with cleisothecia.

Humanized antibody: Genetically engineered antibody designed to transfer the specificity of an animal antibody to a human immunoglobulin by exchange of the complementarity determining regions.

Hybridization: Bringing complementary single strands of nucleic acids together so that they base pair and form a double strand.

Hydrophobin: Highly insoluble protein found in surface of fungal hyphae and spores.

Hymenium: Spore bearing layer of a fruiting body.

Hypersensitive response: Localized plant cell death (necrosis) in infection.

Hypertrophy: Excessive enlargement of plant cells (upon infection).

Hypha (pl., hyphae): Filament composed of highly elongated cells.

Hyphal growth unit: Hyphal length per branch (= hyphal length/number of tips).

Hyphomycetes: Imperfect fungi that produce conidia on a sporodochium or synnema.

Hyphopodium (pl., hyphopodia): Small swelling produced from mature hypha of pathogenic fungi that adheres to the root epidermis before intracellular fungal penetration.

Hypogeous: Growing below the ground.

I

Idiomorph: Genes that occupy similar relative positions in chromosome but have no homology.

Idiophase: Nongrowing phase in which there is no change in mycelial biomass and may actually decline due to lysis.

Incompatible interaction: Interaction between resistant host and avirulent pathogen.

Incubate: To maintain an organism under conditions that will permit its growth.

Indirect germination: Germination of a sporangium by zoospores.

Inducer: Substrates or substrate analogues in whose presence in the environment an enzyme is synthesized in large amounts.

Infection: Process in which a pathogen enters, invades, or penetrates and establishes a parasitic relationship with a host plant.

Inoculation: Transfer of propagules (spores or bit of mycelium) to initiate growth.

in silico **genetics:** Drawing of conclusions about the functions of genes from comparisons between a newly determined DNA sequence and sequences already in the database.

Intercalary: Within a hyphal element.

Internal transcribed spacer: Nonfunctional RNA situated between ribosomal RNAs (rRNA).

Interphase: Time period between cellular divisions in which cellular processes such as protein synthesis are carried out.

Intraradical hyphae: Network of hyphae from mycorrhizal fungi that colonizes the host root tissues.

Introgression: Transfer of a gene from one species into another through backcrossing.

Intron: Noncoding region in gene.

in vitro: Process that takes place under artificial conditions or outside of the living organism; in a test tube.

in vivo: Process that takes place inside the living organism.

Isoelectric focusing: First dimension of separation in two-dimensional gel electrophoresis (2DE) in which proteins are separated by their isoelectric point (pI); proteins are typically separated electrophoretically in gels containing an immobilized pH gradient (IPG).

Isogamous planogametic copulation: Sexual fusion of two similar, motile gametes.

Isolate: Pure microbial culture separated from its natural origin.

Isolation: Process by which an individual fungus is separated from matrix material, such as water, air, soil particles, or eukaryotic tissues for obtaining it in pure culture.

Isozymes: Functionally similar forms of an enzymatic protein produced by different gene loci in the genome or by different alleles at the same locus.

K

Karyotype: Numbers and the morphology of chromosome set of an individual.

Kingdom: Highest category used in classification.

Knockout: Targeted gene disruption or replacement.

Koji: Foodstuff prepared from steamed rice into which a fungus is inoculated.

L

Lichen: Organism that is a combination of a fungus and an alga or cyanobacterium.
Life cycle: Morphological stage or stages of an organism, each serving one or many ecological functions, which appear sequentially as the organisms grows, develops, and reproduces.
Lignicolous: Growing on wood.
Linkage group: All the genes on one chromosome.
Locus: Position on a chromosome where the gene for a particular trait resides.
Lyophilization: Freezing a substance and sublimating the ice in a vacuum; also called freeze-drying.
Lysosomes: Food-digestive organelles in the cell.

M

Macroconidia: Multinuclear spores produced by mitotic division.
Macrocyclic: Rust fungus disease cycle that includes two alternate hosts.
Mat: Flat mass of interwoven hyphae.
Maternal inheritance: Extranuclear inheritance of a character from DNA in mitochondria.
Medium (pl., media): Solid or liquid formulation of materials for culturing a fungus.
Meiosis: Division process in which the number of chromosomes is reduced by half.
Meiotic drive: Distortion of meiosis such that one of a pair of chromosomes in a heterozygote is recovered in greater than half of the progeny.
Melanin: Brown and black pigment that contributes to survival and longevity of fungal propagule.
Metabolic engineering: Transferring entire metabolic pathways for production of chemicals.
Metabolite: Organic compound that is a starting material in an intermediate or an end product of metabolism.
Metagenomics: Analysis of genetic material obtained from environmental samples.
Metaphase: Phase of mitosis in which the chromosome pairs line up at the equator of the cell.
Metula (pl., metulae): Cell below the phialide in species of *Aspergillus* and *Penicillium*.
Microarray: Large set of cloned DNA molecules spotted onto a solid matrix (such as a microscope slide) for use in probing a biological sample to determine gene expression, marker pattern, or nucleotide sequence of DNA/RNA.
Microconidium: Small spore that acts as a fertilizing agent.
Microcosm: Experimental system containing plant and fungus materials that mimics nature.
Microcyclic: Rust fungus disease cycle that includes only one host.
Microcyclic conidiation: Production of conidia directly from a spore or short hypha without the development of a new mycelium.
Micron: One millionth of a meter, symbolized by the Greek letter μ (mu).
Microorganisms: Organisms seen with microscopes.
micro-RNAs (miRNAs): Approximately 21–24 nucleotides-long RNA species implicated in post-transcriptional regulation of biological processes.
Microtubule: Cytoskeleton composed of actin.
Mineralization: Release of mineral nutrients as inorganic ions from breakdown of organic matter.
Minimal medium: Simple medium to which other substances may be added to test their effect upon growth.
Mitochondria: Organelles that produce energy in the cell.
Mitosis: Cellular division that yields two identical cells from one cell through a five-step process.

Model organism: Biological species that is extensively studied to understand a particular biological process.

Mold (mould): Filamentous fungus.

Molecular marker: Any detectable property that identifies a specific region of the genome.

Monoclonal antibodies: Antibodies derived from a single source or clone of cells, all recognizing only one kind of antigen.

Monoecious: Having both male and female reproductive capacities.

MSUD: Acronym for *m*eiotic *s*ilencing of *u*npaired *D*NA, which silences any sequence that does not have a pairing partner at a corresponding position of its homologue.

Mutagen: Agent that is capable of inducing a mutation, such as UV light.

Mutagenesis: Change(s) in the genetic constitution of a cell through alterations to its DNA.

Mutant: Organism or an allele that differs from the wild type in one or more changes in its DNA.

Mutation: Sudden, heritable change appearing in an individual as the result of a change in the structure of a gene.

Mycelium (pl., mycelia): Network of tubular filaments making up the body of a fungus.

Mycobiont: Fungal component of lichen.

Mycoheterotrophy: Dependency of a plant on a fungal partner for carbon and nutrients.

Mycoparasite: Fungus parasitic on other fungi.

Mycoparasitism: Parasitism of one fungus by another fungus.

Mycorrhiza: Fungus growing in symbiosis with plant roots.

Mycotoxins: Organic molecules produced by fungi that evoke a toxic response in higher vertebrates.

Myxomycete: Formerly a group of fungi comprised of cells that can move around and aggregate to form sporangia.

N

Necrosis: Acute tissue injury.

Necrotroph: Fungus that first kills host tissue and then digests the dead tissue.

Niche: Enclosure in nature occupied by an organism.

Nonconventional yeast: Yeast respiring in presence of air.

Northern blot: Technique used to identify and locate mRNA sequences that are complementary to a piece of DNA.

Nuclear membrane: Membrane surrounding the nucleus that controls nuclear traffic.

Nucleolus: Spherical structure within the nucleus that contains RNA.

Nucleotide: Building blocks for DNA or RNA composed of a nitrogenous base, sugar, and phosphate.

Nucleus: Membrane-bound cellular organelle in eukaryotes that contains the genetic material.

O

Obligate parasites: Parasitic fungus that grows only on or in living tissue.

Oogamy: Fertilization in which the oogonium is fertilized by an antheridium.

Oogonium: Female gametangium containing one or more eggs.

Oomycete: Alga-like fungi that have diploid, aseptate hyphae, reproduce sexually by oospores, and have asexual sporangia often containing motile zoopores.

Oospore: Thick-walled spore in the Phycomycotina that develops from fertilization.

Open reading frame: Region of DNA or RNA where a protein could be encoded.

Operon: Set of adjacent genes that are controlled by common regulatory elements.

Optimum temperature (T_{opt}): Temperature at which a species grows best.

Order: Taxonomic classification between class and family.

Organelle: A membrane-enclosed structure within eukaryotic cells.

Orthologue: Same gene in different species.

Ostiole: Opening or pore in an ascocarp or a pycnidium that allows dispersal of conidia.

Outbreeding: Breeding between unrelated individuals.

Oxidative burst: Rapid generation of active oxygen species like superoxide anion (O_2-), hydroxide radical (OH), H_2O_2.

Oxylipins: Oxygenated fatty acid-derived signaling molecules.

P

Palindrome: Sequence of nucleotides that can be read the same way in either direction.

Papillae: Protuberance for discharge of spores.

Paralogous: Genes sharing at least 80% identity in their deduced amino acid sequence derived from a common ancestor by duplication.

Paramutation: Process by which homologous DNA sequences communicate to establish meiotically heritable expression states.

Paraphysis (pl., paraphyses): Basally attached hyphae in the hymenium of perithecia or apothecia.

Parasexual cycle: Genetic recombination without sexual reproduction.

Pathogen: Organism that derives nutrients from a host causing a disease.

Pathogenesis-related protein: Protein that is induced upon infection.

Pathosystem: System that comprises a host and a pathogen.

Pellet: Network of hyphae.

Peloton: Intracellular hyphal coil of some mycorrhizal fungi.

Peridium (pl., peridia): Outside covering or wall of fructification.

Periphysis (pl., periphyses): Short, hair-like cells lining the inside of an ostiole or of a pore in a stroma.

Periplasm: Protoplasm within oogonium.

Perithecium: Closed, rounded, or flask-shaped fruiting body containing spores, with an opening.

pH: Potential of hydrogen ion concentration ranging from 0 (acidic) to 7 (neutral) to 14 (alkaline).

Phenotype: External manifestations of an organism.

Pheromone: Chemical messenger passed between organisms.

Phialide: Cell that produces conidia in basipetal succession.

Photobiont: Photosynthesizing component of a lichen.

Photoperiod: Period during every 24 hours when an organism is exposed to daylight.

Photoreceptor: Light-absorbing molecule with DNA-binding capacity.

Phycomycetes: Group of fungi or fungi-like organisms having coenocytic hyphae and forming motile spores.

Phylogenetic tree: Arrangement of molecular sequences in the form of a branching tree for denoting closeness or differences between different individuals.

Phylum: Taxonomic category beneath the kingdom and above the class; a group of related, similar classes.

Physiological race: Morphologically identical but physiologically different variety that can be distinguished based on pathogenicity reaction on a set of different host cultivars.

Phytoalexins: Antifungal or antimicrobial substances synthesized by plants in response to infection.

Pilot plant: Bioengineering facility for upscaling production of a metabolite.

Plasma membrane: Cell membrane composed of proteins and a phospholipid bilayer.

Plasmid: Small, self-replicating molecule of DNA that is separate from the main chromosome.

Plasmodium: Mass of flowing protoplasm without a cell wall.

Plasmogamy: Fusion of two cells.

Pleomorphic: Production of more than one form or type of spore in the life cycle of a fungus.

Polarisome: Multiprotein complex that regulates formation of polarized growth sites in yeast.

Polarity: Orientation in a given direction.

Polyketide: Natural product containing many ketone groups.

Polymer: Chemical compound formed by polymerization and consisting of repeating structural subunits.

Polymerase chain reaction (PCR): Technique for making an unlimited number of copies of any piece of DNA.

Polymorphism: Naturally occurring variation in sequence between alleles of a gene.

Population: Interbreeding group of organisms belonging to the same species.

Posttranslational modifications: Addition of different chemical groups to a polypeptide after its synthesis on ribosomes.

Primary metabolite: Metabolic product essential for growth and life.

Primary mycelium: Mycelium formed from germination of basidiospores.

Primary structure: Sequence of amino acids in a protein.

Production strain: Strain of microorganisms used in the production of enzymes.

Progametangium: Cell that gives rise to a gametangium.

Prokaryote: Cell without nuclear membrane.

Promoter: DNA sequence preceding a gene that controls its transcription rate gene.

Promycelium: Germ-tube formed by teliospores or chlamydospores that produces sporidia (basidiospores).

Prophase: Phase of mitosis in which the chromatin duplicates itself and thickens into chromosomes, the spindle fibers form, and the nuclear membrane disintegrates.

Protein: Molecule composed of amino acids arranged in a special order determined by the genetic code.

Proteome: Global set of proteins expressed in a cell at a given time and biological state.

Protists: Unicellular eukaryotes.

Protoplast: Cell devoid of a cell wall.

Prototroph: Strain that can grow on minimal medium.

PSR (Phylogenetic Species Recognition): Species recognition based on construction of a phylogenetic tree.

Pycnidium (pl., pycnidia): Closed, fruiting body producing conidia.

Pyrenomycetes: Ascomycotina that form perithecia.

Q

Quelling: Silencing of pink-orange pigmentation in *Neurospora* resulting from a transgene.

R

R genes: Resistance genes in plants active against pathogens.

R proteins: Proteins in host plant that recognize matching pathogen-derived avirulence (*Avr*) proteins.

Ramet: Module capable of independent existence.

Recessive: Phenotype that is expressed in organisms only if it is homozygous for the corresponding allele.

Recombinant DNA: Hybrid DNA molecule produced in the laboratory by joining pieces of DNA from different sources.

Recombinant proteins: Proteins of one species expressed in a host cell, usually of another species.

Recombination: Process in which chromatids exchange parts with other chromatids.

Renewable resource: Living material that can be grown season after season.

Rennin: Enzyme that forms curds as part of any dairy fermentation product; originally from calves' stomachs, now produced by molds and bacteria.

Reporter gene: Gene sequence that is easily observed when it is expressed in a given tissue or at a certain stage of development.

Resistance: Property of host that prevents or impedes disease development.

Restriction enzyme: Enzyme that cuts double-stranded DNA into specific pieces.

Restriction fragment length polymorphism (RFLP): Fragments resulting from restriction-enzyme digestion of DNA.

Retrovirus: Type of virus that can insert its DNA into the genome of its host cell. This ability has been used as a basis for genetic transformation of animal cells.

Reverse genetics: Determination of function of a gene or DNA sequence by altering the expression of the sequence and then identifying the mutant phenotype that is produced.

Rhizobium: Group of bacteria that form symbiotic associations with legume plants and are responsible for fixing atmospheric nitrogen into a form that can be used by plants and animals.

Rhizoids: Short branching root-like hyphae that serve to anchor the unicellular thallus to its substratum and nourish it by digesting and absorbing food.

Rhizomorph: Aggregate of hyphae found in forest soil or on tree trunks.

Rhizosphere: Zone of the soil around roots of plants that contains soil bacteria and fungi.

Ribonucleic acid: Single-stranded nucleic acid that carries the DNA message to parts of the cell where it is interpreted and used.

Ribosomes: Particles of RNA and protein that provide the sites for protein synthesis.

RIP: Process that detects and mutates duplicated DNA sequences during meiosis.

RNA: Ribonucleic acid, a molecule that is a necessary component of the protein synthesis process.

RNA interference (RNAi): Process in which double-stranded RNA triggers the degradation of a homologous messenger RNA.

S

Saprophyte (Saprotroph): Organism that obtains nutrients from dead organic matter.

Scanning electron microscope (SEM): Electron microscope that provides three-dimensional views of the specimen magnified.

Scavenging: Nutrient acquisition strategy wherein nutrients unavailable at some sites are accessed through foraging hyphae.

Sclerotium: Multihyphal structure that remains viable for a long time and develops into new mycelium.

Screening: Method of sorting and selecting microorganisms.

SDS-PAGE: Sodium dodecyl sulfate polyacrylamide gel electrophoresis; electrophoretic technique used to separate proteins according to their molecular weight.

Secondary metabolite: Small chemical molecules produced by the cell that is not essential for maintenance of cellular function or for normal growth of the organism.

Secondary mycelium: Heterokaryotic mycelium formed from the union of two compatible primary mycelia; usually n + n (or dikaryotic).

Secretome: Proteome of the secreted proteins.

Selection: Method to retain specific cells (or clones of cells) expressing a specific trait, such as antibiotic or herbicide resistance, while killing off all other cells that do not express that trait.

Self-activator: Substances in extracts of fungal spores that activate their own germination.

Self-inhibitor: Substances in extracts of fungal spores which inhibit their own germination.

Senescence: Progressive loss of growth potential upon subculturing resulting in death of mycelial colony.

Septum (pl., septa): Cross wall in a hypha.

Sexual reproduction: Reproduction involving nuclear fusion and meiosis.

Sexual spores: Spores generated by meiosis of diploid nuclei.

Shotgun cloning: Large-insert clone for which a full shotgun sequence has been produced.

Shotgun sequencing: Method of determining the sequence of a large DNA fragment.

Signal: Physical or chemical factor that alters gene expression.

Soma: Vegetative body of fungi; mycelium; hyphae (*see* Mycelium, Hyphae).

Somatic cell: Body cells not involved in sexual reproduction (that is, not germ cells).

Sorocarp: Fruiting body of slime molds.

Southern blot: Technique of identifying DNA sequences complementary to DNA.

Speciation: Process by which a single species becomes two species.

Species: Group of organisms capable of interbreeding freely with each other but not with members of other species.

Species diversity: Distribution and abundance of species.

Spermatia: Male sex cells.

Spheroplasts: Cell whose cell wall has been enzymatically removed.

Spindle pole body: Organelle that controls assembly of microtubules in the cell.

Spitzenkörper: Cluster of vesicles in cytoplasm at hyphal apex.

Spontaneous generation: Creation of life from nonliving material.

Sporangiophore: Aerial hypha supporting a sporangium.

Sporangiospore: Asexual spore produced within a sporangium.

Sporangium: Sac-like structure producing asexual spores.

Spore: Reproductive structure formed by either meiosis or mitosis.

Sporidia (sing., sporidium): Basidiospores of *Ustilago maydis*.

Sporocarp: Fruit body.

Sporodochium (pl., sporodochia): Cushion-like fruiting body containing asexual spores.

Sporogenous: Producing or supporting spores.

Sporulate: To produce spores.

Starter culture: Microorganisms that are added to initiate growth.

Stationary phase: The period in growth when the number of cells living equals the number dying.

Sterigma (pl., sterigmata): Small pointed structure upon which a basidiospore forms.

Straminopila: Kingdom of eukaryotic organisms with superficial similarity to fungi.

Stroma (pl., stromata): Compact mass of cells in which fructifications are formed.

Structural gene: Gene that encodes for the synthesis of a protein.

Structural genomics: Characterizing the physical nature of genomes.

Substrate: Chemical substance acted upon by an enzyme.

Substratum: Material medium in which the fungus occurs.

Symbiont: Organism living in beneficial association with another organism.

Symbiosis: Interaction of two organisms to their mutual benefit.

Symptom: Initial reaction of the host during disease development.

Synaptonemal complex: Structure in meiotic nuclei that binds homologous chromosomes together throughout their length.

Synnema (pl., Synnemata): Group of conidiophores forming a stalk.

Synteny: Different species possessing common genome sequence.

Synteny: Conservation of gene order between species.

Syntrophic association: Association of two or more microorganisms for degrading a substrate or substrates that neither can degrade alone.

Systems biology: Interdisciplinary study field of complex interactions in biological systems.

T

Taxonomy (or systematics): The science of classification of organisms (based on their evolutionary relationships).

Teleomorph: Name for the sexual stage of a fungus.

Teliospore: Spore for overseasoning that germinates by producing four haploid basidiospores.

Telophase: Phase of mitosis in which the chromosome pairs have separated and reached opposite poles of the cell as the spindle begins to disintegrate, the nuclear membrane reappears, and the cytoplasm begins to divide.

Tertiary mycelium: Secondary mycelium of a basidiomycete; usually n + n or 2n.

Tetrad: Four products of a single meiotic division.

Tetrad analysis: Isolation and phenotypic analysis of the four products of meiosis of a single meiotic cell.

Tetrapolar: Basidiospores with four different combinations of *A* and *B* genes: *AB*, *Ab*, *aB*, *ab*.

Thallic: Mode of conidial ontogeny where a conidium is formed from a preexisting hyphal segment or cell.

Thallus: Undifferentiated mycelium of a fungus.

Thermophile: Heat-loving fungus whose optimum growth occurs between 45 and 55°C.

Thermotolerance: Acclimation to high temperatures.

Thigmomorphogenesis: Change in morphology in response to contact with surface.

Thigmotropism: Conversion of information received from touch into a growth response.

Tissue culture: Growing cells of multicellular organisms on a nutrient medium.

Toxins: Pathogen molecules that cause plant cell death thereby facilitating colonization by necrotrophic pathogens.

Transcription: Synthesis of messenger RNA on DNA.

Transcription factor: Protein required for initiation of gene transcription.

Transcriptome: Set of mRNA expressed in a cell's particular physiological state; assessed using microarray.

Transformation: Introduction of an exogenous DNA molecule into a cell, causing it to acquire a new phenotype.

Transgenic: Organism that has been transformed with a foreign DNA sequence.

Translation: Synthesis of protein using information contained in a messenger RNA molecule.

Translocation: Transfer of a part of chromosome into a nonhomologous chromosome.

Transposable elements: DNA sequences that insert, relocate, or exit from chromosomes.

Transvection: Phenomenon in which genes are expressed differently when they are paired than when they are unpaired.

Trichogyne: Hypha of the ascogonium that fuses with the male cell.

Trigger: Treatment to break dormancy of spores that is not required by the fungus during its vegetative growth.

Trophocyst: Enlarged cell; swollen portion of the sporangiophore of *Pilobolus*.

Trophophase: Period of active growth (equivalent to the log or exponential phase in a bacterial culture).

U

Urediospores/Urediniospore/Uredospore: Brown-colored spores produced by rust fungi that are liberated through ruptured epidermis of host and are capable of long aerial dissemination and reinfecting same host.

Uredium: Fruiting body of rust fungi that produces urediospores.

V

Vaccine: Preparation of killed or living attenuated microorganisms or part thereof that are administered to a person or animal to produce artificial immunity to a particular disease.

Vacuoles: Membrane-bound organelles in the cytoplasm that are used for storage and digestion.

Variant: Genetically related but distinguishable form of a species.

Vector: DNA molecule, usually a plasmid, that is used to move recombinant DNA molecules from one cell to another.

Vegetative (heterokaryon) incompatibility: Incompatible fusion of hyphal cells.

Vesicle: Swollen cell or a membrane-bound intracellular structure formed for transport of protein between cellular compartments

Virulent: Ability to infect particular host variety.

W

Western blot: Technique used to identify and locate proteins based on their ability to bind to specific antibodies.

White biotechnology: Industrial use of biotechnology to produce biofuel, enzymes for use in detergents, food, and feed.

white collar: Gene responsible for sensing light.

Wild type: Typical form of an organism as found in nature.

Wild-type gene libraries: Wild-type gene library contains a collection of DNA fragments that contain all the genetic information of a particular wild-type organism.

Woronin body: Electron-dense, spherical body found near the septa in hyphae.

X

Xenologous: Homologues that are related by an interspecies (horizontal transfer) of the genetic material.

Y

Yeast: Single-celled species in the Kingdom Fungi that divide by budding.

Z

Zeitgeber: German word for time-giver; light or/and temperature signal.

Zoospore: Fungus spore with flagella, capable of locomotion in water.

Zygomycete: Fungi with aseptate mycelium that produce sexual zygospores and asexual sporangiospores in sporangia.

Zygophores: Hypha in the zygomycetes that develops into a progametangium.

Zygospore: Thick-walled sexual spore resulting from conjugation of two gametes or gametangia.

Zygote: Diploid cell produced by the fusion of two haploid gametes.

Zygotropism: Directed growth of compatible zygophores toward each other.

Appendix: Naming, Defining, and Broadly Classifying Fungi

NAMING OF FUNGI

The scientific name of a fungus follows a binomial system of nomenclature governed by the International Code of Botanical Nomenclature. It is based on the name of the genus and species with both words italicized. For example, the scientific name of wheat stem rust fungus is *Puccinia graminis* Erikss. Here *Puccinia* refers to the genus, *graminis* is the species, and Erikss. is the abbreviation for Jacob Eriksson, a Swedish mycologist (1848–1931), who first founded the species by publishing the description of the fungus. After first use of full binomial name, the genus name is abbreviated to a single capital letter, as *P. graminis*. When applicable, an infraspecies category (*forma speciales*) characterized by physiologic criteria (host adaptation) is added after the species name, for example *P. graminis tritici*, where *tritici* is for the name of the host plant *Triticum* for wheat. For a name to be accepted, a Latin description must be validly published. Nomenclature is based on the priority of publication. If the vernacular name (e.g., wheat stem rust) is used, the organism should also be identified by the scientific name at least once (e.g., *Puccinia graminis*). Some fungi placed in Fungi Anamorphici have two names, one based on the asexual stage (anamorph) and another based on the sexual stage (teleomorph)—for example, *Septoria nodorum* (anamorph) and *Mycosphaerella graminicola* (teleomorph). However, after the connection between the two stages of the same fungus was proven, *Septoria nodorum* is referred to as *M. graminicola*. In writing, the anamorph name should be followed by the teleomorph name.

DEFINING FUNGI

A startling finding from comparisons of sequences of 16S rRNA component of the small subunit of ribosomes found in all organisms was that some species—for example, the long-studied potato late blight fungus *Phytophthora infestans*—are quite distinct from other species of fungi. These species, termed "pseudo fungi," were given the status of kingdom named Straminipila and distinguished from the "true fungi," which are placed in the kingdom Eumycota. Other differences came to light: The vegetative phase in Straminipila is predominantly diploid, whereas in Eumycota it is haploid. Moreover, Straminipila reproduce by means of motile biflagellate zoospores, suggesting that they evolved from an alga that lost chloroplasts (Cavalier-Smith, 2001). Their cell wall is composed of cellulose and not of β-1,3-glucan and/or mannan and chitin as in Eumycota. Despite this reclassification, Oomycetes obviously share a number of phenotypic traits with pathogenic fungi, such as filamentous growth in the vegetative stage, the ability to form spores for both sexual and asexual reproduction, and similar modes of colonization of host plants. For example, Oomycetes produce plant cell-wall degrading enzymes that are very similar to their fungal counterparts.

The Straminipila and Eumycota represent separate domains of life—a situation reminiscent of some forms once included in Bacteria but that were later separated and grouped in the domain Archaea based on distinct DNA sequences and biochemical features. Thus, filamentous pathogenic fungi and Oomycetes are examples of convergent evolution among microbial eukaryotes. Because of these advances in our knowledge of organisms possessing hyphae, a debating issue is how a fungus should be defined: Based on the evolutionary origin or on a unique structure? In this book, the fungi are nonphotosynthetic, generally multicellular and multinucleate, filamentous eukaryotes, encased in a multilayered cell wall. The yeasts, although they are typically unicellular, are included in fungi because their cell wall and reproductive structure (ascus) is very similar to filamentous eukaryotes placed in the phylum Ascomycotina.

NOMENCLATURE RULES

Classification is the placing of an individual in categories. This not only aids in determining whether it is identical or similar to an already known fungus but also in understanding its evolutionary affiliation. Naming and classification go hand in hand. The modern trend is to classify fungi based on genealogy. Some interpretations of relationships based on their evolutionary sequence conflict with the classification schemes that were developed based on morphological characters. A genus is a group of closely related species—for example, *Puccinia coronata*, *P. graminis*, and *P. sorghi*. A family is a group of related genera, an order is a group of related families, a class is a group of related orders, and a phylum is a group of several classes descending from a common lineage. However, no stable definition of these categories has yet been adopted for fungi. In particular, there is no uniformity in the name endings of the higher categories. The phylum or subdivision name ends in -mycotina, the class in -mycetes, the order in -ales, and the family name in -aceae. The classification adopted here is chiefly as given in Burnett (2003). For the black wheat stem rust fungus, its classification and nomenclature are:

Kingdom/Division:	Eumycota
Phylum/Subdivision:	Basidiomycotina
Class:	Urediniomycetes
Order:	Uredinales
Family:	Pucciniaceae
Genus:	*Puccinia*
Species:	*graminis*
Forma specialis (f. sp):	*tritici*
Race:	race 56
Full name:	*Puccinia graminis* Erikss. f. sp. *tritici* race 56

The infraspecies rank of *forma speciales* designates a variant of the species parasitizing a particular host. A race is a category used in pathogenic fungi that is subordinate to forma, which is similar in form but distinguishable on the basis of pathogenic reaction on varieties of a plant. When first used, the full name of the fungus, *Puccinia graminis* Erikss. f. sp. *tritici* race 56, should be given to avoid confusion. For example, the abbreviated binomial *P. graminis* could mean *Polymyxa graminis*.

A BROAD CLASSIFICATION

Kingdom Eumycota

"True" fungi. Predominantly haploid; cell walls with chitin, β-1,3-glucans and/or mannans.

Phylum Chytridiomycotina

Mostly marine forms; single celled with rhizoids; asexual reproduction by motile zoospores; sexual reproduction unknown.

Phylum Zygomycotina

The hyphae are usually nonseptate; sexual reproduction by morphologically undifferentiated gametangia. The fusion of gametangia results in a reproductive structure called a zygospore, which *presumably* germinates by means of a sporangium containing uninucleate spores (Figure A.1). Examples: *Mucor, Rhizopus.*

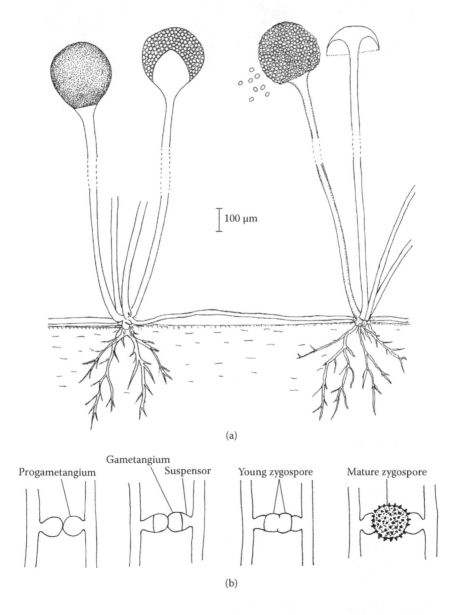

(a)

(b)

Figure A.1 Zygomycotina. *Rhizopus stolonifer,* stolons, rhizoids, young and ripe sporangia. Scale bar = 100 μm. (From Von Arx, J.A., *The Genera of Fungi Sporulating in Pure Culture,* J. Cramer, Vaduz, 1981, with permission.)

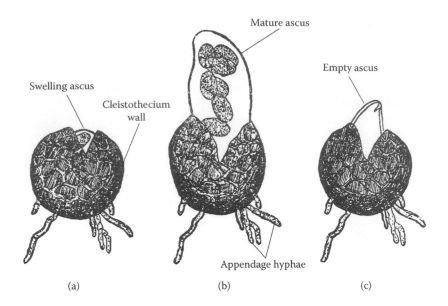

Swelling ascus

Cleistothecium wall

Mature ascus

Empty ascus

Appendage hyphae

(a)　　　　　(b)　　　　　(c)

Figure A.2 Ascomycotina. (a) A swelling ascus just bursting the wall of cleistothecium. (b) Ruptured cleistothecium and ascus. (c) Ascus immediately after discharge of ascospores. (From Ingold, C.T. and Hudson, H.J., *The Biology of Fungi*, Chapman and Hall, London, 1993, with kind permission of Springer Science and Business Media.)

Phylum Glomeromycotina

Here are placed the mycorrhizal fungi, which form symbiotic relationships with plants. Asexual reproduction by chlamydospores. Examples: *Glomus, Acauleospora, Gigaspora, Scutellospora.*

Phylum Ascomycotina

Compartmented hyphae, typically branched and septate with septal pores. Asexual reproduction is by means of uninucleate or multinucleate conidia. The sexual spores are produced in a sac called an ascus, which usually contains four or eight ascospores. Examples: *Neurospora, Sphaerotheca* (Figure A.2).

Phylum Basidiomycotina

Comprise the second biggest group of fungi with some 16,000 species. Compartmented hyphae with characteristic clamp connection at septa. The vegetative mycelium is a dikaryon, each cell containing two sexually compatible, haploid nuclei. The fusion of haploid nuclei produces a transient diploid nucleus in a cell called teliospore or chlamydospores. The diploid nucleus divides to form four meiotic products called basidiospores that, unlike in Ascomycotina, are abjected outside from a tube-like or a club-shaped basidium (Figure A.3). Examples: *Agaricus bisporus* (Figure A.4), *Puccinia graminis* (Chapter 6, Figure 6.11).

Phylum Deuteromycotina (Fungi Anamorphici)

Although a large group of fungi with some 10,000 species, it is a "dust-bin" group in which are placed all those species that have no known sexual stage. The mycelium is septate, multinucleate,

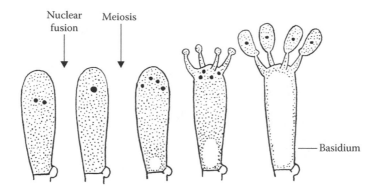

Figure A.3 Basidiomycotina. Diagram of basidium development. (From Ingold, C.T. and Hudson, H.J., *The Biology of Fungi*, Chapman and Hall, London, 1993, with kind permission of Springer Science and Business Media.)

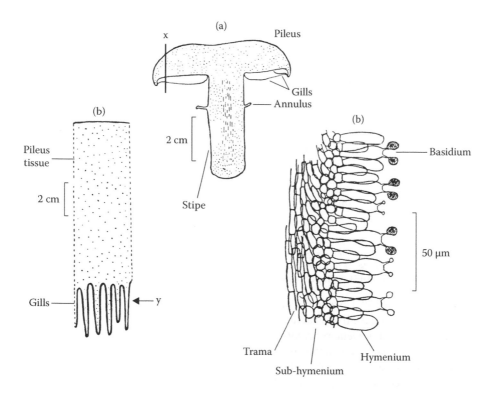

Figure A.4 Basidiomycotina. Fruit body (sporophore) with cap (pileus) and lamellae (gills). (a) Vertical section of cap. (b) Enlarged view of basidium. (From Ingold, C.T. and Hudson, H.J., *The Biology of Fungi*, Chapman and Hall, London, 1993, with kind permission of Springer Science and Business Media.)

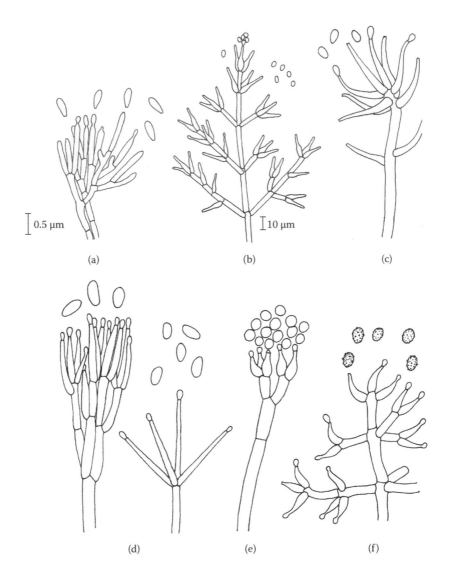

Figure A.5 Deuteromycotina. Conidiogenous structures and conidia. (a) *Myrothecium verrucaria.* (b) *Verticillium tenerum.* (c) *Harziella capitata.* (d) *Gliocladium roseum.* (e) *Gliocladium virens.* (f) *Trichoderma viride.* (From Von Arx, J.A., *The Genera of Fungi Sporulating in Pure Culture,* J. Cramer, Vaduz, 1981, with permission.)

bearing conidia externally on isolated conidiophores (Figure A.5), on a cushion-like mass of hyphae (Figure A.6), or within a flask-like structure (Figure A.7).

Some fungi classified in Deuteromycotina have double names because the sexual stage was discovered after the binomial given on the anamorphic state had become common usage. For example, the sexual (teleomorph) stage of the fungal pathogen of wheat (the glume blotch), *Septoria nodorum*, was recognized 130 years after the fungus had been named on the basis of asexual reproductive structures. This is because the asexual stage (anamorph) *Septoria nodorum* occurs on cereals and grasses during spring and summer in England, whereas the sexual stage *Mycosphaerella graminicola* develops in the wheat stubble and litter in the autumn. Until such time that the connection is established, the same fungus is given two names for the conidial and sexual stages and is classified

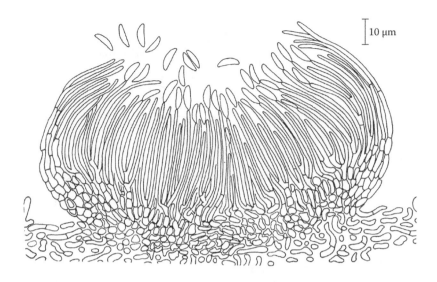

Figure A.6 Deuteromycotina. Conidia produced by a cushion of conidiogenous cells (acervulus). (From Von Arx, J.A., *The Genera of Fungi Sporulating in Pure Culture,* J. Cramer, Vaduz, 1981, with permission.)

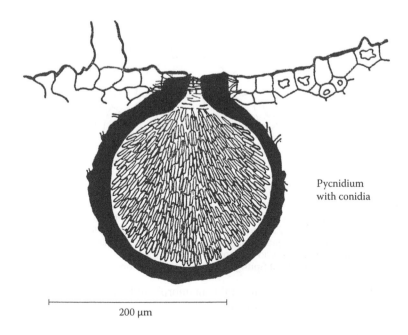

Pycnidium
with conidia

200 μm

Figure A.7 Deuteromycotina. Pycnidium with conidia. (From Ingold, C.T. and Hudson, H.J., *The Biology of Fungi*, Chapman and Hall, London, 1993, with kind permission of Springer Science and Business Media.)

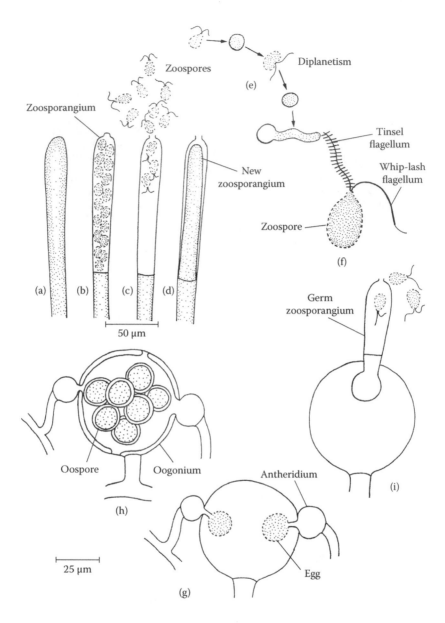

Figure A.8 Straminipila. Diagram of (a)–(c) stages in zoosporangium development of *Saprolegnia*. (d) New sporangium produced by internal proliferation. (e) Encystment and emergence of different form of zoospore (diplanetism). (f) Zoospore with tinsel flagellum directed forward, and whiplash flagellum directed backward. (g) Stages in sexual reproduction. (h) Oogonium with oospore. (i) Germinating oospore. (From Ingold, C.T. and Hudson, H.J., *The Biology of Fungi*, Chapman and Hall, London, 1993, with kind permission of Springer Science and Business Media.)

separately. It is a challenging task to connect the anamorph and teleomorph states, which are produced at different times or in different situations.

Kingdom Straminipila (Stramenopila)

Unicellular or hyphal forms; cell walls composed of cellulose-like β-1,4-glucan; lacking septa; predominantly diploid; asexual reproduction by means of motile spores (Figure A.8). Sexual

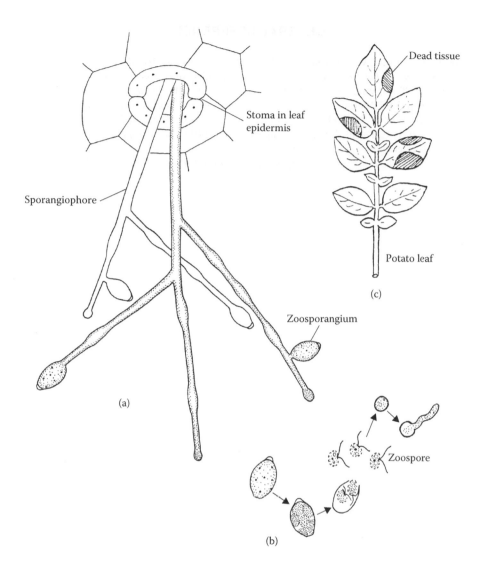

Figure A.9 Straminipila. (a) Sporangiophore of *Phytophthora infestans* emerging through stoma (surface view). (b) Production of flagellated, motile zoospores from sporangium. (c) An infected potato leaf. (From Ingold, C.T. and Hudson, H.J., *The Biology of Fungi*, Chapman and Hall, London, 1993, with kind permission of Springer Science and Business Media.)

reproduction involves gametes of unequal size. Called the "pseudo fungi." Includes *Phytophthora infestans*, the causal agent of late blight of potato (Figure A.9), which provided a great impetus to the study of fungi and contributed to the development of mycology and plant pathology. However, it is now considered not to be a "true fungus" and is placed in the kingdom Straminipila.

GENERAL REFERENCES

Alexopoulos, C. J., Mims, C. W., and Blackwell, M. (1996). *Introductory Mycology*. Wiley, New York, chap. 3.

Burnett, J. (2003). *Fungal Populations and Species*. Oxford University Press, Oxford.

Cavalier-Smith, T. (2000). What are fungi? In *The Mycota*, Vol. VII, *Systematics and Evolution Part A*, pp. 3–37. Edited by D. J. McLaughlin, E. G. McLaughlin, and P. A. Lemke. Berlin: Springer.

Ingold, C. T. and Hudson, H. J. (1993). *The Biology of Fungi*. Chapman and Hall, London.

Kirk, P. M., Cannon, P. F., David, J. C., and Stalpers, J. A. (2001). *Ainsworth and Bisby's Dictionary of the Fungi*. CAB International, Wallingford.

Von Arx, J. A. (1981). *The Genera of Fungi Sporulating in Pure Culture*. J. Cramer, Vaduz.

Index

Printed and bound by CPI Group (UK) Ltd, Croydon, CR0 4YY

21/10/2024

01777040-0009